Marine and Coastal Ecosystem Valuation, Institutions, and Policy in Southeast Asia

Nancy Olewiler • Herminia A. Francisco
Alice Joan G. Ferrer
Editors

Marine and Coastal Ecosystem Valuation, Institutions, and Policy in Southeast Asia

Springer

Editors
Nancy Olewiler
School of Public Policy
Simon Fraser University
Vancouver, Canada

Herminia A. Francisco
Economy & Environment Program
for Southeast Asia
WorldFish
Los Banos, Philippines

Alice Joan G. Ferrer
Division of Social Sciences
University of the Philippines Visayas
Iloilo, Philippines

ISBN 978-981-10-0139-0 ISBN 978-981-10-0141-3 (eBook)
DOI 10.1007/978-981-10-0141-3

Library of Congress Control Number: 2016931420

Springer Singapore Heidelberg New York Dordrecht London

Printed on acid-free paper

Springer Science+Business Media Singapore Pte Ltd. is part of Springer Science+Business Media (www.springer.com)

Foreword

With 30 % of the world's coral reefs and 33 % of the world's mangroves, Southeast Asia (SEA) hosts some of the richest coastal marine systems in the world and is responsible for about 25 % of the global fish production.

However, SEA's highly diverse coastal marine resources are under stress. Overfishing, destructive fishing practices, pollution, and coastal development are all causing resource decline and degradation. Additionally, pressure from population growth and rising income inequality threatens the region's food security and social stability. For millions of Southeast Asians, fish is an important source of dietary protein, and the marine coastal environment provides an important source of livelihood and income. Continuous degradation and decline of coastal marine systems has adversely affected fisher incomes, fisheries employment, revenues, and trade.

Protecting the marine coastal environment and the livelihoods of the people dependent on these resources is a must, but it is not easy. The environmental damages and the impacts on the welfare of the people dependent on these resources must be quantified; appropriate formal and informal institutions for efficient management must be fostered and financing for needed changes must be found.

For the past two decades, the Economy and Environment Program for Southeast Asia (EEPSEA) has captured, shared, and enabled the use of knowledge on the economy and the environment of SEA. This book is a significant addition to body of knowledge EEPSEA has generated. Containing contributions from research projects on coastal resources funded by EEPSEA from 2003 to 2014, the book contains 16 studies from five SEA countries (Indonesia, Malaysia, Philippines, Vietnam, and Thailand). The book's editors and authors have assembled a body of knowledge that provides valuable guidance on how to conserve, protect, and develop coastal marine systems in ways that enhance social welfare. Of benefit to students, researchers, academics, nongovernment organizations (NGOs), resource managers, and policymakers, it is a seminal reference work.

The book groups the studies into three areas: (1) conservation financing, (2) estimating damages to marine coastal resources and impacts on people, and (3) the role and importance of institutions in and policies on coastal resource management. The first set of six studies focuses on valuing key marine ecosystem services in SEA using three valuation methods commonly used to measure ecosystem benefits: travel cost method, contingent valuation methodology, and choice modeling. The next set of four studies offers multiple methods to analyze and quantify the damages brought by pollution and land degradation to marine coastal resources and to determine the best package of policy reforms to reduce emissions and help sustain coastal resources. The last set of six studies focuses on tools and policies to achieve the sustainable management of fisheries and coastal resources such as establishing marine protected areas (MPAs); employing a set of management tools, which may include MPA establishment; establishing artificial reefs; and setting fishing effort controls, including individual transferable quotas.

WorldFish greatly appreciates and welcomes this book because it will help us better achieve our mission to reduce poverty and hunger by improving fisheries and aquaculture.

My personal hope for the future is for protected, better managed, and more sustainable coastal marine systems that are able to sustain fish supplies and deliver food and nutrition security, income, and livelihood benefits to those that need them most, especially poor women and young adults. The chapters of this book provide important insights on what can be done to achieve such a state.

For anyone interested in coastal marine resource conservation and protection and the application of existing knowledge in this area toward promoting sustainable coastal systems, food security, and poverty alleviation in SEA—be it through the efforts of the government, NGOs, civil society, or academia—this book is a must.

WorldFish, Penang, Malaysia — Stephen Hall

Preface

This book is one of several publications of the Economy and Environment Program for Southeast Asia (EEPSEA). EEPSEA's mandate is to support capacity building in the field of environmental economics in Southeast Asia (SEA). The Program was set up by the Canadian International Development Research Centre in 1993, with co-funding provided by the Canadian International Development Agency (CIDA) and the Swedish International Development Cooperation Agency (SIDA). In November 2012, IDRC and SIDA devolved EEPSEA's management to WorldFish, with funding support for another 4 years.

EEPSEA's end goal is to develop in each country the needed expertise to support SEA governments to improve the management of their environment and natural resources. The Program relies on a pool of international and regional experts who provide researchers with mentoring support and who serve as resource persons in its various training courses.

EEPSEA has supported more than 400 research projects on various topics related to the economics of environment and natural resource management. The results of these projects are published as Research Reports, with some accompanied by Policy Briefs that summarize key policy-relevant findings. The various studies produced through EEPSEA research grants are grouped into the following areas: Ecosystem Valuation, Pollution and Waste Management, Climate Change Issues, Natural Resource Management, and Fisheries and Coastal Resources.

To make the results of these various studies more accessible to end users, EEPSEA is now compiling studies on common themes into books. We believe that the materials in these books will continue to contribute to capacity development of students who are enrolled in resource and environmental economics. They can also serve as source materials of environmental service values that could be used by natural resource managers to analyze environmental problems and evaluate alternative solutions to address those problems.

This book is the second that EEPSEA is producing under this book project. A collection of studies on fisheries and coastal resources, the book demonstrates how ecosystem values of marine resources can be derived, shows how to value the damages to coastal and marine ecosystems as a result of pollution and development projects, and evaluates institutions and policies to manage overfishing in marine resources.

The writing of this book took off in 2014 when the editors had a week-long writing workshop in Manila, Philippines. We are appreciative of the hard work of all the study authors who have responded to our multiple requests in the process of putting together this book. More importantly, this book will not be possible without their contributions through their research projects. As teachers of environmental economics courses, they too hope that students will learn from their work. Ms. Noor Aini Zakaria, who patiently coordinated with the authors and also helped in making sure that the publication format is compliant with the Springer requirements, deserves our gratitude. Our thanks also go to Ms. Julienne Bariuan-Elca who liaised with Springer to get this book published and Ms. Kei Cuevas who also helped to get this book done. Finally, our appreciation goes to the three external reviewers who shared their precious time in reviewing various parts of the book. We continue to be responsible for whatever errors may remain in the book.

Vancouver, BC, Canada — Nancy Olewiler
Los Banos, Philippines — Herminia A. Francisco
Iloilo, Philippines — Alice Joan G. Ferrer

Contents

Contributors

Dessy Anggraeni Sustainable Fisheries Partnership (SFP), Bogor, Indonesia

Rizal Bahtiar Department of Resource and Environmental Economics, Faculty of Economics and Management, Bogor Agricultural University (IPB), Bogor, Indonesia

Raul G. Bradecina Partido State University, Goa, Camarines Sur, Philippines

Chantip Chandee Faculty of Economics, Prince of Songkla University, Songkhla, Thailand

Pathomwat Chantarasap Faculty of Economics, Prince of Songkla University, Songkhla, Thailand

Sukampon Chongwilaikasaem Faculty of Economics, Prince of Songkla University, Songkhla, Thailand

Alice Joan G. Ferrer Division of Social Sciences, University of the Philippines Visayas, Iloilo, Philippines

Karen Lou A. Francisco Private Consultant, Bay, Laguna, Philippines

Herminia A. Francisco Economy & Environment Program for Southeast Asia, WorldFish, Los Banos, Philippines

Asuncion B. de Guzman Gaia Resource and Environmental Consultancy Services, Naawan, Misamis Oriental, Philippines

Dieldre Harder Department of Economics, College of Economics & Management, University of the Philippines Los Baños, Laguna, Philippines

Wittawat Hemtanon Faculty of Economics, Prince of Songkla University, Songkhla, Thailand

Nia Kurniawati Hidayat Department of Resource and Environmental Economics, Faculty of Economics and Management, Bogor Agricultural University (IPB), Bogor, Indonesia

Jinky C. Hopanda University of the Philippines Visayas Foundation, Inc., Iloilo, Philippines

Anabeth L. Indab Resources, Environment and Economic Center for Studies, San Juan City, Philippines

Gazi Md Nurul Islam School of Economics, Finance and Banking, College of Business, Universiti Utara Malaysia, Kedah, Malaysia

Esmyra P. Javier Part of an MSc Thesis, University of the Philippines, Los Banos, Laguna, Philippines

Jin Jianjun College of Resources Science & Technology, Beijing Normal University, Beijing, China

Resources, Environmental & Economic Center for Studies, Beijing, China

Sakchai Kiripat Faculty of Economics, Prince of Songkla University, Songkhla, Thailand

Lourdes O. Montenegro Lee Kuan Yew School of Public Policy, National University of Singapore, Singapore, Singapore

Alan Dino E. Moscoso School of Technology, University of the Philippines Visayas, Iloilo, Philippines

Orapan Nabangchang-Srisawalak School of Economics, Sukhothai Thammatirat Open University, Bangkok, Thailand

Pham Khanh Nam School of Economics, University of Economics Ho Chi Minh City, Ho Chi Minh City, Vietnam

Kusairi Mohd Noh Institute of Agricultural and Food Policy Studies, Universiti Putra Malaysia, Serdang, Malaysia

Aswani Farhana Mohd Noh Institute of Agricultural and Food Policy Studies, Universiti Putra Malaysia, Serdang, Malaysia

Nuva Department of Resource and Environmental Economics, Faculty of Economics and Management, Bogor Agricultural University (IPB), Bogor, Indonesia

Nancy Olewiler School of Public Policy, Simon Fraser University, Vancouver, BC, Canada

Michael Q. Orquejo University of the Philippines Visayas Foundation, Inc., Iloilo, Philippines

Mildren Peñales One Global Place, World Bank (Philippines), Taguig, Metro Manila, Philippines

Kunlayanee Pornpinatepong Faculty of Economics, Prince of Songkla University, Songkhla, Thailand

Chotima Pornsawang Faculty of Economics, Prince of Songkla University, Songkhla, Thailand

Resurreccion B. Sadaba Division of Biological Sciences, College of Arts and Sciences, University of the Philippines Visayas, Iloilo, Philippines

Papitchaya Saelim Faculty of Economics, Prince of Songkla University, Songkhla, Thailand

Udomsak Seenprachawong School of Development Economics, National Institute of Development Administration, Bangkok, Thailand

Jumtip Seneerattanaprayul Faculty of Economics, Prince of Songkla University, Songkhla, Thailand

Shaufique Fahmi Sidique Institute of Agricultural and Food Policy Studies, Universiti Putra Malaysia, Serdang, Malaysia

Tran Vo Hung Son School of Economics, University of Economics Ho Chi Minh City, Ho Chi Minh City, Vietnam

Rodelio F. Subade Division of Social Sciences, University of the Philippines Visayas, Iloilo, Philippines

Zenaida M. Sumalde Department of Economics, College of Economics and Management, University of the Philippines, Los Baños, Laguna, Philippines

Truong Dang Thuy Faculty of Development Economics, Environmental & Natural Resources Unit, University of Economics, Ho Chi Minh City, Vietnam

Sinad Treewanchai Faculty of Economics, Prince of Songkla University, Songkhla, Thailand

Abbreviations

ADB	Asian Development Bank
AFMA-DMEQCMA	Agriculture and Fisheries Modernization Act—Database on Monitoring of Environmental Quality for Coastal Management and Aquaculture
AIMS	Apo Island Marine Sanctuary
ANOV	Annual net operating values
APRI	Indonesia Crab Processors Association
AR	Artificial reefs
ARMNP	Apo Reef Marine Natural Park
AS	Activated sludge
AUU	Abandoned, underdeveloped, and undeveloped
B-C	Benefit-cost
BDFI	Batanes Development Foundation Inc.
BFAR	Bureau of Fisheries and Aquatic Resources
BFAR-FRQD	Bureau of Fisheries and Aquatic Resources-Fisheries and Quarantine Division
BIPLAS	Batanes Islands Protected Landscape and Seascape
BLCC	Biri-Larosa Coastal Community
BOD	Biological oxygen demand
BPLS	Baliangao Protected Landscape and Seascape
BSC	Blue swimming crab
CAC	Command and control
CBA	Cost-benefit analysis
CBRMP	Community-Based Resource Management Program
CDP	Coastal Development Plan
CEA	Cost-effectiveness analyses
CEP	Coastal Environmental Program
CF	Conservation fee
CIDA	Canadian International Development Agency
CM	Choice modeling
CPPAP	Conservation of Priority Protected Areas Project

CPR	Common pool resources
CPUE	Catch per unit of effort
CRM	Coastal resource management
CRP	Cordova Reclamation Project
CS	Consumer surplus
CSRP	Cebu South Reclamation Project
CVM	Contingent valuation method
DA	Department of Agriculture
DAO	Department of Environment and Natural Resources Administrative Order
DB-REMO	Danao Bay Resource Management Organization
DENR	Department of Environment and Natural Resources
DENR-CMMO	Department of Environment and Natural Resources—Coastal and Marine Management Office
DLA	Department of Local Administration
DO	Dissolved oxygen
DOF	Department of Fisheries
DOSTE	Department of Science, Technology and Environment
DR	Discount rates
DTCP	Department of Public Works and Town and Country Planning
EC	Emission charges
ECNQA	Enhancement and Conservation of National Quality Act (1992)
ECS	Emission Charge System
EEPSEA	Economy and Environment Program for Southeast Asia
FADs	Fish aggregating devices
FAO	Food and Agriculture Organization
FARMC	Fisheries and Aquatic Resources Management Council
FGD	Focus Group Discussion
FIES	Family Income and Expenditure Survey
FIRR	Financial internal rate of return
FLA	Fishpond Lease Agreement
GEF	Global Environment Facility
GNI	Gross national income
GSECI	Ground Structures Engineering Consultants, Inc
GSO	General Statistical Office
HCMC	Ho Chi Minh City
HH	Households
IADP	Itbayat Integrated Area Development Program
IDRC	International Development Research Centre
IEC	Information and education
IID	Independently and identically distributed
INTAQ	Integrated aquaculture
IPAF	Integrated Protected Area Fund
IQ	Individual quotas

IRR	Internal rates of return
ITCM	Individual travel cost method
ITQ	Individual transferable quotas
ITWG	Interagency technical working group
IUCN	International Union for Conservation of Nature
JICA	Japan International Cooperation Agency
KII	Key informant interviews
LGC	Local Government Code
LGCAMP	Lingayen Gulf Coastal Area Management Plan
LGU	Local Government Unit
LKIM	Fisheries Development Authority Malaysia
MARINA	Maritime Industry Authority
MEA	Millennium Ecosystem Assessment
MEY	Maximum economic yield
MFO	Municipal Fishery Ordinance
MFRMP	Marine Fisheries Resource Management Program
MFRS	Marine Fishery Reserve-Sanctuary
MIIC	Malayan Integrated Industries Corporation
MLS	Minimum legal size
MMAF	Ministry of Marine Affairs and Fisheries
MMC	Marine Management Committee
MMEA	Malaysian Maritime Enforcement Agency
MOA	Memorandum of agreement
MOAC	The Ministry of Agriculture and Co-operatives
MOI	Ministry of Industry
MOINT	Ministry of Interior
MONRE	Ministry of Natural Resources and Environment
MOP	Macao pataca
MOPH	Ministry of Public Health
MPA	Marine protected area
MSY	Maximum sustainable yield
MWTP	Mean willingncss to pay
NAV	Net annual valucs
NEM	Northeast monsoon
NESDB	National Economic and Social Development Board
NFI	National Fisheries Institute
NFRDI	National Fisheries Research and Development Institute
NGOs	Nongovernment Organizations
NIACDEV	Northern Iloilo Alliance for Coastal Development
NICA	National Institute of Coastal Aquaculture
NIPAS	National Integrated Protected Areas System
NOAA	The National Oceanic and Atmospheric Administration
NPV	Net present values
NSO	National Statistics Office

OA	Open access
OECF	Overseas Economic Cooperation Fund
OLS	Ordinary least square
ONEB	Office of the National Environment Board of Thailand
ONEP	Office of Nature and Environmental Policy and Planning
ORRAF	Office of the Rubber Replanting Aid Fund
PR	Pag-ro
PAMB	Protected Area Management Board
PASU	Protected Area Supervising Unit
PDI	Philippine Daily Inquirer
PEMSEA	Partnerships in Environmental Management for the Seas of East Asia
PENRO	Provincial Environment and Natural Resources Officer
PES	Payment for environmental services
PHP	Philippines peso
PI	Panggangan Island
PO	People's organization
SACRED	Samar Center for Rural Education and Development
SEA	Southeast Asia
SEAFDEC	South East Asian Fisheries Development Center
SEC	Securities and Exchange Commission
SEDF	Surigao Economic Development Foundation Inc.
SIDA	Development Cooperation Agency
SIPLAS	Siargao Islands Protected Landscape and Seascape
SMR	Sagay Marine Reserve
SMU	Special management unit
SSDP	Southern Seaboard Development Project
SWM	Southwest monsoon
TAC	Total allowable catch
TB	Total benefit
TCM	Travel cost method
TDP	Tradable discharge permit
TEV	Total economic value
THB	Thai baht
THIPA	Turtle Islands Heritage Protected Area
TKN	Total Kjeldahl nitrogen
TRMNP	Tubbataha Reef Marine National Park
TSS	Total suspended solids
UASB	Upflow anaerobic sludge blanket
UNDP	United Nations Development Programme
UNEP	United Nations Environment Programme
UP–MSI	University of the Philippines–Marine Science Institute
USAID	United States Agency for International Development
USC	University of San Carlos

USD	US dollar
USEPAERL	US Environmental Protection Agency's Environmental Research Laboratory
UT	U-Tapao
VISSEA	Visayan Sea Coastal Resources and Fisheries Management
VND	Vietnamese dong
WB	World Bank
WEPA	Water Environment Partnership in Asia
WQCI	Water Quality Composite Index
WTP	Willingness to pay
WWF	World Wildlife Fund
ZTCM	Zonal travel cost method

Chapter 1
Introduction: Marine and Coastal Ecosystem Valuation, Institutions, and Policy in Southeast Asia

Herminia A. Francisco

Southeast Asia (SEA) contains 30 % of the world's reefs (Chua 2009), 33 % of the world's mangrove resources, and at least 20 % of its seagrasses (Fortes 2010). It also accounts for a quarter of global fish production (Garces et al. 2008) supporting the livelihood of 60 % of the people residing within 60 km of the coast (Chua 2009). According to Komatsu (2013), the leading producer of fish in the region is Indonesia (33.8 %), followed by the Philippines (16.3 %), Vietnam (15 %), Myanmar (13.8 %), and Thailand (10.9 %).

Fish production from coastal areas in Asia has been declining. It was estimated that coastal fish stocks in South Asia and SEA have declined by 5–30 % over the last five decades, negatively impacting fisher incomes, fisheries employment, revenues, trade, and social stability (ADB 2014). The decline in fish production is attributed to overfishing and use of destructive fishing methods, among other human activities (PEMSEA 2003; UNEP 2004; Burke et al. 2002), and climate change (Fortes 2010; Hoegh-Guldberg 1999).

Sustaining fish production to support the economy and the lives of those who depend on this resource requires that coastal resource management is improved. Toward this end, natural resource managers will require better information on the value of marine ecosystem services and goods, how these values are affected by economic activities, and how the ecosystem services could be enhanced through better institutions and policies. Research helps generate this information as inputs to decisions on how to reduce damaging activities and enhance those that protect coastal resources.

The Economy and Environment Program for Southeast Asia (EEPSEA) has supported numerous research projects related to coastal and marine resources over the years. This book is a collection of 16 studies carried out from 2003 to

H.A. Francisco (✉)
Economy & Environment Program for Southeast Asia, WorldFish, Los Banos, Philippines
e-mail: H.Francisco@cgiar.org; hermi@eepsea.net

N. Olewiler et al. (eds.), *Marine and Coastal Ecosystem Valuation, Institutions, and Policy in Southeast Asia*, DOI 10.1007/978-981-10-0141-3_1

2014 that cover the following themes: Part I, *Valuing Ecosystem Services and Conservation Financing*; Part II, *Valuing Damages to Coastal and Marine Resources*; and Part III, *Institutions and Policies for Fisheries Management*. The featured studies are intended for both natural resource managers tasked with coastal resource management and for students of environment and natural resource economics. The collection of studies covers a wide range of environmental problems that many coastal resource managers face. It is our hope that the studies will help in understanding these problems and how to evaluate the options to deal with them.

Part I (*Valuing Ecosystem Services and Conservation Financing*) includes six studies that value key marine ecosystem services in SEA and China so that their importance can be better appreciated and understood. The papers highlight both the valuation methods used in the study, the findings, and the policy relevance of those environmental values. The studies showcase how researchers interested in the analysis of environmental problems can use resource and environmental economics tools in valuing ecosystem services. Three valuation methods commonly used to measure ecosystem benefits (i.e., travel cost method, contingent valuation methodology, and choice modeling) are thus discussed extensively (with survey instruments included) in some of the papers.

Part II (*Valuing Damages to Coastal and Marine Resources*) features the threats to marine resources and how the damages that they bring about can be valued. Among the threats featured are the pollution caused by economic activities and by fish farms' overstocking and excessive feed use and damages to coastal resources from a land reclamation project. A good understanding of the costs associated with economic activities is crucial in determining appropriate charges to impose on those damaging activities. The numbers are also crucial in evaluating the options to deal with the problem; they represent benefits (cost avoided) that will have to be weighed with the cost of implementing the corrective/pollution control options. The fourth paper looks into ways to restore mangrove resources that have been abandoned from unsuccessful aquaculture development by exploring the institutional aspect of making this land use reversion.

Part III (*Institutions and Policies in Fisheries Management*) is made up of five studies. Among those presented here is a discussion on the problems and management reforms needed in marine protected area (MPA) in the Philippines. While the discussion is location specific, the issues presented are not limited only to the Philippines as MPA establishment is also popular in other countries. The study in Malaysia discusses the benefits brought about by artificial reef enhancement measures, a management tool that is also popular in other parts of the region. Finally, the study in Thailand evaluates the use of individual transferable quotas (ITQs) in controlling fishing efforts. The study finds that ITQs can potentially improve coastal fishery resource management only if institutions to effect better collaboration among stakeholders are in place.

Each of the three parts of the book has an overview section. The overview brings attention to the main issues on the theme and provides a summary of the various papers included in that theme. The papers in that collection are then presented to the readers. The book ends with a summary that highlights the main lessons of policy significance derived from the various studies. It also presents what the editors think are the remaining research gaps toward improving marine and coastal management in SEA.

References

ADB (Asian Development Bank) (2014) Economics of fisheries and aquaculture in the coral triangle. Asian Development Bank, Mandaluyong City

Burke L, Selig E, Spalding M (2002) Reefs at risk in southeast Asia. World Resources Institute, Washington, DC

Chua T (2009) Status of coral reefs: Singapore and beyond. 5 Feb 2009. http://wildshores.blogspot.com/2009/02/status-of-coral-reefs-singapore-and.html#.VIpX1tKUdgA. Retrieved 15 Jun 2015

Fortes MD (2010) The seagrass-mangrove connection as a climate change mitigation and adaptation factor in east Asian coasts. Paper presented at the international symposium on integrated coastal management for marine biodiversity in Asia, 14–15 Jan 2010, Kyoto

Garces LR, Pido MD, Pomeroy RS (2008) Fisheries. In: Pandya A, Laipson E (eds) Southeast Asia: challenges and opportunities in transnational trends: middle Eastern and Asian views. The Henry L. Stimson Center, Washington

Hoegh-Guldberg O (1999) Climate change, coral bleaching and the future of the world's coral reefs. Mar Freshw Res 50(8):839–866

Komatsu M (2013) Special study on sustainable fisheries management and international trade in the southeast Asia and Pacific region, ADBI working paper 438. Asian Development Bank Institute, Tokyo

PEMSEA (Partnerships in Environmental Management for the Seas of East Asia) (2003) Sustainability development strategy for the seas of South of East Asia. Regional implementation of the world summit on sustainable development requirements for the coasts and oceans. PEMSEA, Quezon City

UNEP (United Nations Environment Programme) (2004) Seagrass in the South China Sea. United Nations Environment Programme, Nairobi

Part I
Coastal Marine Ecosystem Services in Southeast Asia: Support for Conservation Initiatives

Herminia A. Francisco

Introduction: Coastal Marine Ecosystems of Southeast Asia, the Big Picture

Coastal marine ecosystems, comprised of mangroves, coral reefs, sea grass beds, and sandy beaches, provide goods and services that support human populations. MEA (2005) classified these goods and services as provisioning, regulating, supporting, and cultural values. Most countries in Southeast Asia (SEA) are endowed with highly diverse and rich coastal marine ecosystems as two major marine ecosystems (the South China Sea and the Sulawesi Sea) surround the region. The SEA countries with the richest coastal marine resources are Indonesia, the Philippines, and Malaysia.

The region accounts for less than 10 % of the global area but contains 30 % of the world's reefs (Chua 2009). Fortes (2010) cited that the region has 33 % of the world's mangrove resources and at least 20 % of its sea grasses. Garces et al. (2008) attributed a quarter of the global fish production, about 19 million metric tons of fish, to the region, with Asia accounting for 60 % of the global total.

The region is also highly dependent upon its coastal marine ecosystems for economic activities. With 60 % of its people residing within 60 km of the coast (Chua 2009), there is a large population dependent on these ecosystems for livelihood and nutrition. These ecosystems are also the economic base for many industries (e.g., shipping, transport, ecotourism, etc.) but are subjected to expanding settlement and infrastructure development.

All of these put pressure on coastal marine ecosystems such that their health and condition are in serious decline (PEMSEA 2003; UNEP 2004; Burke et al. 2002). In

H.A. Francisco (✉)
Economy & Environment Program for Southeast Asia, WorldFish, Los Banos, Philippines
e-mail: H.Francisco@cgiar.org; hermi@eepsea.net

addition, climate change, through coral bleaching and acidification, is damaging the marine environment (Fortes 2010; Hoegh-Guldberg 1999). Specific human activities that cause serious harm to these ecosystems include the conversion of mangroves to fishponds, overfishing, coral reef harvesting, pollution and sedimentation that destroy sea grass, the quality of coastal environment, and illegal harvesting of endangered marine species (Clark 1992).

The destruction of coastal marine ecosystems means the loss of both the consumptive products (e.g., fishery and mangrove wood products) and the nonconsumptive services (e.g., recreation, regulating function, supporting function, and cultural function) that they provide. The consumptive or provisioning function of these ecosystems is well recognized and abused through excessive extraction, but their other crucial ecological functions have gained global recognition only in the last decade or two.

For instance, besides serving as spawning ground and fish habitat, mangroves also provide shoreline protection. Coral reefs are associated with fish production[1] and ecotourism, but they too are important in shoreline protection. Sea grasses are food to dugong, green turtles, and manatees but also serve as nursery grounds for fish and shellfishes (Short et al. 2011). Also, like coral reefs and mangroves, they protect the shoreline, providing erosion control by trapping sediments. Of late, the realization about the important function of sea grasses and mangroves in sequestering carbon and providing natural protection against climate change (i.e., temperature increase, sea level rise, increased frequency of storms, and strong waves) provides even stronger justification for their protection (Fortes 2010; Hoegh-Guldberg et al. 2009).

Given the important roles various marine ecosystems perform for mankind, there is an urgent need to protect what remains of them by regulating their use through property right's reform and by implementing conservation and protection measures and policies. The most common approach to managing these resources is by declaring them as marine protected areas (MPAs). However, this is only a first step, as laws governing their use or access need to be strengthened. A crucial element is earmarking a reasonable amount of government funds to support conservation initiatives. In addition, engaging local governments and communities, partnering with private/business sectors, and setting economic incentives to encourage the sustainable use of these ecosystem services are needed. While all of these reforms are important, the lack of funding to support conservation and protection work in coastal marine areas is a big hurdle. Conservation financing is thus one area that is of interest to researchers in this field.

This chapter will discuss the potential to mobilize conservation finance from the beneficiaries of ecosystem services, as inferred from how much they value the ecosystem benefits they receive. This chapter includes seven studies that were

[1] The World Fisheries and Aquaculture Atlas of 2002 (FAO 2003) estimated that about 35 million people, mostly from developing countries, rely on marine fisheries and aquaculture for livelihood.

funded by the Economy and Environment Program for Southeast Asia (EEPSEA)[2] to value key marine ecosystem services in SEA and China. A short discussion of the pressures faced in each of the coastal ecosystem study sites is also provided. The three valuation methods commonly used to measure ecosystem benefits – travel cost method, contingent valuation methodology, and choice modeling – are discussed extensively in some of the papers. Researchers and students who are interested in undertaking similar research in the future are encouraged to use these materials as reference.

Ecosystem Values and Their Policy Significance

The ecosystem values of interest, and their corresponding valuation studies as included in this chapter, are seascape beauty, by Bradecina (2014); reef-based recreation, by Seenprachawong (2001) and Pham and Tran (2001); quality improvement of a bay ecosystem, by Seenprachawong (2002); and endangered marine species habitats, by Indab (2007) and Nabanchang et al. (2008).

In all six papers, ecosystem service valuation is a means to either (a) evaluate the potential for capturing a portion of the consumer surplus from the recreation value of the marine ecosystem (e.g., for rent capture) or (b) serve as inputs in analyzing the potential impacts of development projects in coastal areas (e.g., for trade-off analysis).

A detailed discussion on the application of each methodology is presented in some of the papers in this chapter. Readers should refer to Pham and Tran (2001) study on Hon Mun Islands for theory and application of travel cost method (TCM). For information on the contingent valuation method (CVM), Indab's (2007) study is a useful reference. Finally, Seenprachawong's (2002) study provides a good discussion on the application of choice modeling (CM) in valuating attribute/ecosystem benefits from Phang Nga Bay, Thailand.

Seascape Beauty of Caramoan, Philippines: How Much Is It Worth?

Caramoan is a municipality of Camarines Sur Province in the Philippines. It has a 71 km, irregular coastline surrounded by bay, seas, swamps, and ocean. The Caramoan beachscape is considered unique, with fine white sand, beautiful beach

[2] EEPSEA is a program of WorldFish with funding from the International Development Research Centre (IDRC) and the Swedish International Cooperation Development Agency (SIDA). Its mission is capacity building for environmental economic research in Southeast Asia.

sceneries, and clear blue waters; its coral resources, habitat to diverse fishes, are a diver's paradise.

The threats to this beachscape come from unregulated fishing activities, domestic wastes from the growing tourist population and the local community, and siltation from upland farming. To protect the condition of the area, the local government of Caramoan established Marine Protected Areas (also known as MPAs). A stronger solid waste management program and environmental awareness campaign were also planned as part of the MPA program, but these activities require funding beyond what the local government can provide.

As such, Bradecina (2014) assessed whether visitors would support the abovementioned programs, if asked. He set up a hypothetical scenario and asked visitors if they would be willing to pay a given level of entrance fee that would support Caramoan beachscape coastal resource management and protection activities. Several entrance fee levels were used. The study found out that a fee of around PHP500 (USD12) would be acceptable to a majority of the visitors. The mean willingness to pay (MWTP) was estimated at PHP563 (USD14).

The researcher presented his results to the Caramoan local government. He was subsequently invited to be part of the team that develops the tourism development master plan for the area. The results of his study were used in the cost-benefit analysis of the said master plan, as well as in writing a more detailed plan on how to institute a user fee system for Caramoan.

Recreation Benefit and User Fee to Access the Coral Reefs and Diverse Fishes of Phi Phi Islands, Thailand

Phi Phi Islands' (in the Andaman Sea, off the west coast of Southern Thailand) reefs have a rich biodiversity of resources. The islands are surrounded by hard coral gardens that are home to a wide variety of marine creatures. The islands' beauty has made them world famous. Hundreds of visitors land on Phi Phi's shores every day. The tourism industry is well developed, with numerous operators running boat tours and offering services for snorkeling and diving.

The rapid growth in tourist numbers, accompanied by infrastructure development to meet tourist needs, has resulted in high levels of wastes (e.g., wastewater, solid wastes, and erosion from construction projects). This reduces the quality of the island shoreline ecosystems. While Phi Phi Islands remain a popular ecotourism destination, there is a need to put in place a management regime that will slow down, if not arrest, the degradation of their coastal marine ecosystems.

Seenprachawong (2001) measured the recreational benefit derived from the islands' coral reef ecosystems by both domestic and international visitors using the travel cost method (TCM). Since TCM captures only part of the use value of the reef ecosystem, the contingent valuation methodology (CVM) was also employed to estimate both its use and nonuse values.

Results indicate that the net recreational value (accounting for the cost of travel) for both domestic and international visitors (also called consumer surplus) is USD205.41 million/year. The CVM measures of both use and nonuse values of Phi Phi Islands' reef ecosystems indicate a higher value for the reef of USD497.38 million/year.

These numbers are potential revenues that agencies responsible for island management and conservation could tap, either through an entrance fee or a user fee added to marketed recreational goods like snorkeling and diving. The author recommended charging an entrance fee of up to USD7 per visit based on the computed consumer surplus per visit to the site. For areas with reefs in critical condition, additional fees set even higher should be imposed to reduce the pressure on these coral ecosystems.

Coral Reef-Based Recreation Benefits and Port Expansion: A Trade-Off Analysis

The Hon Mun Islands, located in the south of Nha Trang Bay, have the richest coral biodiversity in Vietnam. The islands support a variety of habitats and ecosystems that include fringing coral reefs, mangroves, and sea grass. These ecosystems give rise to a growing local fishing industry. The coastal water also supports a thriving shipping industry. The rich coral biodiversity of the islands and their close proximity to Nha Trang City, which is a 30–60 min boat ride away, make these islands a favorite tourist destination for both domestic and international visitors.

The rapid economic development in Nha Trang City was accompanied by infrastructure expansion along the coastal areas, as well as the rise in pollution and sedimentation discharged into water bodies. Overfishing and the use of destructive fishing practices (e.g., blast and poison fishing) also contributed to coral reef destruction. These translate to lower recreational benefits for tourists who visit the area. The planned Nha Trang Port expansion could reduce these recreational benefits still further.

Pham and Tran (2001) saw an opportunity to contribute research to the policy dialogue on the proposed port expansion. They estimated the recreational benefits derived by both domestic and international visitors to the islands. The port expansion is one such development project, but there could be others. The generated values will serve as inputs in evaluating any development project that will impact the condition of the islands' coral reefs.

Using the zonal travel cost method (ZTCM), the researchers estimated the value of Hon Mun Islands' coral reef-based recreational benefits to be USD17.98 million. This figure includes the visitor consumer surplus (USD3.1 million) with the rest being revenue to transportation companies and service providers such as hotels, restaurants, tourist agencies, and other services patronized by tourists while in Nha Trang. The consumer surplus reflects the amount that visitors are willing to pay beyond what they have spent to enjoy the diversity of coral reefs and fish species, as well as the scenic beauty of the marine landscape.

ZTCM captures only the use value (recreational benefits) derived from coastal ecosystems, which also produce many other nonuse types of benefits. However, even considering only the use value of a given resource in evaluating a proposed marine development project could lead to a different decision. For instance, Pham and Tran (2001) compared the recreational benefits derived from Hon Mun Islands with the revenue projected for the planned port expansion, which is projected at USD3.1 million/year (equivalent to the consumer surplus portion of the recreational value). The comparison revealed that the foregone recreational benefit from the proposed port expansion – projected to be equivalent to a 20 % loss of recreational value – is greater than the port's projected annual revenue. The researchers thus recommended that the local government should revisit the port expansion proposal.[3]

Valuing Individual Attributes/Services of Coastal Ecosystems in Phang Nga Bay, Thailand

Phang Nga Bay is a large bay in the Andaman Sea and is part of the 400 km^2 Ao Phang Nga National Park. The bay's long coastline is found in Thailand's Phuket, Phang Nga, and Krabi provinces. Its dominant coastal ecosystems include mangroves, coral reefs, and sea grass beds. It is also habitat to numerous rare and endangered aquatic and bird species. Its islands of limestone mountains add to the bay's attractions. With this rich and diverse marine environment, the bay supports a thriving ecotourism industry. Its rich marine resources and wood products from mangroves also support the livelihood of local communities with.

Both the tourism industry and the harvesting of marine and other mangrove products contribute to the decline in the quality of the bay's marine coastal ecosystems. In particular, the infrastructure development that comes with a growing tourism industry has led to vegetation removal and land use changes. In addition, the growing number of tourists increases wastes that pollute the bay, and unregulated activities of some tourists have also reduced the quality of the coral reefs.

Seenprachawong (2002) examined the value of the bay's ecosystem services using choice modeling (CM). Specifically, four coastal ecosystem attributes/services were considered: (a) its diversity in flora and fauna, which relates to its recreation benefits; (b) its being a source of local livelihood, which relates to its

[3] It is not clear to what extent the decision-makers used the findings of the study. Port expansion pushed through but at a smaller scale than originally planned. It must also be mentioned that, around the time of the study, there were ongoing discussions about declaring the area a marine protected area (MPA): it was declared an MPA in January 2001.

provisioning benefits; (c) its flood protection function, an ecological function; and (d) its being a habitat for rare and endangered species. Three choice sets, representing the status quo condition and two alternative conditions, each having different combinations of the ecosystem attributes and payment level were presented to 300 Thai respondents (aged 18–75 years old and living in or traveling to Phang Nga Bay).

The results showed that among the four attributes, the most important is Phang Nga Bay's diversity in flora and fauna, which provides recreational and tourism benefits. The welfare estimate for this service was 1133 THB/person/year (USD28). The bay's provisioning and ecological functions were rated equally important with both attributes earning a welfare value of USD13/person/year. The bay's habitat function is perceived as the least important attribute with WTP for this ecosystem benefit estimated to be USD3/person/year. These four attributes translate to a value of 2263 THB/person/year (USD 57) to improve the Phang Nga Bay ecosystem. Aggregating across all visitors to the bay, the value of improving the condition of the bay's marine coastal ecosystems from the status quo level was estimated to be 5784 million THB/year (USD144.6 million). This number could be compared to the cost of implementing the proposed conservation program, but that is beyond the scope of the paper.

Seenprachawong (2002) discussed ways to capture a portion of the economic benefits from the bay's coastal marine ecosystems. He suggested that consumers could pay a user fee on activities that make use of marine resources such as snorkeling, boating, and diving operations. The fee could be added as part of the accommodation fee paid to service providers. Alternatively, or in addition to, the producers or service providers who earn revenue through the use of the marine ecosystem could be taxed. For instance, fishermen, tour operators, and water sport operators could be charged for their operations and those payments put toward maintaining the resource base that supports their livelihoods or businesses.

Protection of an Endangered Species at the Local Level: The Whale Shark

The whale shark (*Rhincodon typus*) is the world's largest fish. It is migratory and is declared a vulnerable species by the International Union for Conservation of Nature (IUCN 2002). The *Urban Times* in 2014 and *Time Magazine* in 2004 named Donsol, Sorsogon Province, Philippines, the best place for an encounter with the whale shark; Donsol is known as the "whale shark capital" of the world.

The local government of Donsol is aware of the need to protect whale sharks in the area. However, it does not have sufficient resources to support a more comprehensive conservation program. As such, Indab (2007) assessed whether the residents of the province of Sorsogon would be willing to contribute to fund such a program. The study asked people directly about their WTP, using the dichotomous contingent methodology.

The results of the study show that there is a high level of awareness and concern for environmental issues, including the protection of the whale sharks of Sorsogon. They are not, however, able or willing to pay for the implementation of a whale shark conservation program due to lack of money and the need to prioritize economic concerns over environmental issues. The study revealed a positive but very low MWTP value ranging from PHP17/household/month (USD0.34) to PHP35 (USD0.70).

The main conclusion from this study is that there is very limited capacity to pay for conservation programs in local communities of many developing countries like the Philippines, particularly in areas where income is low and where livelihood opportunities are limited, as is the case for Indab's (2007) study area. Financing conservation programs, specifically those that contribute to biodiversity conservation or endangered species protection, would have to rely on financing from other sources.

Protection of a Shared Endangered Species Through a Regional Conservation Program

Marine turtles are another migratory species under threat. They are harvested for a variety of uses (e.g., for their meat, shells, skin, and eggs or to serve as pets), but unregulated harvesting results in the loss of their ecological functions. They are considered "ecosystem engineers" as they consume sea grasses, sponges, and jelly fish and thereby return marine nutrients into the terrestrial ecosystems (Teelucksingh et al. 2010). They are indicators of the health of a marine ecosystem (Perrine 2003), and with four of the seven species of marine turtles declared critically endangered and two in the at risk category (IUCN 2002), the future of this endangered species and their habitat is bleak.

Efforts to protect marine turtles need to cross national borders as this species also migrates across countries. Southeast Asian (SEA) Nations, the Philippines, Cambodia, Thailand, Malaysia, Vietnam, Indonesia, and China need to work together to form a more coordinated marine turtle conservation program if efforts to protect this species are to succeed.

A group of SEA researchers (Nabanchang et al. 2008) worked together to carry out cross-border research using CVM to ascertain whether citizens of the Philippines, Vietnam, Thailand, and China would be willing to pay for a regional conservation program. The study sites were located in key cities in the four countries since they represent populations with a higher ability to pay for environmental goods (i.e., protection of endangered species). A regional conservation program was proposed in the scenario, consisting of improved monitoring capacity, investment in breeding facilities, and research, among other activities.

The results show that there is low willingness to support the proposed regional marine turtle conservation program. If a referendum were launched to find whether a majority of the citizens would vote to support such a program – which would require a monthly household contribution over a 5-year period – it would pass only at the lowest payment (bid) level of USD 0.02/month. Considering the responses for the five bid levels proposed in the study, the mean WTP/household/month was estimated at USD 1.41 for Thailand, USD 1.16 for China, USD 0.83 for Vietnam, and USD 0.17 for the Philippines. These numbers are really small but are comparable to the values derived by other studies on valuing endangered species conservation in Asia. It was noted that there is a high level of awareness about the plight of marine turtles and people agree that they should be protected. However, the need to address short-term economic concerns receives higher priority everywhere.

On the positive side, about a third of the 1250 respondents are willing to pay to support a regional marine turtle conservation program. Even if only this group of respondents could be convinced to support marine turtle conservation in the region, the potential revenue could be substantial. The challenge now lies in developing a fund-raising campaign that could mobilize resources from this group.[4] On this front, there is a need to improve the trustworthiness of government agencies responsible for endangered species conservation, to enlist the help of conservation NGOs, and to implement fund-raising strategies.

Conclusions: Ecosystem Values, Conservation Financing, and Valuation Research

The benefits derived from coastal marine ecosystem services are many and varied. The value of products harvested from this resource (e.g., wood, fish, and other marine resources) is huge and generates income and feeds a large coastal population in the region. The resource also supports various industries, for example, fisheries, shipping, and ecotourism – ecotourism is seen as the most aligned with ecosystem conservation and protection goals (UNEP 2008). Ecotourism provides local communities with alternative sources of income. Its development is thus seen as important to livelihood support. However, when coastal areas are declared MPAs, this source of livelihood may suffer at the outset (Gucu and Gucu 2003).

The papers included in this chapter showcase research that value ecosystem services, such as recreation benefits, existence value for biodiversity in coral reefs, ecological functions like flood control, and habitat functions for endangered marine species. All the valuation studies included in this book are policy driven – researchers were driven by the need to determine the potential to raise funds from consumers or beneficiaries of the ecosystem services, who in most cases are tourists. The premise was that if the benefits derived by consumers of the ecosystem

[4] The lead researcher was in fact invited by one hotel operator in Thailand to a gala dinner to raise funds to support marine turtle conservation. The funds raised that night were given to the national agency responsible for marine turtle protection.

services can be estimated, then a fund-raising campaign or appropriate economic instruments (e.g., user fee, entrance fee, surcharges, or taxes) can be devised to collect part of the ecosystem values. Likewise, if these values are available, decisions on whether to proceed with development projects in coastal areas can be made with sufficient information about the associated trade-offs in terms of loss of ecosystem services.

Listed below are the lessons derived from the seven EEPSEA-funded studies on coastal ecosystem valuation in SEA.

1. The region has rich marine ecosystems, and these support ecotourism in countries like Thailand, Vietnam, the Philippines, Malaysia, and Indonesia. Most of these coastal destinations are open access with minimal or no entrance fees charged. The rich biodiversity and open access to the resource encouraged rapid development of tourism that brings about problems related to pollution and degradation of the reef conditions. This situation calls for measures to conserve and protect the natural beauty and resources of the marine environment. There is currently limited funding available to undertake these measures.
2. Tourists are often willing to pay more than the current entrance fee on top of the travel cost incurred by visiting the site – a reflection that tourists value the benefits derived from these ecosystems beyond the actual costs. This excess amount constitutes a consumer surplus that could be collected on through rent capture to help support conservation activities. All of the papers that valuate tourism-related benefits (e.g., seascape, coral reefs, and biodiversity) pointed out the potential to raise funds for conservation activities from tourists.
3. Several papers suggested ways to capture part of the consumer surplus of tourists. One way is to increase entrance fees at sites where they are set at very low levels. Where it is difficult to impose such fees (e.g., when a marine park is huge with several entry points or without well-defined entrance locations), the user charge can be imposed on ecosystem-based activities currently marketed (e.g., snorkeling, diving, and boating). A fee could also be imposed as a surcharge on tourism operators like hotel owners, boat owners, or water sports operators, who, in turn, would pass this added cost to consumers. Such recommendations are useful to natural resource managers, who are in the position to respond to or implement fee payment schemes.
4. Researchers generally pass along study recommendations to relevant authorities (e.g., local governments and park managers), which is commendable. Some of the studies cited managed to bear fruit in terms of concrete actions. For example, Bradecina's (2014) study results were used to formulate a tourism development plan in Caramoan, Philippines. Likewise, Nabangchang et al.'s (2008) study led to a fund-raising activity for marine turtles in Thailand.
5. In general, however, efforts to create impactful outcomes appear limited. Part of the problem is that almost all the recreation-valuation papers do not include the evaluation of alternative economic instruments to capture rent. While they do discuss possible options, it should be noted that for coastal resource managers to take on their recommendations, the analysis should be extended to either a policy analysis of the implementation of economic instrument *X* or a

comparative policy analysis of the implementation of several economic instruments. Such a policy analysis should consider the political dimension, social acceptability, administrative feasibility, and costs of implementation of the proposed policy measure.

Perhaps researchers who are driven to investigate this policy issue should pursue such an analysis as an ongoing activity. This would assist natural resource managers in assembling and synthesizing all the factors that they need in future considerations on whether to recommend raising user fees or resource use charges.

6. The research study goals to estimate environmental values that could later be used to evaluate the viability of any proposed development project in coastal areas is indeed a good justification for conducting ecosystem valuation research. Often when decisions need to be made, natural resource managers do not have the luxury of time and resources to implement research to support decision-making. The usefulness of these studies may take time to be realized, but having these numbers readily available is no doubt a better scenario than assuming these numbers are zero simply because there are no values available to measure ecosystem services.
7. The two studies on endangered species conservation show that SEA nationals attach little importance to these species. This is probably not surprising for most areas where poverty is prevalent. However, even in urban cities, where people with higher income levels live, the willingness to support such conservation programs is also low. This implies that this good has to be provided by governments or agencies tasked with protection of global goods like biodiversity. It could also be a signal that fund-raising campaigns to support the plight of these endangered species need to be directed only toward those individuals with a strong affinity for endangered species protection; a study to identify such people or to determine ways to elevate the concern for endangered species could prove useful.

References

Bradecina RG (2014) Economic valuation of Caramoan, Camarines Sur beachscape in the Philippines: toward establishing an equitable payment for environmental services scheme for sustainable ecotourism. EEPSEA research report no 2014-RR3. Economy and Environment Program for Southeast Asia, Los Baños

Burke L, Selig E, Spalding M (2002) Reefs at risk in Southeast Asia. World Resources Institute, Washington, DC

Chua T (2009) Status of coral reefs: Singapore and beyond. 5 Feb 2009. http://wildshores.blogspot.com/2009/02/status-of-coral-reefs-singapore-and.html#.VIpX1tKUdgA. Retrieved 15 Jun 2015

Clark JR (1992) Integrated management of coastal zones. FAO Fisheries technical paper no 327. Food and Agriculture Organization, Rome

FAO (Food and Agriculture Organization) (2003) World fisheries and aquaculture atlas, 2nd edn. FAO, Rome

Fortes MD (2010) The seagrass-mangrove connection as a climate change mitigation and adaptation factor in East Asian Coasts. Paper presented at the international symposium on integrated coastal management for marine biodiversity in Asia, Kyoto, 14–15 Jan 2010

Garces LR, Pido MD, Pomeroy RS (2008) Fisheries in Southeast Asia: challenges and opportunities. In: Pandya A, Laipson E (eds) Transnational trends: Middle Eastern and Asian views. The Henry L. Stimson Center, Washington, DC

Gucu G, Gucu AC (2003) Is ecotourism an appropriate tool to ensure sustainable Mediterranean monk seal conservation in the Cilician Basin, Turkey? Evaluation report of the experimental eco-tourism application in Bozyazi – Mersin, Middle East Technical University Institute of Marine Sciences

Hoegh-Guldberg O (1999) Climate change, coral bleaching and the future of the world's coral reefs. Mar Freshw Res 50(8):839–866

Hoegh-Guldberg O, Hoegh-Guldberg H, Veron JEN, Green A, Gomez ED, Lough J, King M, Ambariyanto, Hansen L, Cinner J, Dews G, Russ G, Schuttenberg HZ, Peñaflor EL, Eakin CM, Christensen TRL, Abbey M, Areki F, Kosaka RA, Tewfi A, Oliver J (2009) The coral triangle and climate change: ecosystems, people and societies at risk. WWF-Australia, Brisbane

Indab AL (2007) Willingness to pay for whale shark conservation in Sorsogon, Philippines. EEPSEA research report. Economy and Environment Program for Southeast Asia, Singapore

IUCN (International Union for Conservation of Nature and Natural Resources) (2002) IUCN red list of threatened species. IUCN, Gland

MEA (Millennium Ecosystem Assessment) (2005) Millennium Ecosystem Assessment report. http://www.unep.org/maweb/en/About.aspx. Retrieved 14 June 2015

Nabanchang O, Jianjun J, Indab A, Thuy TD, Subade RF (2008) Mobilizing resources for marine turtle conservation in Asia a cross-country perspective. ASEAN Econ Bull 25(1). The environment, climate change and natural resources in Southeast Asia: issues and challenges: 60–69

PEMSEA (Partnerships in Environmental Management for the Seas of East Asia) (2003) Sustainability development strategy for the seas of South of East Asia. Regional implementation of the world summit on sustainable development requirements for the coasts and oceans. PEMSEA, Quezon City

Perrine D (2003) Sea turtles of the world. Voyageur Press, Stillwater

Pham KN, Tran VHS (2001) Analysis of the recreational value of the coral surrounded Hon Mun Islands in Vietnam. EEPSEA research report. Economy and Environment Program for Southeast Asia, Singapore

Seenprachawong U (2001) An economic analysis of coral reefs in the Andaman Sea of Thailand. EEPSEA research report. Economy and Environment Program for Southeast Asia, Singapore

Seenprachawong U (2002) An economic valuation of coastal ecosystems in Phang Nga Bay, Thailand. EEPSEA research report. Economy and Environment Program for Southeast Asia, Singapore

Short FT, Polidoro B, Livingstone SR, Carpenter KE, Bandeira S, Bujang JS, Hilconida P, Calumpong HP et al (2011) Extinction risk assessment of the world's seagrass species. Biol Conserv 144(7):1961–1971

Teelucksingh SS, Eckert S, Nunes PALD (2010) Marine turtles, ecosystem services and human welfare in the marine ecosystems of the Caribbean Sea: a discussion of key methodologies. Etude Caribeenes: Ressources Marine, no 15, pp 2–14

UNEP (United Nations Environment Programme) (2004) Seagrass in the South China Sea. United Nations Environment Programme, Nairobi

UNEP (United Nations Environment Programme) (2008) Vital water graphics – an overview of the state of the world's fresh and marine waters, 2nd edn. UNEP, Nairobi

Chapter 2
Economic Valuation of the Philippine's Caramoan Beachscape

Raul G. Bradecina

Abstract This study determined the tourists' willingness to pay (WTP) for the conservation of Caramoan beachscape in Camarines Sur, Philippines, as inputs in establishing the appropriate payment for environmental services for the sustainable development of ecotourism and coastal resource management of the area. The study employed the contingent valuation method (CVM) using four analytical models.

Based on the results, majority of the Caramoan tourists are younger, adventurous, without familial obligations, and highly educated and earn a relatively high income. One half of the visitors are willing to pay to promote the sustainability of the beachscapes. The visitors' WTP in the general model was estimated to be PHP 897. The average monthly visitors were estimated to be 1,000 tourists per month. Using these data, the total economic value of conserving the beachscapes in Caramoan was estimated to be PHP 10.76 million annually. Bid amount, age, and income were the only factors that consistently correlated with WTP in all of the analytical models. This indicates that younger visitors and those who have higher income are more likely to be willing to pay for conservation.

The estimated economic value of conserving the Caramoan beachscape justifies the relevance of investing public funds to pursue sustainable beachscape ecotourism development in Caramoan. This study is an attempt to contextualize PES for beachscape. It highlighted the priority concerns for sustainable source of fund for conservation and harmonized institutional arrangements for beachscape tourism and coastal resource management. It supported the potential of implementing PES within a community-based coastal resource management framework under a marine fishery reserve-sanctuary setting comanaged by the community and the Caramoan LGU or within a natural protected area framework managed by the Caramoan Natural Park Protected Area Management Board.

Keywords Beachscape valuation • Coastal resource management • Payment for environmental services • Caramoan

R.G. Bradecina (✉)
Partido State University, Goa, Camarines Sur, Philippines
e-mail: rgbradecina@yahoo.com

N. Olewiler et al. (eds.), *Marine and Coastal Ecosystem Valuation, Institutions, and Policy in Southeast Asia*, DOI 10.1007/978-981-10-0141-3_2

2.1 Introduction

The beachscapes of Caramoan are part of the Lagonoy Gulf coast, the largest fishing ground in the Philippines' Bicol Region. Like other coastal municipalities along the gulf, the municipality of Caramoan shares in the problems common to the gulf's coastal resource utilization and governance, such as overfishing, degraded coastal habitat due to illegal fishing, and weak enforcement of fishery laws (Pelea et al. 2005). The establishment of marine protected areas (MPAs)—used interchangeably with Marine Fishery Reserve-Sanctuary (MFRS)—as a tool for management has been seen to address some of these problems through comanagement with fisher folk and the local government units (LGUs) or municipal governments. As a community-based coastal resource management tool, the MFRS consists of a core no-take zone called a "sanctuary" and a buffer zone called a "reserve" where fishing using sustainable gear is allowed. Many of these MPAs and coastal resource management (CRM) activities in Lagonoy Gulf are financed from the LGUs' general fund and compete with various public service and investment priorities. The lack of sustainable funding and the dependency of CRM-oriented projects on the usually deficient LGU coffers have typically been credited for the failures of these conservation initiatives.

Several studies have recommended the development of non-extractive resource utilization in the Lagonoy Gulf area, such as ecotourism, as an alternative option to generate income for fisher folk and to fund conservation of the coastal resources (Pelea et al. 2005), but its potential has yet to be tapped by most LGUs.

According to Rosales (2003), the lack of proper economic valuation is one of the weaknesses in the management of coastal resources. If corrected, it could encourage more investment and better management and conservation because comprehensive economic valuation involves identifying the various types of resources used and attaching monetary values to each. Rosales (2003) further stressed that recreation—while highly undervalued—has been identified as one of the more important uses of beaches, particularly those located in MPAs.

Conservation fees (CFs), or user or entrance fees, offer a potential source of funds. CFs are charged for access to areas that offer natural attractions (e.g., beaches, dive sites, lakes, caves, and forests). CF schemes are based either on user-pay or cost-recovery principles (Padilla et al. 2005a).

The parameters of LGU collection of user fees from the recreational use of beachscapes are found in the Local Government Code of 1991. The code gives LGUs the power to generate and mobilize revenue through taxes, fees, or charges to enable efficient and effective governance. Moreover, the Philippine Fisheries Code of 1998 bestows upon the LGUs the primary mandate to manage municipal waters in the coastal zone that are not covered by the NIPAS Act,[1] specifically municipal waters that extend to a distance of up to 15 km from the shoreline. The NIPAS Act

[1] Republic Act 7586, otherwise known as the National Integrated Protected Areas System (NIPAS) Act, provides for the establishment and management of protected areas in the Philippines.

and Forestry Code[2] are other significant laws that provide for fee collection from the users of environmental goods and services.

The use of economic valuation in computing the fees is mandated in DENR Administrative Order (DAO) 2000–2051, *Guidelines and principles in determining fees for access to and sustainable use of resources in protected areas*. In setting fees, the guidelines adopt the willingness-to-pay (WTP) principle, in which fees are determined from "WTP estimates of the visitors based on appropriate surveys (Sect. 8.1.2)," and the cost-recovery principle, in that "collected revenues shall cover, as much as possible, a reasonable proportion of all costs incurred in protecting, maintaining and enhancing the natural attractions of the protected area (Sect. 8.1.1)."

The Caramoan LGU is fortunate to have been gifted with white sand beaches, limestone rock formations, islands, and bays. Four years ago, Caramoan was unknown in tourism communities. In 2008, the *Philippine Daily Inquirer* reported that the area had generated local and international media attention when the international series of the hit reality show "Survivor" was alternately filmed in the area by French, Israel, and Bulgarian television from 2008 to 2012; since then, local and foreign travelers have started to explore its beachscapes. In 2010, the Caramoan local tourism office recorded an average of 1,000 visitors per month. This development also showcased the potential of non-extractive resource use e.g., coastal ecotourism as a viable option in providing livelihood and in generating conservation fees for promoting sustainable tourism and CRM in Caramoan.

The recent shift in resource-use patterns in the coastal area—from mainly fishing to recreation tourism—has spawned resource-use conflict and displacement of some sectors that traditionally benefitted from the resource as a fishing ground. In addition, the open access nature of the Caramoan beaches, as in most common pool resources, will likely lead to the degradation of the coastal tourism assets in the long run if not properly managed. These developing issues underscore the need to estimate WTP values to establish a payment for environment services (PES) scheme, through means such as user fees, for conserving the Caramoan beachscapes. Revenue could fund CRM efforts, thereby ensuring the sustainable livelihood of stakeholders who may be disproportionately impacted by the shift in resource-use pattern in Caramoan. This study estimates for the conservation of Caramoan beachscape in Camarines Sur, Philippines, as input for establishing a PES system for sustainable ecotourism and coastal resource management in the area.

[2] Republic Act 7161, or the Revised Forestry Code of the Philippines, indicates forest charges on timber and other forest products.

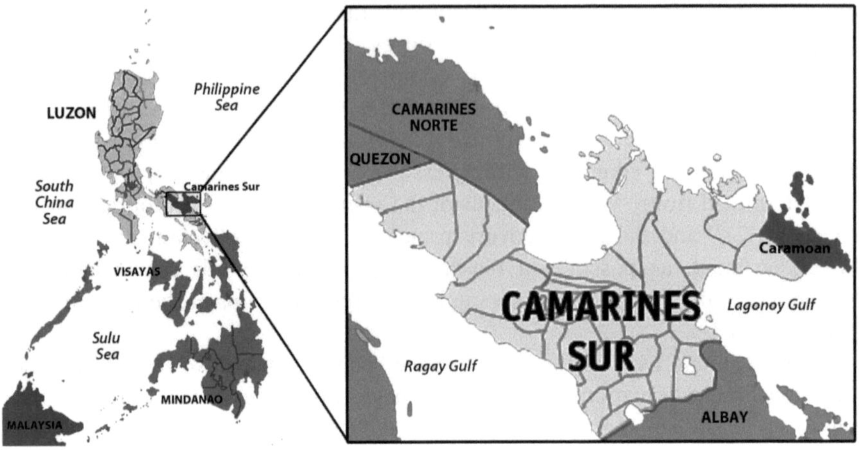

Fig. 2.1 Location of the study area in the Philippines

2.2 The Site

The small town of Caramoan is located at the tip of the Caramoan Peninsula, a rugged stretch of land extending into the waters of the Maqueda Channel on the north and east and Lagonoy Gulf on the south (Fig. 2.1). It covers approximately 27,741 km^2 with approximately 71 km of irregular coastline surrounded by the vast ocean, bay, seas, and swamps.

The municipality is home to 34.7 km^2 Caramoan National Park, declared via Proclamation No. 291 on 20 July 1938, and is currently a natural protected area. It is a hilly peninsula, with deep gorges and a rough, rocky terrain. The forests are mainly lowland in type. Habitats in Caramoan National Park include mangrove forests, sand dunes, and beaches. The natural features of this park include caves, limestone formations, white sandy beaches, a lake let, and a subterranean river. Its marine ecosystem is part of the Ticao-San Bernardino Strait-Lagonoy Gulf MPA whose marine waters are adjacent to Maqueda Bay, the mouth of the Lagonoy Gulf. This protected natural marine area ranks extremely high in terms of conservation priority.

2.3 Research Methods

2.3.1 Activities

The contingent valuation method was used in this study; several activities were done to achieve the objectives. First, exploratory visits and key informant interviews were conducted to gather information on the status of tourism management in the area as well as to seek affirmation from the municipal government officials. Second, a survey was prepared and pretested. The information collected included respondent perceptions on national and local governance of the coastal environment, attitudes toward conservation, willingness to pay (WTP) for conservation of the beachscapes, and socioeconomic status. The enumerators were trained through classroom lectures and hands-on experience in conducting the survey in the field. The questionnaire was pretested on 50 respondents using preliminary bid amounts of PHP100, PHP250, PHP500 (11.9USD), PHP750 (17.85USD), and PHP1,000 (23.80USD).

The questionnaire was finalized based on insights generated from the pretest. The WTP scenario was revised and a separate brochure was developed. The brochure contained concise information on the status of the beachscapes and coastal area, the threats to the beachscapes, a description of the worst-case scenario, current conservation initiatives implemented by the LGU to prevent the worst-case scenario from happening, and the hypothetical fee collection mechanism. In the bid elicitation part, final bid amounts of PHP100 (23.8 USD), PHP450 (10.71 USD), PHP950 (22.61 USD), PHP1,250 (29.76 USD), and PHP1,800 (42.85 USD) were adopted. Finally, the survey was conducted using the finalized questionnaire and WTP scenario brochure.

Two intercept strategies were adopted to engage with participants, namely, survey drop-off and face-to-face interviews. In the drop-off intercept strategy, the enumerators left the questionnaire and the WTP scenario brochure with the respondents after having explained them. On occasions when respondents requested more time to carefully answer the questionnaire, they arranged to have the completed questionnaires picked up the next day. The respondents in the study were local tourists who had stayed for at least 1 day in Caramoan and were employed. Sample respondents were randomly assigned bid amounts. A total of 1,000 tourists were interviewed between January and April 2010, representing the late (and lean) and the pre-peak tourist seasons; there was a 100 % return rate.

2.3.2 *Scenario Design*

The scenario was conveyed to the respondents by way of a colorful brochure showing the beachscapes of Caramoan, the current status of its attributes, and the projected changes to these attributes if ecotourism were left unmanaged. The brochure read as follows:

> The beachscape beauty of Caramoan is unique among the beaches of the Philippines because of the following features: high biodiversity of nearby Karst forest; fine white sand; beautiful beach sceneries; good coral cover teeming with diverse fishes; and private and secluded clear blue waters.
>
> However, these are threatened by: 1) excessive and unregulated fishing, 2) siltation from upland areas, 3) growing tourist population, and 4) domestic wastes.

In ten years, we believe that if these threats will not be addressed, Caramoan will suffer from a decline in biodiversity and reduced coral cover, filthy sands and degraded beach beauty, polluted waters, and crowded beaches.

The Caramoan Local Government has established Marine Fishery Reserve-Sanctuary areas, and is currently implementing solid waste management and an environmental awareness campaign.

To raise funds to implement other programs to protect and preserve the beauty of Caramoan beaches and its ecosystem, we are thinking of creating a sustainable source of revenue by establishing entrance fees. The fund will be managed by the Tourism Council and the Caramoan local government unit. The fund will go to coastal resource management, biodiversity conservation and fisher folk livelihood.

Considering the above information about the Caramoan beachscape beauty and ecosystem conservation and management fund to be set up, let us suppose that tourists will be asked to pay entrance fees /or environment services fees. The entrance fee will be collected by authorized personnel in a booth at Guijalo Port, the sole port of entry to Caramoan mainland.

2.3.3 Elicitation Format

The elicitation format chosen for this study was the dichotomous choice format, stated as follows:

> Let me take you back to the time that you were thinking of going to Caramoan. Would you have come if you knew that there was a PHP__ entrance fee and that this fee were to be managed by the Caramoan local government unit and the Tourism Management Council and used solely for the purpose of protecting the beauty and ecosystems of Caramoan beachscapes? Please be reminded that there is no right or wrong answer to this question.

2.3.4 Data Analysis

The analysis adopted the mean WTP formula from Hanemann (1984 as cited in Calderon et al. 2005; Launio et al. 2011) whose utility model assumes that if there exists a representative consumer who has an indirect utility function $V(P,M,Q,S)$, then the level of the consumer utility function depends on price (P), income (M), socio-characteristics (S), and quality (Q). When asked if he/she would pay to help protect the beachscape beauty in Caramoan at a given price (P), the respondent will say yes if:

$$V(M-P, Q^1, S) > V(M-0, Q^0, S) \tag{2.1}$$

Equation 2.1 shows that the respondent will answer yes if the utility derived from improved quality of the beachscape (Q^1) and paying the price (P) is higher than that derived from having no improvement in the beachscape quality (Q^0) and not paying the price ($P = 0$). If V (P,M,Q,S) is the observable component of the utility, the probability of the respondent saying yes is

$$\text{Prob(yes)} = Prob[V(M-P, Q^1, S) + \varepsilon_1 > V(M-0, Q^0, S) + \varepsilon_0] \qquad (2.2)$$

where ε_i is an unobservable component of the utility. Assuming that the random variable ε_i follows a logistic probability distribution, one can write:

$$Prob(yes) = \frac{1}{1 + e^{-\Delta V}} \qquad (2.3)$$

where $-\Delta \text{V} = V(M-P, Q^1, S) > V(M-0, Q^0, S)$

The recreational benefit of the hypothetical market to conserve beachscape is measured as WTP and is defined as

$$V(M - WTP, Q^1, S) > V(M-0, Q^0) \qquad (2.4)$$

Hanemann shows that if $V(M - P,Q,S)$ is linearly specified, then the probability of the respondent saying yes is

$$\text{Log}\left[\frac{Prob(yes)}{1 - Prob(yes)}\right] = \propto_0 - \beta_1 P + \beta_2 Q + \sum \beta_i S_i \qquad (2.5)$$

Parameters α_0 and β_i will be estimated parametrically. The mean maximum WTP for the conservation of beachscape can be calculated using Eq. 2.6:

$$\text{Mean maximum WTP} = \frac{1}{\beta_1}\left[ln\left(1 + e^{\alpha_0 + \beta_2 Q + \sum \beta_i S_i}\right]\right. \qquad (2.6)$$

Equations 2.5 and 2.6 were estimated using the routine of STATA®.

2.4 Results and Analysis

2.4.1 *Awareness of National Issues Related to Coastal Environment*

Among the national issues related to coastal environment, water pollution, degradation of coastal habitats (i.e., coral reefs, sea grass beds, and mangroves), and the *decline of biodiversity* are the top three of which the respondents are most aware (Table 2.1). These issues are commonly attributable to institutional weakness and

Table 2.1 Respondent's ranking of their awareness of national issues related to coastal environment

National issues	Frequencies by rank			Weighted mean	Rank
	1	2	3		
Water pollution	413	187	151	1.651	1
Degradation of critical coastal habitats	144	217	152	2.016	2
Decline of biodiversity	49	43	62	2.084	3
Typhoon	18	9	35	2.274	4
Deforestation	37	98	105	2.283	5
Garbage disposal	37	297	203	2.309	6
Overcrowding of settlements in coastal areas	83	116	234	2.349	7
Coastal erosion	13	31	50	2.394	8

Table 2.2 Respondents' ranking of their awareness of local coastal environment and environment governance issues in Caramoan

Local environmental and governance issues	Frequencies by rank			Weighted rank	Rank
	1	2	3		
Littering on the beach	456	187	153	1.619	1
Blast fishing	56	56	56	2.000	2
Sand quarrying	54	77	58	2.021	3
Unplanned, uncontrolled proliferation of houses, resorts, and infrastructures	147	286	243	2.142	4
Compressor fishing	6	20	12	2.158	5
Siltation	7	9	12	2.179	6
Overcrowding of tourists	36	74	86	2.255	7
Unregulated overfishing	34	86	96	2.287	8
Cutting of mangrove trees and vegetation	70	99	168	2.291	9
Small-scale mining	14	46	49	2.321	10
Cyanide fishing	0	57	63	2.525	11

the increasing pressures on the environment and natural resource base from uncontrolled population growth in the coastal areas.

2.4.2 *Awareness of Local Coastal Environment and Environment Governance Issues in Caramoan*

Among the coastal and environmental governance issues in the study area, littering on the beach, blast fishing, and sand quarrying are the top three issues of which respondents are most aware (Table 2.2). This reflects the area's weak solid waste management, as well as the presence of unsustainable economic activities, which exert pressure on the area's resource base.

2.4.3 *Institutional and Management Issues Impacting Caramoan Beachscapes*

Among the institutional and management issues impacting Caramoan beachscapes, the respondents consider *sustainable source of fund for conservation*, *political issues*, and *harmonized institutional arrangement for beachscape tourism management* as the top three issues that will impact the long-term sustainability of Caramoan beachscapes (Table 2.3). These issues highlight the importance of determining the value of protecting the beachscapes to visitors as key sources of information to justify public investments in their management. The information would also be useful in determining the level of rent that could be captured from visitors, should the local government impose one.

A concern was also expressed on the complexity of managing the Caramoan beachscapes since it falls within the jurisdiction of two governing institutions, the local government and the protected area management board (PAMB), which is coordinated by the provincial government with the Department of Environment and Natural Resources (DENR). The park portion of the Caramoan beachscape is under the jurisdiction of PAMB, while the portion that falls outside of the park is under the control of the Caramoan Municipal Government, in accordance with the Local Government Code (LGC). The same code declares coastal waters within 15 km of the municipality as municipal waters and places the responsibility of managing these resources in the hands of the municipal government.

Table 2.3 Respondents' rankings of institutional and management issues impacting the ecological sustainability of the Caramoan beachscape

	Frequencies by rank				
Issues	1	2	3	Weighted rank	Rank
Sustainable source of fund for conservation and protection	254	102	1	1.696	1
Political issues	119	44	54	1.700	2
Harmonized institutional arrangement and mechanisms for beachscape tourism management	118	110	107	1.967	3
Enforcement of environment- and fisheries-related laws	97	164	94	1,992	4
Geographic and political jurisdictions between local and provincial governments	56	105	68	2.052	5
Comprehensive management plan for ecotourism and conservation of beachscape	163	194	192	2.053	6
Technical capability for management	78	62	102	2.099	7
Awareness of conservation	59	138	154	2.271	8
Institutional frameworks for ecosystem governance	42	77	107	2.288	9

Table 2.4 Respondents' attitude toward Caramoan beachscape conservation

	Attitude toward beachscape conservation			
	Agree		Disagree	
Bid amount	N	%	n	%
All	981	98	19	2
100	192	96	2	1
450	194	97	6	3
950	198	99	2	1
1,250	197	99	3	2
1,800	200	100	0	0

Table 2.5 Respondents' willingness to pay for Caramoan beachscape conservation

	Willingness to pay for conservation			
	Yes		No	
Bid Amount	N	%	n	%
100	172	86.0	28	14.0
450	116	58.0	84	42.0
950	79	39.5	121	60.5
1,250	64	32.0	136	68.0
1,800	62	31.0	138	69.0

2.4.4 *Attitude Toward Caramoan Beachscape Conservation*

Almost all (98 %) respondents support the protection of the Caramoan beachscape (Table 2.4) independent of the bid level assigned to them.

2.4.5 *Respondents' Willingness to Pay for Caramoan Beachscape Conservation*

The number of respondents willing to pay for Caramoan beachscape conservation decreases as the bid level increased, as seen in Table 2.5. Any fee higher than PHP450 had a majority of the respondents no longer supporting the collection amount. It is safe to say that a fee around PHP 500 would be acceptable to a majority of tourists.

2.4.6 *Reasons for Willingness and Unwillingness to Pay for Caramoan Beachscape Conservation*

Among the reasons cited by respondents for positive responses to the hypothetical payment plan for beachscape beauty conservation, the highest percentage went to

Table 2.6 Reasons why respondents are willing or unwilling to pay

Reasons	Freq.	%
Willing to pay		
It will promote the sustainable use of the beachscape	183	37
It will provide a steady source of fund for ecotourism management	95	19
It will lessen the threats to coral reefs, mangrove, and sea grass beds	64	13
It will maintain the cleanliness of fine white sand	55	11
It will promote equitable sharing of benefits	34	7
It will provide livelihood	31	6
It will sustain the good quality of water	19	4
It will improve law enforcement	13	3
Unwilling to pay		
I do not believe that the money I will pay will be used for the conservation of beachscape beauty	196	39
I cannot afford to pay for the program	129	25
Only fishers and resort owners who directly benefit from its beaches should pay	83	16
Majority of the poor will be affected	62	12
Only the rich should pay for this	25	5
I prefer to give money to humanitarian causes instead	6	1
I do not think conservation of beachscape beauty is necessary	5	1

wanting to improve the sustainability of the beachscape. This is followed by the following reasons: to provide a steady source of funds for ecotourism management (19 %) and to reduce the threats to coral reefs, mangrove, and sea grass beds (13 %). Improving law enforcement received the lowest support (3 %) for reasons respondents are willing to pay into a beachscape conservation program.

On the other hand, distrust or refusal to believe that the money paid would actually be used for beachscape conservation was the most frequently cited reason for unwillingness to pay (39 %). This is followed by respondents reporting an inability to afford the fees under the hypothetical conservation payment scheme (25 %) and the belief that only fishers and resort owners, who directly benefit from the beachscape, should pay (16 %; Table 2.6).

2.4.7 Parametric Logistic Regressions

The coefficients and probability values of the factors affecting WTP resulting from the logistic regressions of the CV model using the data set are presented in Table 2.7. Results showed that the bid amount (*BID*), age (*AGE*), income (*INCOME*), and attitude toward conservation (*ATTITUDE*) were significantly correlated with WTP. Consistent with economic theories, the bid amount was negatively correlated to WTP, while the variables on age, income, and attitude were found positively correlated with WTP.

Table 2.7 Significant variables affecting willingness to pay

Variables	Model with adjustment for Certainty
CONSTANT	−0.0814624
BID	−0.0016287
	(0.000)**
GENDER	0.0461732
	(0.754)
AGE	−0.0433655
	(0.000)**
CIVIL STATUS	0.16748
	(0.922)
EDUCATION	0.0673712
	(0.276)
INCOME	7.90e-07
	(0.007)**
ATTITUDE toward conservation	1.212579
	(0.025)**
No. of observations	975
LR chi-square	192.84
Prob> chi-square	0.000
Pseudo R square	0.1484

**Significant at p(z) 5 % level. Values in parenthesis are p-values

The negative vectors of BID and GENDER and the positive vector of INCOME are consistent with the study hypothesis. The negative and significant correlation of bid amount to the WTP implies that the probability of the respondents' willingness to pay for Caramoan beachscape conservation decreases as the bid amount increases. The negative and significant relationship of age to WTP indicates that the younger respondents are more likely to be willing to pay for beachscape conservation and protection in Caramoan. This is consistent with the a priori expectation that since environmental conservation is an investment in the future, only those who stand to benefit from returns on their investment in the future—depending on how long they expect to enjoy these benefits—will be willing to pay. Because younger people are likely to outlive older people and are more likely to see and enjoy the fruits of protecting these assets over the long term, they are more likely to invest. Gratification of one's need for future recreational enjoyment serves as the sole driver for the hypothetical willingness to pay for the conservation of Caramoan beachscapes and its ecosystems.

The positive and significant relationship of income to WTP indicates that the respondents with higher incomes are more willing to pay for beachscape beauty conservation in Caramoan. High-income respondents put premiums on environmental improvements compared with their lower-income counterparts, whose limited income is a major constraint in valuing natural resources and environmental services (Seenprachawong 2001).

The positive and significant relationship of attitude to WTP indicates that respondents with higher appreciation for environment conservation are more willing to pay for it. The absence of a causal relationship between attitude and WTP in the protest and uncertainty plus protest models strengthens this observation.

There was no significant relation between WTP and the variables of gender, civil status, and education. This suggests that these factors do not impact respondent acceptance or refusal to pay for beachscape conservation.

The parametric mean WTP derived in this study (PHP563) is comparable to values generated elsewhere: PHP552 for Anilao, Batangas, by Padilla et al. (2005b) and PHP543 for the Tubbataha seascape by Subade (2005).

2.5 Recommendations

The visitor imputation of economic value of the Caramoan beachscape beauty provides the strong justification for the investment of public funds for its conservation. The economic benefits of conserving beachscapes represented by the visitors' aggregate WTP reflect the extent of the value of the beachscape beauty at risk of being lost in the long run if conservation efforts are inadequate. This can help drive the Caramoan municipal government to set up a payment scheme to collect such fees and allocate a larger annual budget for the management and conservation of beachscapes and coastal resources.

Most of the local environmental issues known to the visitors are associated with weak enforcement of laws brought about by the relevant institutions' scarcity of funds and the poor economic condition of a larger proportion of the fishing community. In Caramoan, as in the rest of coastal municipalities bordering Lagonoy Gulf, the lack of employment opportunities and the open access nature of coastal resources spell poverty in the coastal fishery. This situation forces people to discount future benefits from coastal resources, resulting in resource degradation. To stop this trend of coastal resource degradation, there is a need to provide alternative livelihood activities for the community and to invest in coastal environment rehabilitation efforts; these are key challenges to the Caramoan LGU to promote sustainable coastal ecotourism.

Also, the three top ranking institutional and management issues perceived by respondents as having impact on the long-term sustainability of Caramoan beachscapes (i.e., not having sustainable sources of funding for conservation, political issues, and being without harmonized institutional arrangements for beachscape tourism management) highlight the need for sustainable funding, transparency, and sustainable resource governance in the area. The lack of monetary incentives to key personnel of coastal resource management institutions (e.g., *bantay dagat*, MFRS management body) and resource users to enforce laws and adopt sustainable resource utilization practices, respectively, compromise effective coastal law enforcement and resource management. The establishment of PES or its form (e.g., user fees) as a scheme to finance conservation and development sustainably is an opportunity for the Caramoan LGU to incentivize the positive behavior of resource management personnel and resource users.

The Local Government Code and the Fisheries Code of the Philippines, which provide LGUs with the authority to impose taxes, and fees for the use of the resources and environmental assets within the municipal waters, augur well toward exploring the potential of PES in the future. The current impetus for determining the viability of granting Territorial Use Rights to fishing communities in the use of common pool resources in coastal areas also provides a ray of hope.

References

Calderon M, Camacho LD, Carandang,MG, Dizon JT, Rebugio LL, Tolentino, NL (2005) A water user fee for households in Metro Manila. EEPSEA research report. Economy and Environment Program for Southeast Asia. Singapore

Hanemann WM (1984) Welfare evaluations in contingent valuation experiments with discrete response data: comment. Am J Agric Econ 71:1053–1056

Launio CC, Shinbo T, Morooka Y (2011) Island villagers' willingness to work or pay for sustainability of marine fishery reserve: case of San Miguel island, Philippines. Coast Manag 39(5):459–477

Padilla JF, Ansula AD, Tolosa MO (2005a) Getting users to pay for conservation: a guide to site-based sustainable user fee schemes. World Wildlife Fund, Philippines

Padilla JF, Tongson EE, Lasco RD (2005b) PES: sustainable financing for conservation and development. Proceedings from the national conference-workshop on payments for environmental services: direct incentives for biodiversity conservation and poverty alleviation. Manila, 1–2 Mar 2005

Pelea NR, Borbe SG, Pelea MJ (2005) Socioeconomic status of fisherfolk in Lagonoy gulf. In: Soliman VS, Pelea NR, Dioneda RR Sr (ed) Lagonoy gulf post-resource and socio-economic assessment (terminal report). Report Submitted to the Fishery Resource Management Project, Department of Agriculture-Bureau of Fisheries and Aquatic Resources. Bicol Small Business Institute Foundation, Inc., Legaspi City. 1–97

Rosales RMP (2003) A survey to estimate the recreational value of selected MPAs: Moalboal-Cebu, Siquijor and Pamilacan Island-Bohol. Marine ProtectE Areas Project Coastal Conservation and Education Foundation, Inc. (CCE Foundation, Inc.), Cebu

Seenprachawong S (2001) An economic analysis of coral reefs in the Andaman Sea of Thailand. EEPSEA research report. Economy and Environment Program for Southeast Asia. Singapore

Subade RF (2005) Valuing biodiversity conservation in a world heritage site: citizens' non-use values for Tubbataha reefs National Marine Park. EEPSEA resource report. Economy and Environment Program for Southeast Asia. Singapore

Chapter 3
An Economic Analysis of Coral Reefs in the Andaman Sea of Thailand

Udomsak Seenprachawong

Abstract This study valued the benefits of coral reefs in the Andaman Sea in a proposed project development area known as the Southern Seaboard Development Project. The specific area of interest was the Phi Phi Islands, which is a highly valued group of islands both in terms of its use values and non-use values. Accordingly, this project calculated the recreational values of coral reefs in the Andaman Sea of Thailand. The specific objectives were (1) to estimate the recreational benefits of Phi Phi Islands, (2) to estimate consumer willingness to pay for improved coral reef quality at Phi Phi Islands and (3) to use these values to determine an appropriate entrance fee for visiting the reef sites at Phi Phi Islands. This study employed both travel cost method and contingent valuation method to generate estimates of the reef value at Phi Phi. The consumer surplus associated with visits to Phi Phi represents an annual value of THB 8216.4 million (USD 205.41 million).

Keywords Economic analysis • Coral reef • Consumer surplus • Entrance fee • Thailand

3.1 Introduction

The southern coast of Thailand is of great significance, as it marks the boundary between two of the region's major oceans: to the east, the Gulf of Thailand, contained within the Pacific Ocean, and to the west, the Andaman Sea connecting to the Indian Ocean. Because of this strategic location, Thailand's seas are regarded as one of the country's major food sources, supported by rich coral reef resources. The Andaman Sea coast also has mangrove forest distributed along the coastal belt, with a major portion concentrated in Phang Nga coast.

The country has plans to implement a massive project to spur economic development in the area. This is known as the Southern Seaboard Development Project (SSDP). The SSDP will include an area along the west coast at the Andaman Sea

U. Seenprachawong (✉)
School of Development Economics, National Institute of Development Administration, Bangkok, Thailand
e-mail: udomsak.s@nida.ac.th

N. Olewiler et al. (eds.), *Marine and Coastal Ecosystem Valuation, Institutions, and Policy in Southeast Asia*, DOI 10.1007/978-981-10-0141-3_3

coast in Krabi and Surat Thani provinces and the Gulf of Thailand in Nakhon Si Thammarat province. The project will entail building an east-west toll highway and the construction of two commercial port developments. These ports will be built at the Tublamu in Phang Nga province, which is home to pristine coral reefs found around the Surin and Similan Islands.

This research intends to value the benefits of coral reefs in the Andaman Sea in the proposed project development area. Preservation of the unique marine ecosystem in the Andaman Sea requires maintaining the coral reefs intact. The specific area of interest is the Phi Phi islands. Phi Phi Islands are highly valuable, both in terms of use values (e.g. recreation, tourism, educational opportunities and scientific research) and non-use values (e.g. genetic resources and future uses of ecological functions, both known and unknown). It is a well-known site with relatively easy access and is popular among tourists.

This study will discuss how results from Phi Phi may be transferred to Similan Island, or to other coral reef sites being considered as sites for development projects, such as a thermal power plant.

The general objective of this project is to calculate the recreational values of coral reefs in the Andaman Sea of Thailand. The specific objectives are:

1. To estimate the recreational benefits (consumer surplus) of Phi Phi Islands
2. To estimate consumer willingness to pay (WTP) for improved coral reef quality at Phi Phi Islands
3. To use these values to determine an appropriate entrance fee for visiting the reef sites at Phi Phi Islands

3.2 The Study Site

3.2.1 Description of the Phi Phi Islands

Over the past few years, Ko (Islands) Phi Phi has grown from a peaceful little Muslim fishing village to one of the busiest tourist destinations in the country. Located about 45 km east of Phuket, the group of Phi Phi Islands is in Krabi province and is composed of the Ko Phi Phi Don (Hilly Island), Ko Phi Phi Lae (Phi Phi in the Sea), Ko Yung (Mosquito Island) and Ko Mai Pai (Bamboo Island).

Phi Phi Islands are fringed with hard-coral gardens, which are home to a wide assortment of brilliant tropical creatures, including five species of sea anemones. In most areas, the coral growth and marine life are prolific and most of the fish species around the Similan Islands can also be found around Ko Phi Phi. The beauty of Phi Phi Islands has made them world famous. Hundreds of visitors charter boats to visit these from either Krabi or Phuket provinces every day to go snorkelling around the archipelago's coral reefs or climbing the karst limestone.

3.2.2 *Visitors and Facilities of the Islands*

Visitors have to travel first to either Phuket or Krabi and then take a boat to get to the Islands. Most visitors choose to take a public boat from the pier at Phuket or Krabi. The transit cost for a return trip from Phuket to the Islands is about 400 THB (100 USD). Currently, there is a 20 THB entrance fee to visit the Islands. In addition, any tourist agency can arrange day trips. For example, the cost of travelling from Phuket Town to the Islands ranges between 500 and 1200 THB, including lunch and bus transportation to the boat pier. Generally, the cost depends on how luxurious a boat the visitor chooses, as well as whether snorkelling or viewing the coral reefs is included. Once on Phi Phi, visitors can rent a private boat for 1000 THB a day to explore the Islands.

3.3 Methodology

This study employed both travel cost method (TCM) and contingent valuation method to generate estimates of the reef value at Phi Phi. It used the TCM to estimate the consumer surplus for both domestic and international tourists to Phi Phi Islands. Travel cost data was collected from visitors who had had a recreational experience of the coral reefs (such as diving, snorkelling and fishing). In addition, the CVM study was used to estimate the non-use values (option, existence and bequest values) of coral reefs at Phi Phi from domestic non-users.

Travel from a visitor's home could be by public transport or private car. Costs incurred by private means are based on the amount of fuel consumed in making a return trip from the visitor's home to Phuket or Krabi. The cost of fuel per trip is divided by 4 (the seating capacity is assumed to be 4) to derive the cost per visitor using private transport means. Costs incurred by using public transport are twofold: firstly, the fare from the individual traveller's home area to the bus terminal (or the airport) in Phuket or Krabi and then that from the bus terminal (or the airport) to the boat pier.

3.3.1 *Travel Cost Method*

The TCM measures the demand function for visits to a site. A demand function is an empirical relationship between the price of a goods item and the quantity purchased:

$$Q = \mathrm{f}(P, X) \tag{3.1}$$

where Q is the quantity purchased, P is the price and X represents a number of demographic variables which might shift the demand function, such as income or age. Economic theory suggests that demand curves slope downward ($dQ/dP < 0$), meaning as its price rises, people purchase less of an item. The travel cost demand function is a specific application of this general tool to recreational trips. It describes how many times people purchase trips depending on the price of each trip. As with market goods, theory predicts that the higher the price of a trip, the less often people tend to visit; and so the travel cost demand function should likewise be downward sloping.

The TCM uses a survey technique based on interviews whereby visitors of recreational sites are invited to provide information on their trips (such as the cost, length, purpose of trip and other sites visited) and on other demographic characteristics (including income, age and gender). The fundamental principle that drives this model is that if a consumer wants to use the recreational services of a site, he has to visit it. The travel cost to reach the site is considered to be the implicit or surrogate price of the visit, and changes in the travel cost will cause a variation in the quantity of visits. Determination of these variations across individuals will contribute to the estimation of demand functions and the value of the site.

Refining the TCM, this study uses the individual travel cost method (ITCM); the demand curve in this model relates an individual's annual visits to the costs of those visits. That is:

$$V_i = \mathrm{f}(P_i, X_i) \tag{3.2}$$

where V_i = the number of visits made per year by individual i, P_i = the visit cost faced by individual i and X_i = all other factors determining individual i's visits (income, age and other demographic characteristics). The total cost is the sum of expenditures made on fuel, opportunity cost of time for travelling and for visits on-site by all individuals.

The functional form relating the dependent variable (visits per year) and independent variables (travel cost and socio-economic variables) is either linear or a double log. To obtain a more accurate demand curve, this study uses the double-log demand function:

$$V_i = e^{\alpha_0 + \sum_{c=1}^{l} \alpha_c D_{ci} + \varepsilon_i} \bullet \prod_{j=1}^{k} X_{ji}^{\beta_j} \bullet P_i^{\beta_P} \tag{3.3}$$

V_i = number of visits of individual i
D_{ci} = dummy variables referring to individual i
X_{ji} = socio-economic features of individual i and other variables referring to i
P_i = price paid by individual i; P_i is integration variable
$i = 1,\ldots,n$ index of observations
$c = 1,\ldots,l$ index of additive dummy variables
$j = 1,\ldots,k$ index of socio-economic variables

α_o = constant
α_c = coefficients of the additive dummy variables
β_j = coefficients of socio-economic variables
β_p = coefficient of the price variable
ε_i = error term

Once estimated, the model is expressed in the following form:

$$V = e^{\alpha_0 + \sum_{c=1}^{l} \alpha_c D_c} \bullet \prod_{j=1}^{k} X_j^{\beta_j} \bullet P^b \tag{3.4}$$

For each single individual, the consumer surplus (CS) is the integral of the demand function V with respect to the price p, between the lower bound p_{li} and the choke price, or the upper bound p_{ui}. The choke price is the price that leads to a demand equal to zero. The indefinite integral of the demand function is:

$$\int vdp = e^{\alpha_0 + \sum_{c=1}^{l} \alpha_c D_c} \bullet \prod_{j=1}^{k} X_j^{\beta_j} \bullet \frac{p^{b+1}}{b+1} \tag{3.5}$$

The integral between p_l and p_u is:

$$CS = \frac{e^{\alpha_0 + \sum_{c=1}^{l} \alpha_c D_c}}{b+1} \bullet \prod_{j=1}^{k} X_j^{\beta_j} \bullet \left(p_u^{b+1} - p_l^{b+1}\right) \tag{3.6}$$

For each individual, consumer surplus is computed by plugging the values for each individual dummy variable D_{ci} (the travel cost p_{li}, the choke price p_{ui} and the value of the explanatory variables X_{ji}) into the general formula in Eq. 3.6:

$$CS_i = \frac{e^{\alpha_0 + \sum_{c=1}^{l} \alpha_c D_{ci}}}{b+1} \bullet \prod_{j=1}^{k} X_j^{\beta_{ji}} \bullet \left(p_{ui}^{b+1} - p_{li}^{b+1}\right) \tag{3.7}$$

The annual consumer surplus per individual can be computed by summing up the consumer surplus estimates from all observed consumers (N) and dividing by N:

$$CS \text{ per individual} = \frac{1}{N} \sum_{i=1}^{N} CS_i \tag{3.8}$$

The annual consumer surplus per visit is calculated by dividing the annual consumer surplus per individual by the annual sample average number of visits:

$$CS \text{ per visit} = \frac{CS \text{ per individual}}{\text{Sample average visits per year}} \tag{3.9}$$

The *CS* per visit is then multiplied by the total number of visitors to Phi Phi during the year to obtain the annual total benefit of Phi Phi:

$$\text{Total benefit}(TB) = CS \text{ per visit} \times \text{Total visitors} \tag{3.10}$$

Loss of the site usually means loss of all future recreational opportunities, not just the current annual value. The entire future stream of annual recreational values must therefore be included. Because they happen in the future, economic theory suggests this stream of benefits be discounted to make them comparable with the present. Assuming that the annual value of recreation is constant over time, the present value of the stream of future benefits can be calculated simply by the following formula:

$$PV = \sum_{t=1}^{T} \frac{TB}{(1+r)^t} \tag{3.11}$$

where $r =$ discounting rate

3.3.2 Contingent Valuation Method (CVM)

The CVM is a technique that allows the value of environmental goods and services to be estimated by asking people directly, usually by means of a survey, their willingness to pay (WTP) for a change in the availability of such environmental goods and services. The individual maximum WTP for an environmental change is assumed to be the value the individual attaches to such a change. The major advantage of this approach, compared with the reveal preference methods, is that the CVM can elicit both use and non-use values.

The elicitation format chosen in this study is the dichotomous choice format. This means that respondents were asked whether or not they were willing (yes/no answer) to pay a predetermined contribution to the trust fund (PRICE) to restore coral reefs at Phi Phi Islands. The PRICE[1] was randomly assigned among respondents so as to generate price variation. The price range used in this study was based on a pretest survey carried out in the open-ended elicitation format.

[1] The values for each subsample are 50, 100, 300, 500, 700, 1000, 1200, 1500, 1800 and 2000 THB for domestic respondents; USD1, 3, 8, 15, 20, 25, 30, 40, 45 and 50 for international respondents.

3.3.2.1 Econometric Specification

Hanemann (1984) shows that if there exists a representative consumer who has an indirect utility function $V(P, M, Q, S)$, then the level of the consumer depends on price (P), income (M), socio-characteristics (S) and the quality (Q). The respondent is asked if he would pay to help restore the coral reefs at Phi Phi Islands at the given price, P. The respondent will say yes if

$$V(M - P, Q^1, S) > V(M - 0, Q^0, S) \tag{3.12}$$

Equation 3.11 shows that the respondent will answer yes if his/her utility, derived from improved reef quality (Q^1) and the price paid (P), is higher than not having improved reef quality (Q^0) and not paying ($P = 0$). If $V(P, M, Q, S)$ is the observable component of the utility, the probability of the respondent saying yes is:

$$\text{Prob(yes)} = \text{Prob}\left[V(M - P, Q^1, S) + \varepsilon_1 > V(M - 0, Q^0, S) + \varepsilon_0\right] \tag{3.13}$$

where ε_i is an unobservable component of the utility. Assuming that the random variable ε_i follows a logistic probability distribution, one can write:

$$\text{Prob(yes)} = \frac{1}{1 + e^{-\Delta V}} \tag{3.14}$$

where $-\Delta \text{V} = V(M - P, Q^1, S) > V(M - 0, Q^0, S)$

The recreational benefit of the hypothetical market to improve the coral reefs at Phi Phi Islands is measured as WTP and is defined as

$$V(M - WTP, Q^1, S) > V(M - 0, Q^0, S) \tag{3.15}$$

Hanemann shows that if $V(M - P, Q, S)$ is linearly specified, then the probability of the respondent saying yes is

$$\text{Log}\left[\frac{\text{Prob}(yes)}{1 - \text{Prob}(yes)}\right] = \alpha_0 - \beta_1 P + \beta_2 Q + \sum_i \beta_i S_i \tag{3.16}$$

Parameters α_0 and β_i will be estimated parametrically. The mean maximum WTP for coral reef restoration can be calculated using formula in Eq. 3.16:

$$\text{Mean maximum WTP} = \frac{1}{\beta_1}\left[\ln\left(1 + e^{\alpha_0 + \beta_2 Q + \sum \beta_i S_i}\right)\right] \tag{3.17}$$

The data set used in this paper was collected through questionnaires to acquire information on travel cost or to elicit the amount of WTP. The questionnaire had three parts: travel cost data, socio-economic data and WTP questions.

3.3.2.2 Scenario Design

The hypothetical market is stated as follows:

> There are many ways of measuring marine biodiversity. One simple indicator is coral abundance, in terms of area covered. An abundance of zero (ABU = 0) would mean that all of the coral has disappeared. An abundance of 100 (ABU = 100%) means that the reef is in its natural pristine state. The reef at Phi Phi Islands, according to the best scientific evidence, is about one-quarter degraded: at ABU = 75%. If we 'do nothing', scientists estimate that it will fall to a value of ABU = 60% in about 20 years. Current ongoing management of the area will maintain the level of biodiversity at a stable level of abundance. This level corresponds to a 75% abundance on the index just explained. A trust fund will be established to help the ecological restoration of Phi Phi's coral reef system if contributions are adequate. The trust fund will be held by the Phi Phi Islands Committee for exclusive use on projects to increase the biodiversity at Phi Phi Islands from the current 75% ABU to a 100% ABU. Examples of the projects proposed include treatment of sewage to a high standard, a new drainage system for storm water, planting mangroves and coastal plants to reduce the impacts of run-off, and establishing monitoring of fish, plant life and mangroves.

3.3.2.3 Elicitation Format

The elicitation format chosen in this study was the dichotomous choice format. This means that respondents were asked whether or not they were willing to pay a predetermined contribution amount to the trust fund:

> Would you be willing to pay USD__________ per year for the next 5 years to a trust fund to help restore the coral reefs of Phi Phi Islands from their current level of 75% ABU to 100% ABU?

3.4 Empirical Results

3.4.1 Estimation of Benefits Based on the Travel Cost Method

Table 3.1 reports domestic visitors' total travel costs and average visits by region, and Table 3.2 reports those of international visitors. In both tables, total travel costs consist of three components: cost of transport, time cost and food, lodging and other expenses. The tables indicate that transportation costs increase with distance, however note that the value of travel time used to calculate the price of a trip is somewhat controversial. Some (for example, Pearce et al. 1989) have suggested using a full hourly wage as a measure of the value of recreational travel time. Yet, other estimates approximate the value to be one-third of the full wage rate or even

Table 3.1 Total travel costs and visits by region (domestic visitors)

	North ($n = 14$)	Northeast ($n = 6$)	Central ($n = 194$)	South ($n = 416$)	Total ($n = 630$)
Round-trip transportation costs (THB)	4475	3491	3344	898	1755
Time costs (full wage rate) (THB)	2325	1496	1528	899	1130
Time costs (1/3 wage rate) (THB)	767	494	504	297	373
Other costs (food, lodging) (THB)	1681	408	1206	1439	1363
Total costs (full wage rate) (THB)	8482	5396	6080	3323	4246
Total costs (1/3 wage rate) (THB)	6924	4394	5056	2631	3490
Visits	1.07	1.00	1.15	2.34	1.93

Table 3.2 Total travel costs and visits by region (international visitors)

	America ($n = 20$)	Europe ($n = 75$)	Australia ($n = 7$)	Asia-Pacific ($n = 26$)	Total ($n = 128$)
Round-trip transportation costs (USD)	106	189	218	131	166
Time costs (full wage rate) (USD)	748	770	489	498	696
Time costs (1/3 wage rate) (USD)	247	254	162	164	230
Other costs (food, lodging) (USD)	56	72	47	78	69
Total costs (full wage rate) (USD)	911	1031	755	707	931
Total costs (1/3 wage rate) (USD)	409	515	427	373	465
Visits	1.2	1.1	1	1.8	1.2

10 % of the wage (Farber 1988). Both the full wage rate and one-third wage rate have been presented in the tables below.

Table 3.3 gives the definition of each variable used in the travel cost modelling and in the calculation of the consumer surplus, while Table 3.4 summarises the results of the ITCM estimation of the double-log demand function.

The demand for recreation was estimated using a double-log functional form; the natural logarithm of the number of visits (LNVISIT) to the site over a period of 12 months is related to selected model explanatory variables. These results are shown in Table 3.4.

The annual consumer surplus (CS) with the double-log demand function was calculated using Eq. 3.7. The upper limit of integration (the choke price) was set at

Table 3.3 Definition of variables used in travel cost modelling and consumer surplus calculations

Variable name	Definition
Again	=1 if the respondent will be back to Phi Phi again
Age	The respondent's age
HHNUM	The respondent's household size
INC	The respondent's income
Male	=1 if the respondent is male
Member	=1 if the respondent is a member of an environmental organisation
Numed	The respondent's number of years at school
Osite	=1 if the respondent visits other site(s) than Phi Phi
Own	=1 if the respondent owns a house
Single	=1 if the respondent is single
TC1	TRAV + time cost (full wage rate) + other costs (food, lodging, others)
TC2	TRAV + time cost (1/3 wage rate) + other costs (food, lodging, others)
TRAV	Round-trip transportation costs
Visit	Number of visits per year

Table 3.4 ITCM re-estimation of double-log demand function parameters

Model	Coefficient	Std. error	T-ratios
Dependent variable: LNVISIT			
Domestic: $n=630$, $R^2=0.207$, $F=55.727$			
(Constant)	2.799	0.349	8.01
Member	0.201	0.100	2.01
LNNUMED	−0.219	0.126	−1.73
LNTRAV	−0.266	0.072	−12.15
Dependent variable: LNVISIT			
International: $n=128$, $R^2=0.119$, $F=4.16$			
(Constant)	−0.836	0.477	−1.753
Male	0.132	0.064	2.057
Single	0.146	0.085	1.723
LNAGE	0.334	0.125	2.665
LNTRAV	−0.076	0.031	−2.499

99 % of the round-trip transportation costs, which equals 7353 THB (USD183.82) for domestic visitors and 80,400 THB (USD 2010) for international visitors. An annual CS per individual was computed using Eq. 3.8 at 6568.86 THB (USD164.22) for domestic visitors and 75,920 THB (USD1898) for international visitors. These numbers were then divided by the average number of visits per individual, which equals 1.93 for domestic visitors and 1.27 for international visitors. This results in an annual CS per visit of 3403.55 THB (USD85) for domestic visitors and 59,760 THB (USD1494) for international visitors.

The total benefits from the recreational services of Phi Phi are estimated to be about 69.90 million THB (USD1.75 million) a year for domestic visitors and 8146.4

Table 3.5 Distribution of non-users' responses by bid amount

Price (THB)	Pay	Not pay	Total
50	15	5	20
100	15	5	20
300	17	3	20
500	10	10	20
700	4	16	20
1000	3	17	20
1200	2	18	20
1500	3	17	20
1800	2	18	20
2000	2	18	20

million THB (USD203.66 million) a year for international visitors[2]; added together, the total benefit of Phi Phi (in terms of recreational value) being 8216.4 million THB (USD205.41 million) a year. Therefore, the value of Phi Phi is around 249,720 THB (USD6243) per hectare per year.[3] Assuming the real value of this recreational value of 8216.4 million THB (USD205.41 million) per year remains the same over 30 years, and using a real interest rate of 5 %, the present value of recreational benefits from Phi Phi is 126,280 million THB (USD3157 million).

3.4.2 *Estimation of Non-use Value of Coral Reef Restoration*

The study used the CVM to estimate the non-use value of coral reefs at Phi Phi to domestic vicarious users (urban Thai population in the labour force). Table 3.5 shows the distribution of 200 respondents to the bid levels used. The results of the logit estimation of the linear utility model by maximum likelihood are reported in Table 3.6. To calculate the mean maximum WTP, a new model was estimated by dropping all insignificant variables. The econometric results of the new model are reported in Table 3.7 with the mean maximum WTP calculated using the formula in Eq. 3.16 that is:

$$\begin{aligned}\text{Mean maximum WTP} &= \frac{1}{0.0022}\left[\ln\left(1+e^{-1.54+0.18(15.2)}\right)\right]\\ &= 634\ \text{THB (USD 15.85) per person}\end{aligned}$$

Multiplying the mean WTP of 634 THB (USD 15.85) per person by the total labour force in Thailand of 31.3 million yields the non-use value of Phi Phi's coral reefs of

[2] Based on the 1998 figures of total domestic and international visitors to Phi Phi of 20,540 and 136,277 visitors, respectively

[3] The reef area at Phi Phi is estimated to be 32,900 ha.

Table 3.6 Parameter estimates of the logit model by maximum likelihood

Model	Coefficient	T-ratios
(Constant)	0.16	0.083
Inc	0.000014	0.664
Price[a]	0.0022	−6.549
Age[a]	−0.0651	−1.859
Male	0.554	1.328
Single	−0.266	−0.578
Numed[a]	0.185	1.654

[a]Significant at 0.10

Table 3.7 Re-estimation of parameter estimates of the logit model by maximum likelihood

Variable	Coefficient	T-ratios
(Constant)	−1.54	−1.00
Numed[a]	0.18	1.80
Price[a]	−0.0022	−6.66

[a]Significant at 0.10

19,840 million THB (USD 496 million) a year. Therefore, the benefit values (use and non-use) of coral reefs at Phi Phi were estimated to be 19,895 million THB (USD 497.38 million) a year, averaging 604,720 THB (USD 15,118) per hectare per year.

The travel cost method reveals that domestic and international recreation alone represents an annual value of 8216.4 million THB (USD 205.41 million). Given that the Park comprises 32,900 ha, its recreational value averages about 249,720 THB (USD 6243) per hectare per year. The contingent valuation method indicates that together the use and non-use values of Phi Phi's coral reefs represent an annual value of 19,895 million THB (USD 497.38 million), averaging about 604,720 THB (USD 15,118) per hectare per year. These figures were used to calculate coral reef values at other sites at risk, which are shown in Table 3.8. Nine marine national parks in Thailand include significant reef areas. Most of the parks containing reefs are in the Andaman Sea, with only three sites in the Gulf of Thailand. Together with the Fisheries Protected Areas, approximately 60 % of Thailand's coral reefs are located within a protected area.

3.5 Conclusions and Policy Recommendations

The focus of this study is to valuate the coral reefs and develop ways this information could be used to improve planning for coral reef management in Thailand. The site analysed, Phi Phi Islands, is rich in reef systems. Phi Phi is representative of many coastal areas in Thailand that have potentially rich coral reefs in need of improved management so that economic and other benefits might be restored and

Table 3.8 Economic values of coral reefs in marine national parks of Thailand

Marine national park	Sea area (ha)	Region	Recreational values million THB/year (USD million/year)	Total values million THB/year (USD million/year)
Tarutao	126,000	Andaman	31,466.8 (786.67)	76,194 (1904.85)
Mu Ko Ang Thong	8400	West Gulf	2083.2 (52.08)	5040 (126.00)
Mu Ko Surin	10,205	Andaman	2530.8 (63.27)	6122.8 (153.07)
Hat Nai Yang	8000	Andaman	1984 (49.60)	4800 (120.00)
Khao Laem Ya-Mu Ko Samet	12,000	East Gulf	2976 (74.4)	7200 (180.00)
Mu Ko Similan	9300	Andaman	2306.4 (57.66)	5580 (139.50)
Mu Ko Chang	4480	East Gulf	1110.8 (27.77)	2688 (67.20)
Mu Ko Phi Phi	32,900	Andaman	8216.4 (205.41)	19,895 (497.38)
Mu Ko Lanta	10,850	Andaman	2690.8 (67.27)	6510 (162.75)

Reef areas protected in marine national parks are taken from ONEB (1991)

enhanced. Phi Phi generates large ecotourism benefits. The consumer surplus associated with visits to Phi Phi represents an annual value of 8216.4 million THB (USD 205.41 million). Local and national levels of government in Thailand can justify larger annual budget allocations for the management of coastal resources.

The data from Phi Phi indicate that it may be rapidly approaching a point at which increased dive tourism results in measurable degradation of the marine environment. The beauty of Phi Phi Islands has made them world famous. A number of private tour operators have been running boat tours to the islands for tourists, for whom snorkelling and fish feeding have become popular activities. This increased use of the area has resulted in demands for improved facilities to serve the tourists. The consensus with regard to Phi Phi is that the rapid growth in tourist numbers, together with the infrastructure established to service their needs, have produced a severe reduction in the quality of the islands' ecosystem, particularly in the heavily used shallow sea areas.

Phi Phi, however, remains a popular place for ecotourism but to continue doing so would require developing a management regime for marine tourism to manage pressure on the ecosystem resources.

One strategy is to use economic instruments to modify human behaviour. An example of this type of strategy is the use of higher entry fees to facilities during peak-use times in an attempt to spread visitor density. Alternatively, permits could be auctioned to commercial tourist operators to restrict their numbers. Another example is use of a regulation combined with an economic disincentive, such as imposing fines for littering, taking undersized fish, or other inappropriate behaviour. Discounts on access fees to a marine park could be provided if groups undertook a clean-up project or assisted with research during their visits. Given the increasing financial pressure under which many of these public management agencies find themselves, taking the opportunity to utilise economic instruments to raise funds and accomplish management objectives may be worthwhile.

When using economic instruments to capture the net benefit values of Phi Phi, there are two general options to consider: to target the producers or the consumers directly. If the tourists (consumers) are to be charged, instruments could be applied to those activities that physically use the environment, such as offshore water sports (specifically, snorkelling boats and dive operations), swimming and beach activities. However, such activities are dispersed among many individual operators providing different services and are geographically spread throughout the Islands. The most obvious complementary service utilised by all tourists at Phi Phi is the accommodation sector. Therefore, a charge levied through the use of accommodation would effectively target this consumer group and at the same time facilitate the administration and enforcement of the charge.

The second option is to charge producers (namely fishers and water-sport operators) an annual user fee or resource use charge. However, there are notable problems with setting the fee at an appropriate amount and enforcing the use of the resource—for example, ensuring that only those licenced are the exclusive users and monitoring to ensure that licensee usage does not rise above specified or reported levels (Huber et al. 1998). Without an effective ability to control access to the resource, licenced users would be reluctant to pay the associated fee if their exclusive rights to the resource could not be upheld.

Finally, physical approaches (technical interventions) for mitigating the ‘trampling effects’ that have been successful elsewhere could be implemented at Phi Phi; physical structures have successfully been used in a number of situations to control tourist access while at sea. A typical example is to set up mooring buoys for vessels to reduce anchor damage to coral reefs. In addition, regulations could restrict the type of vessel permitted in an area, such as by allowing only electrically powered boats in an area sensitive to noise disturbance. Thus, the negative impacts of tourist activities can be mitigated by a combination of regulatory and physical approaches.

This study utilised the CVM to estimate also the utility values associated with coral reef biodiversity at Phi Phi. The 400 domestic survey respondents were asked whether they would contribute toward a trust fund that would be managed by the Phi Phi Islands Committee to increase biodiversity. The payment would be made on a per-annum basis over 5 years and would lead to a 25 % increase in coral reef cover. At the sample means, a consumer’s WTP toward increasing biodiversity was estimated as 287 THB (USD 7.18) per visit. Based on this figure, the results of this study suggest a basic entrance fee of 40 THB (USD1) per person per visit to Phi Phi, which is twice as much as the current rate of 20 THB (USD 0.50) per person. It would also seem reasonable for the Park to impose charges for tourists visiting certain special and environmentally vulnerable recreational sites. Both of these measures could help relieve the negative pressure on the delicate marine environment.

References

Farber S (1988) The value of coastal wetland for recreation: an application of travel cost and contingent valuation methodologies. J Environ Manage 26:299–312, Academic Press

Hanemann M (1984) Welfare evaluations in contingent valuation experiments with discrete responses. Am J Agric Econ 66:332–341

Huber R, Ruitenbeek HJ, Seroa Da Motta R (1998) Market-based instruments for environmental policymaking in Latin America and the Caribbean: lessons from eleven countries. The World Bank, Washington, DC

ONEB (Office of the National Environment Board of Thailand) (1991) A national coral reef strategy for Thailand, Vol. 1, Statement of need, Thailand Coastal Resources Management Project, Office of the National Environment Board, The University of Rhode Island and USAID

Pearce DW, Markandya A, Barde JP (1989) Environmental policy benefits: monetary valuation. Organisation for Economic Co-operation and Development, Paris

Chapter 4
Conservation Versus Development: Valuation of Coral Reefs Questions Port Expansion Plan in Vietnam

Pham Khanh Nam and Tran Vo Hung Son

Abstract Coral reefs in Nha Trang City have been threatened by human activities such as use of dynamite, illegal harvesting, shipping, and pollution from the nearby port development project. How much recreational benefit from coral reefs is lost if these activities are not held in check? Using travel cost method, this study estimated the recreational value of coral reefs in Nha Trang to be VND 259.8 billion (USD18.3 million) annually, of which Hon Mun's consumer surplus was estimated to be VND 45.4 billion (USD3.2 million). If this coral reef benefits were to be reduced by 20 % due to increased pollution created by the nearby expanded port, the resulting decrease in the recreational value would be more than the projected annual revenue of the port. The proposed port expansion plan needs to be seriously reconsidered.

Keywords Coral reefs • Recreational benefit • Travel cost method • Vietnam

4.1 Introduction

Making trade-offs between development and biodiversity conservation is challenging for policy-makers. It requires explicit evidences on losses, gains, costs, and benefits. This study estimates the most important benefit of Hon Mun Islands' coral reefs in Nha Trang City, recreational values, and challenges the Nha Trang Port expansion plan.

Coral reefs that dwell in the shallow depths of coastal water are essential for diverse economic activities of local population, as well as related coastal ecosystems. Coral reefs provide habitat, spawning, and nursery grounds for fish species; prevent coastline erosion; protect coastline from storm; provide the beauty for tourism; and support marine biodiversity.

P.K. Nam (✉) • T.V.H. Son
School of Economics, University of Economics Ho Chi Minh City, Ho Chi Minh City, Vietnam
e-mail: khanhnam@ueh.edu.vn; sontran@ueh.edu.vn

N. Olewiler et al. (eds.), *Marine and Coastal Ecosystem Valuation, Institutions, and Policy in Southeast Asia*, DOI 10.1007/978-981-10-0141-3_4

However, coral reefs in Vietnam have been facing increasing anthropological threats. Many coral reef ecosystems along the coasts have been overexploited, and marine biodiversity has decreased dramatically. Public recreational marine areas, such as the Ha Long Bay, the Son Tra Peninsula of Da Nang Province, or the Hon Mun Islands of the city of Nha Trang, have contributed significantly to the economy, but these areas have been polluted and overexploited by development projects, illegal harvesting, and overfishing.

Nha Trang City is situated 450 km from Ho Chi Minh City and 1280 km from Hanoi (GSO 1998). Nha Trang, with its attractive marine features like coral reefs and bird nests, is one of the most important tourism sites in Vietnam. With an airport and a seaport, Nha Trang is also strategically located along National Route 1 and the railway route linking the North and the South.

The Hon Mun Islands are located in the south of Nha Trang Bay. The islands support a variety of habitats and ecosystems, including fringing coral reefs, mangroves, and sea grasses, with an adjacent deepwater upwelling that supports the local fishing industry. The Vietnamese government designated the Hon Mun Islands as a pilot marine protected area (MPA) (Vo 1998). The aim is to maintain biodiversity, protect coral reefs, improve fisheries, control pollution, manage tourism, and create new jobs for local people hired to manage the MPA.

In recent years of increasing economic development, the marine environment of Nha Trang City, especially in the Hon Mun Islands, has faced increased exploitation. Coral reefs have been destroyed by many factors, mainly human activities. Destructive exploitation by shipping, usage of dynamite, coral harvesting, and marine tourism has decreased marine biodiversity and precious genetic resources, such as hawksbill sea turtles, false killer whales, and leatherback sea turtles from the South China Sea. These losses in turn diminish the benefits reaped from tourism in the islands. The question is, "How much recreational benefit is lost if these activities are not held in check?" There is a plan to expand Nha Trang Port. If this plan becomes a reality, the quality of water in the Hon Mun area will deteriorate due to the increase in port traffic, affecting marine ecosystems and recreational activities. Policy-makers will have to choose between recreational activities and marine biodiversity or the port. So far, there has been no decision made by the government. The port expansion proposal is facing opposition especially from the Department of Science, Technology, and Environment (DOSTE 1996).

The potential threat posed by the port expansion project necessitates estimation of the recreational value of the islands, so that decision-makers can compare this value with that projected for the proposed port expansion. The estimated recreational value of the islands could be used to assess the economic impact of expansion of the port and to devise future recreational development plans for the islands. Policy-makers would need to know the advantages tourism has over other activities (e.g., fishing and collecting bird nest) in order to decide how to allocate resources among the competing uses.

4.2 Objectives of the Study

The overall objective of the research is to estimate recreational value of the Hon Mun Islands and suggest whether policy-makers should expand the Nha Trang Port. Specifically, the objectives are as follows:

1. To determine the factors that affect recreational demand for the Hon Mun Islands
2. To estimate the annual recreational value of the islands using the zonal travel cost model
3. To compare the estimated recreational value of the islands with revenue of the proposed port expansion

4.3 Methodology

4.3.1 Valuation Methodology: The Zonal Travel Cost Model (ZTCM)

The Hon Mun area is a public site with neither an admission fee nor a fee to use the site's resources for fishing, aquaculture, and recreation. As such, it is impossible to use market prices directly to value the site. Instead, this research project applies the zonal travel cost model (ZTCM) to estimate the recreational value of the publicly managed Hon Mun Islands in Nha Trang.

The ZTCM is characterized by the limited number of visits made by a tourist in a specific time period (e.g., in a year), coming from a variety of points of departure. Tourists who travel to Hon Mun area come from various places in Vietnam, and one-third of all visitors are foreign travelers who frequently visit a specific site less often. The individual travel cost model (ITCM) was not used as it is more suitable for urban parks where local residents could make frequent visit for recreational purposes (Georgiou et al. 1997).[1]

According to the Khanh Hoa Department of Tourism, most tourists visit the Hon Mun Islands one to three times each year. The Hon Mun Islands are about 8 km from the port. It is difficult even for visitors living in Nha Trang City to frequently spend a holiday at the islands because their willingness to travel by boat to the islands depends very much on the weather and tour expenses. This is different from the case of a park or a lake. In the case of a park like Khao Yai National Park (DeShazo 1997) or various city parks, local residents visit the park several times a week for recreation, so it is possible to use the ITCM to estimate the recreational value in such cases.

[1] Georgiou et al. (1997, p. 38) wrote that "...(ITCM) requires that there is variation in the number of trips individuals make to the recreational site in order to estimate the demand function." However, this drawback of the ITCM is not a problem for the ZTCM, which uses the number of trips per capita from each zone as a function of the travel cost.

4.3.2 *Zonal Travel Cost Model (ZTCM)*

The ZTCM divides the area surrounding the site into ten zones, and the unit of observation is the zone. The number of visits per capita from each zone is a function of the travel cost. The ten zones fan out with increasing distance from Hon Mun. The first zone, and closest zone, is Nha Trang and the last, and farthest, is Hanoi (The number of zones used can be quite large). In a zone, the inhabitants must have similar preferences, and each zone can contain one administrative area or a group of several administrative areas. Table 4.1 shows the zoning structure.

As in the Khanh Hoa Tourism Report (So Du Lich Khanh Hoa 1998), the origins of foreign visitors are divided into two regions, namely, Asia and Oceania (Australia and New Zealand) and North America and Europe. Visitation rates were calculated for both of these regions.

The trip generating function for the zonal model is

$$V_i = V(C_i, POP_i, S_i) \tag{4.1}$$

where

V_i represents visits from zone i to the Hon Mun Islands
POP_i represents the population of zone i and
S_i represents socioeconomic variables such as the average income for each zone.

In this research, the dependent variable is expressed in terms of the visitation rate (V_i/POP_i). The linear and semilog specification for the demand function is common in the economic literature on demand for recreation (Garrod and Willis 1999; Ward and Beal 2000; Mwebaze and MacLeod 2013). Given the demand function for visits to the islands, consumer surplus and recreational value can be estimated; then consumer surplus is calculated by the integral calculus of the demand function

Table 4.1 Zones of origin (GSO 1998 with a population average growth rate of 1.65 %)

Zone	Distance (km)	Administrative district	Population
1	5	Nha Trang	341,000
2	33.3	Dien Khanh, Ninh Hoa, Cam Ranh, and Van Ninh	647,700
3	110	Phan Rang and Tuy Hoa	350,200
4	217	Da Lat and Buon Ma Thuot	786,200
5	250	Phan Thiet and Binh Dinh	545,900
6	441	HCMC	5,155,700
7	497	Long An, Tay Ninh, Vung Tau, and Dong Nai	925,600
8	516	Da Nang and Hue	1,112,600
		Quang Nam and Quang Ngai	
9	677	An Giang, Can Tho, Ca Mau, and Tien Giang	1,456,000
10	1140	Hanoi, Haiphong, Nam Dinh, Thanh Hoa, and Nghe An	5,050,500

When this table was established, the data on population from the General Statistical Office was only available for 1999. The data in this table was estimated for the year 2000 based on the average growth rate of the population with respect to the travel cost between the price paid and the "choke price." (The "choke price" is the price at which demand is "choked off" or zero.)

One issue in the calculation of travel cost is the multiple destination and multiple purpose trips. Tourists generally visit not only the Hon Mun Islands but also various places in Nha Trang City and the neighboring areas. Although coral is the unique characteristic of the Hon Mun Islands, a few tourists make a trip from their home to Hon Mun only for the benefit of admiring coral unless they live in Nha Trang City. This argument is supported by the fact that tourism is still a luxury commodity in Vietnam and that no foreign tourists come to Vietnam to visit only one site unless their journey is for some special purpose, for example, meetings or research. Information collected in the questionnaire covered their travel expenditure for the whole trip to Nha Trang, not exclusively the Hon Mun Islands. In this study, the travel costs for only Hon Mun have been elicited basing on two factors: (1) the respondent's satisfaction with the Hon Mun Islands in comparison with other recreational sites in Nha Trang and (2) the time the respondent spent on the Hon Mun Islands out of the total time spent in Nha Trang.

The problem was accentuated for international tourists: Foreigners do not visit just Nha Trang, but they also travel to various other sites in Vietnam such as Da Lat, Hoi An, Hue, Ha Long Bay, Hanoi, Sa Pa, and Mai Chau—Hon Mun is just a short stopover. In sites that attract an insignificant number of international tourists, the tourism value derived from them may typically be omitted. However, for sites like Hon Mun where foreign visitors make up about a third of the total, inclusion of their behavior is compulsory. In this study, travel costs to both domestic visitors and foreign visitors to the Hon Mun Islands were extracted and included in the estimation calculations. It should be noted, however, that it was not possible to accurately isolate the travel costs for Hon Mun; only rough estimates, those sufficient for the purposes of this study, were extracted.

The information on transportation costs from the questionnaire covered the cost of a visitor's whole trip, not just the trip to the Hon Mun Islands. In order to estimate the recreational value of the islands, the costs for traveling to the islands were identified, calculated, and extracted from the total costs of the trip.

4.3.3 Data Collection Techniques

4.3.3.1 Collection of Secondary Data

1. Demographic information, such as population and income, was obtained mainly from statistics by the General Statistical Office.
2. The Institute of Oceanography in Nha Trang and the Department of Science, Technology and Environment in Khanh Hoa provided information on environmental aspects.
3. The Department of Tourism in Khanh Hoa provided information on tourism activities.

4. The feasibility study on the port expansion and other related information were obtained from the Nha Trang Port Authority.

4.3.3.2 Collection of Primary Data

Primary data were collected on visitor experience and socioeconomic characteristics. Specifically, the questionnaire was designed to collect information on:

1. On-site and off-site recreational behavior
2. Travel experiences and trip costs
3. Socioeconomic factors

4.3.3.3 Sampling

In the survey, systematic sampling was employed.[2] Because survey data from nonresidents of Nha Trang City could not be obtained, this survey only concentrated on the users' group. Individual visitors were chosen as respondents for the interviews. A "visitor" was defined as one who used the Hon Mun Islands for recreation. Clearly, villagers who lived within the range of the islands were not included in the survey. Samples were taken using two approaches. The first approach was by directly interviewing visitors to the islands. The interviewer was required to speak to every given number of visitors encountered (e.g., the interviewer would interview every fifth or sixth visitor.) The second approach was by handing the questionnaire to visitors on boat trips and getting them to complete the forms. A pretest survey was conducted to test the validity of the questions and their relevance to the planned analysis.

The number of samples was 390, of which 180 are domestic visitors and 210 are foreign visitors. The research population covered the urban population of Vietnam because most Vietnamese tourists are people from urban areas. Vietnamese in rural areas are too poor to afford the luxury of touring.

4.3.4 Characteristics of the Study Area: Resources and Threats

The Hon Mun Islands are a group of small islands, namely, Hon Mot, Hon Tam, Hon Mieu, Hon Mun, and part of Hon Tre, that lie about 8 km from the Nha Trang Port.

[2] Scheaffer et al. (1996) states that, "A systematic sample is generally spread more uniformly over the entire population and thus may provide more information about the population than an amount of data contained in a simple random sample."

The Hon Mun Islands support a very lucrative edible bird nest colony. According to the Vietnam Biodiversity Action Plan (Tran 1998), the area has the highest level of marine biodiversity in Vietnam. The Institute of Oceanography (1998) reports Nha Trang as having the second highest rating for marine biodiversity in the region, with only slightly less diversity (65 genera) than the Indo-Pacific center of diversity (70 genera).

Besides coral, the islands contain other rich resources in the form of fisheries and bird nests. Fishery is one of the main sources of livelihood for the villagers of Hon Mun Islands. Tran (1998) estimated that the value of fishery in the Hon Mun Islands was around USD 6,123,200 (VND 73.5 billion) in 1997. Tran (1998) also estimated that the average individual earnings from lobster culture were around USD 52 per month (VND 625,000/per month). According to Vo (1998), fishing anchovy and cuttlefish and catching ornamental fishes are the main activities around the islands which provide many benefits to the villagers.

The Hon Mun Islands have a very valuable, but scarce, resource: bird nests (or "yen sao"), which are in the eight caves found on Hon Mun and Hon Noc Islands. According to the Khanh Hoa Statistics Agency, the production of bird nests on these two islands is about 100–120 kg/year. With a market price of USD 2000/kg (VND 28.4 million/kg), the bird nest business of the Hon Mun Islands brings approximately USD 200,000–240,000/year (VND 2.84–3.40 billion) to the region.

Finally, the area is of considerable value to research and monitoring, as it contains high genetic diversity due to a combination of being home to various reef types and being close to the edge of the continental shelf and upwelling. The National Institute of Oceanography of Nha Trang has conducted significant research into the biodiversity, biology, and ecology of the area's living coastal resources, aquaculture and restoration, biochemistry, hydrochemistry, and marine physics and geology. The Institute of Oceanography is also part of the Global Coral Reef Monitoring Network, and the area is an important area for field research for the Nha Trang University of Fishery (located within 10 km of Nha Trang Port).

The most immediate source of threat to the area comes from shipping activities in Nha Trang Port—about 3 km from the nearest point of the islands—which could directly affect fishing operations and tourism in the Hon Mun Islands. Nha Trang Port is presently the most important seaport of Khanh Hoa Province and processes 640,000 metric tons of goods and 18,000 passengers annually. Its activities have a definite bearing on the management of the Hon Mun Islands.

4.4 The Tourism Industry in Nha Thrang

The number of tourists to Nha Trang City, as well as the revenue from tourism, has increased considerably over the recent years; Table 4.2 presents the revenue and number of tourists from 1994 to 2000.

The number of visits to Nha Trang City in 2000 was about 397,000, of which 70 % visited the islands for recreation. Recreational activities at the islands include

Table 4.2 Revenue from and number of tourists to Nha Trang City (1994–2000) (So Du Lich Khanh Hoa 1997, 1998, 1999 and 2000)

	Unit	1994	1995	1996	1997	1998	1999	2000
Revenue	VND million (USD million)	60,661 (5.8)	85,110 (7.7)	115,000 (10.0)	115,200 (9.6)	147,700 (10.9)	150,500 (10.8)	197,200 (13.9)
Total tourists	Person	260,000	317,000	390,000	315,500	331,400	370,000	397,000
Foreign tourists	Person	73,500	91,500	109,000	105,000	99,600	130,000	118,700

snorkeling, scuba diving, boating, jet skiing, sunbathing, swimming, and visiting fishing villages.

Coral reefs and ornamental fishes are features peculiar to the Hon Mun Islands compared to other recreational sites, but they only attract about 10 % of the visitors (those who scuba dive), and other water sports like boating, sailing, and jet skiing are not very popular either.

4.4.1 Socioeconomic Characteristics of Tourists

The average visitor visited the Hon Mun Islands 1.7 times over 1 year. Over 50 % of domestic visitors visited the Hon Mun Islands for the first time. This can have two implications: (1) holidaying in Vietnam is not an annual activity, and/or (2) the Hon Mun Islands are not that attractive to the Vietnamese. Many visitors visited Hon Mun only once in the analyzed year, even local residents.

The average income of visitors is around VND 1,300,000/month (USD 92), which is higher than the national average level. The educational level of the visitors averages 13.7 school years, which is higher than the national average. The age structure showed that most visitors fell into the working age group, with the average age of 32.2 years, and 70 % of the respondents are male. The detailed information is available in Table 4.3.

Table 4.4 showed that the average number of visits by foreigners is 1.17/year—lower than that of domestic visitors. Foreign tourists have to pay a large amount of money to visit Hon Mun, so it is reasonable to expect that the frequency of their visits in any given year would be less than that of domestic visitors.

Table 4.3 Statistical data on the socioeconomic characteristics of Vietnamese visitors to the Hon Mun Islands ($n = 180$)

Characteristics	Mean	Standard deviation	Median	Minimum	Maximum
Number of visits	1.7	1.19	1.00	1.00	5.00
Distance (km)	401	345	385	5	1140
Travel time (days)	4.35	3.41	4.00	1.00	30.00
Group (persons)	15.00	25.14	8.00	1.00	160.00
Income (VND)	1,325,556	683,739	1,200,000	300,000	3,000,000
Age (years)	32.2	10.02	30.00	11.00	60.00
Education (schooling years)	13.68	2.57	14.00	5.00	18.00
Sex (male = 1; female = 0)	0.69	0.46	1.00	0.00	1.00
Marital status (married = 1; not married = 0)	0.51	0.50	0.00	0.00	1.00

Table 4.4 Statistical data on the socioeconomic characteristics of foreign visitors to the Hon Mun Islands ($n = 210$)

Characteristics	Mean	Standard deviation	Median	Minimum	Maximum
Number of visits	1.17	0.65	1.00	1.00	6.00
Travel time (days)	2.4	1.15	2.00	1.00	10.00
Group (persons)	5.29	4.03	3.00	1.00	18.00
Income (USD)	3642	2604	3000	500	10,000
Age (years)	32.5	10.78	30.00	12.00	68.00
Education (schooling years)	15.17	2.4	16.00	5.00	22.00
Sex (male = 1; female = 0)	0.52	0.50	1.00	0.00	1.00
Marital status (married = 1; not married = 0)	0.34	0.47	0.00	0.00	1.00

The foreign visitors' socioeconomic features show that their average monthly income is USD 3642/month (VND 51.7 million). Their average number of schooling years was 15.1, considerably higher than that of Vietnamese visitors, and their average age is 32.5 years, similar to Vietnamese visitors. However, there was a difference in the gender figure. Fifty-two percent of foreign respondents were male compared to 70 % for Vietnamese respondents.

4.5 Analysis of the Recreational Value of the Hon Mun Islands

4.5.1 Visitors' Travel Cost Structure

Table 4.5 presents detailed expenditures of domestic and foreign visitors to Hon Mun. It is worth mentioning that these expenditures have been adjusted from the whole trip costs to reflect money paid for Hon Mun only (i.e., addressing the issues of multiple destination and purpose trips). A very small part of the recreational value contributes to the local economy; this consists of expenditure on food and accommodation in Nha Trang, tourist boat tickets, and services on the islands.

Visitor costs make up Nha Trang's gross income from tourism in 2000 received through boat trip tours, boat rental owners, diving services, villagers in the fishing village (Lang Chai[3]), and other tourism service suppliers. This gross income was estimated to be VND 48,994 million/year (USD 3.4 million), of which the contribution of foreign tourists was less than that of the domestic tourists. However, the

[3] There are some small fishing villages on the Hon Mun Islands. Lang Chai is the biggest and the island tourists visit most commonly.

Table 4.5 Detailed expenditure of tourists to the Hon Mun Islands (based on the survey data; Unit: VND 1000)

	Domestic tourists		Foreign tourists	
Costs	All	Per capita	All	Per capita
Transportation costs	19,936,849 (USD 1.4 million)	127 (USD 8.9)	150,832,903 (USD 10.6 million)	1587 (USD 112)
Hotel costs in Nha Trang	14,026,320 (USD 1.0 million)	72 (USD 5.1)	6,841,728 (USD 0.48 million)	72 (USD 5.1)
Time costs	986,548 (USD 0.7 million)	6.6 (USD 0.5)	14,502,599 (USD 1.0 million)	152 (USD 10.7)
On-site costs	14,805,580 (USD 1.1 million)	89 (USD 6.3)	13,322,251 (USD 0.9 million)	140 (USD 9.8)
Total	35,728,977 (USD 2.5 million)	219 (USD 15)	178,657,753 (USD 12,6 million)	1880 (USD 132)

contribution of international tourists to the local economy is greater per capita compared with domestic tourists considering that the number of foreigners to Hon Mun makes up only one-third of the total number of visitors. It is worth making a comparison here: According to a report by the Khanh Hoa Tourism Department (December 2000), the total tourism revenue for Khanh Hoa in 2000 was estimated at VND 197.2 billion (USD 13.9 million). Assuming that over an average of 3 days of recreation in Nha Trang, tourists use 1 day to visit Hon Mun, and weighing the tourism value of Hon Mun by a factor 1/3, then the revenue gained from tourism in Hon Mun is VND 197.2 billion × 1/3 = VND 65.7 billion (USD 4.6 million).

The greatest part of visitor expenditures lies in transportation costs. For domestic visitors, these costs make up over half of their total outlay. For foreign tourists, this figure was about 85 % of their total expenditure. The major part of these costs are acquired by airline companies and complementary service suppliers.

4.5.2 *The Zonal Travel Cost Model*

4.5.2.1 Domestic Visitors

Visitation rates for zones are calculated and presented in Table 4.6.

The visitation rates decrease drastically with distance, from 63.48 per 1000 of the population in the innermost zone to 3.46 per 1000 of the population in the outermost zone. Zone 1 (the Nha Trang area) has the highest visitation rate. This is understandable because this zone contains the second largest number of samples (respondents) and a small population. The visitation rate of Zone 6 (Ho Chi Minh

Table 4.6 Visitation rate per 1000 of the population per year for all zones (based on the survey data)

Zone	Population	Sample		Visitation rate/1000
		Persons	%	
1	341,000	20	11.1	63.48
2	647,700	7	4.0	11.70
3	350,200	8	4.4	24.70
4	786,200	15	8.3	20.65
5	545,900	6	3.3	13.88
6	5,155,700	85	47.2	17.48
7	925,600	8	4.4	9.35
8	1,112,600	7	4.0	6.81
9	1,456,000	6	3.3	4.46
10	5,050,500	18	10.0	3.86
Total	16,371,400	180	100	

City) highlights some specific and interesting elements. Samples from this zone make up approximately half of the total. There are reasons for this:

1. Firstly, the population of Ho Chi Minh is about five million (nearly one-third of the population represented in this study).
2. Secondly, just like Vung Tau and Da Lat, Nha Trang[4] is a popular recreational site in the south of Vietnam.
3. Thirdly, Ho Chi Minh City is Vietnam's largest city. It is the economic center of the country, and its residents can afford to (and do) take holidays.
4. Fourthly, transportation facilities—air, train, and coach—between Ho Chi Minh and Nha Trang are readily available. The most popular form of transportation for tourists is by train, which is a 10-h, overnight trip. Zone 2 (districts in Khanh Hoa Province) is near Hon Mun, but it being a rural area means the number of visitors from here is small.

Demand Curve The visitation rate variable that was calculated violated the econometric assumption of normal distribution. As such, the log of the visitation rate was used as a dependent variable in the demand function (Fig. 4.1). Table 4.7 shows results from the ordinary least square (OLS) regressions calculated for zonal demand functions.

Equation 4.3 shows that both income variable and cost of substitute site variable have a relation to the cost variable because income was used to calculate time cost and travel cost was used to calculate substitute price. The coefficients of cost and

[4] Ho Chi Minh City's residents often choose Da Lat, Vung Tau, or Nha Trang to take holidays.

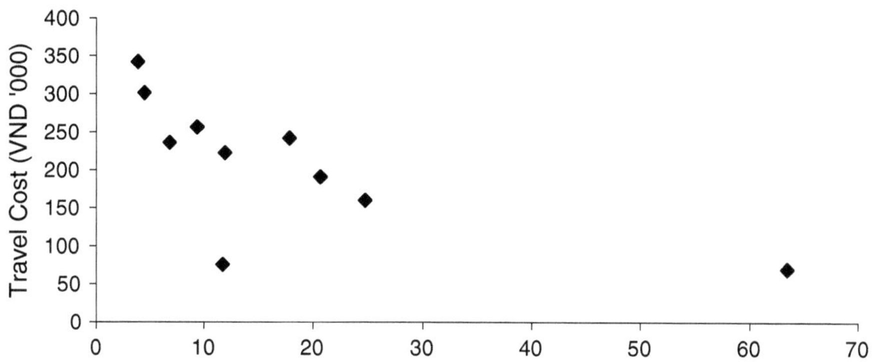

Fig. 4.1 Graphical relationship between the visitation rate and travel cost

Table 4.7 Domestic demand for visits to the Hon Mun Islands

LN(VISIT) = 4.163 − 0.007 COST (8.54)(−3.55) R − squared = 0.61	(4.2)
LN(VISIT) = 3.408 − 0.01 COST + 0.001 INCOME + 0.002 SUBSTITUTE PRICE (3.94)(−3.34)(0.99)(0.45) R − squared = 0.69	(4.3)

The t-statistics are in parenthesis. The number of observations (zones) is 10

income have the expected signs. Although Function 3 results in a higher R-squared value, it has multicollinearity[5] problems.

Figure 4.2 shows the user demand curve for Hon Mun visits in the year 2000. The curve was drawn based on Function 2. The user demand or marginal WTP curve for Hon Mun's recreational resources reflects a way of summarizing consumption attitudes and capabilities for such resources. This user demand curve is curvilinear and convex to the origin, that is, relatively flat at low prices and steep at higher prices. At low travel costs and high rates of visitation, relatively small increases in travel prices will lead to substantial reductions in the number of visits to Hon Mun. At high travel costs and low visitation rates, however, travel price increases have much smaller effects, resulting in much smaller reductions in the number of visits.

[5] Multicollinearity refers to where two or more explanatory variables in the regression model are highly correlated, making it difficult or impossible to isolate their individual effects on the dependent variable.

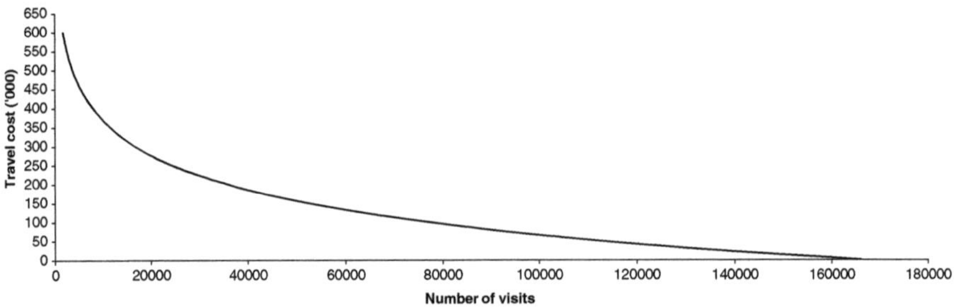

Fig. 4.2 Demand curve for visits to Hon Mun

Table 4.8 Consumer surplus and price paid for Hon Mun visits in 2000 (calculated from survey data)

Zone	Number of visits	Consumer surplus VND 1000 (USD million)	Price paid VND 1000 (USD million)
1	12,811	1,672,058 (0.118)	897,503 (USD 0.063 million)
2	23,414	3,055,923 (0.215)	1,757,994 (USD 0.124 million)
3	6612	863,008 (0.061)	1,056,982(0.074)
4	11,707	1,527,937 (0.108)	2,234,320 (0.157)
5	6385	833,395 (0.059)	1,419,963 (0.100)
6	51,865	6,769,170 (0.477)	12,553,297 (0.884)
7	8359	1,090,989 (0.077)	2,140,915 (0.151)
8	11,739	1,532,049 (0.108)	2,768,215 (0.195)
9	9320	1,216,453 (0.086)	2,805,743 (0.198)
10	23,695	3,092,591 (0.218)	8,094,045 (0.570)
Total	165,910	21,653,574 (1.52)	35,728,978 (2.52)

Consumer Surplus and Recreational Value Table 4.8 presents calculations of consumer surplus and expenditure of domestic tourists from the ten zones. In all zones, consumer surplus is significantly larger than zero, meaning visitor welfare increased by visiting Hon Mun. Those from the adjacent areas, i.e., zones 1 and 2, obtained the biggest surpluses when compared with travel costs they paid for Hon Mun. The ratio of consumer surplus to price paid tends to decrease as the distance from home to Hon Mun increases. Overall, local tourists visit Hon Mun 166,000 times a year, spend VND 35.7 billion (USD 2.5 million), and enjoy a net welfare value of VND 21.6 billion (USD 1.5 million).

4.5.2.2 Foreign Visitors

Unlike Vietnam, where samples of tourist populations were restricted to urban areas, statistical populations of foreigners were based on regional populations.

Table 4.9 Visitation rates and travel costs of foreign tourists by region (calculated from survey data)

Region	Number of samples	Travel cost VND million (USD)	Visitation rate (visits/ 1000 pop.)
1 (Asia and Oceania)	117	1.623 (115)	0.276
2 (Europe and North America)	93	2.203 (155)	0.065
Total	210		

Region 2 (Europe and North America) has a much larger population but has fewer samples than Region 1 (Asia and Oceania); therefore, it has a smaller visitation rate.

Based on the minimum requirement of two observations to estimate a demand curve, the linear demand function is as follows (calculated from the data given in Table 4.9):

$$P = 2.381 - 2.737 \times Q \tag{4.4}$$

where P is the travel costs (in VND million) and Q is visits per 1000 of the population.

Given a linear demand curve, the annual consumer surplus (CS) per visitor is the choke price minus the actual price paid, divided by two:

$$\begin{aligned} \text{Total CS} &= \tfrac{1}{2} \times (\text{number of visit}) \times (\text{choke price} - \text{price paid}) \\ \text{per visitor CS} &= \tfrac{1}{2} \times (\text{choke price} - \text{price paid}) \\ (\text{since CS} &= \text{Total CS}/\text{number of visit}) \end{aligned} \tag{4.5}$$

For Region 1 (Asia and Oceania), values from Table 4.5 substituted into Eq. 4.5 result in a consumer surplus (CS) of

$$\text{CS} - \tfrac{1}{2} \times (2.381 - 1.623) = \text{VND } 379{,}000$$
$$\text{*Choke price} = 2.381;\ \text{*price paid} = 1.623\ (\text{Table 4.9})$$

For Region 2 (Europe and North America), it is

$$\text{CS} = \tfrac{1}{2} \times (2.381 - 2.203) = \text{VND } 89{,}000$$
$$\text{*Choke price} = 2.381;\ \text{*price paid} = 2.203\ (\text{Table 4.9})$$

The weighted average consumer surplus is about VND 250,000 (USD 17.6). The average recreational value for foreigners is estimated to be VND 2.13 million. This is derived by summing up the average consumer surplus and average travel cost to the islands.

Table 4.10 Recreational value of the Hon Mun Islands in the year 2000 (Unit: VND million) (calculated from survey data)

	Consumer surplus		Price paid		Recreational value	
	All visitors (USD mil.)	Per visitor (USD)	All visitors (USD mil.)	Per visitor (USD)	All visitors (USD mil.)	Per visitor (USD)
Domestic visitors	21,654 (1.52)	0.13 (9.2)	35,728 (2.52)	0.22 (15.5)	57,382 (4.04)	0.33 (23.24)
Foreign visitors	23,810 (1.68)	0.25 (17.61)	178,657 (12.58)	1.88 (132.4)	202,467 (14.3)	2.13 (150)
Total	45,464 (3.20)		214,385 (15.10)		259,849 (18.3)	

Total Recreational Value The total recreational value equals the total consumer surplus plus the total price paid. These have been calculated for domestic and foreign visitors and summarized in Table 4.10.

The annual monetary recreational value of the Hon Mun Islands contributes annually to the economy is about VND 259.8 billion (USD 18.3 million). However, this revenue does not go directly to Hon Mun. This value is distributed, firstly, in the form of the consumer surplus of visitors who have gained recreational benefit at Hon Mun and then in terms of prices paid, to transportation companies and agents for service providers, such as hotels, restaurants, and tourist agencies, among others. A very small part of the estimated recreational value of Hon Mun is given to the local economy through expenditures on food and accommodation in Nha Trang, tourist boat tickets, and services on the islands.

Reflecting the annual recreational benefit of the Hon Mun Islands, the consumer surplus was estimated to be VND 45.4 billion (approximately USD 3.2 million). This figure is the value of the benefit that visitors gained by visiting the Hon Mun Islands. It also reflects the amount that visitors are willing to pay to enjoy the natural resources of the islands, such as the air, sea, scenic beauty, coral, and fish. This figure, however, does not reflect the nonuse value of Hon Mun. With fewer visits, international tourists received more surplus than domestic tourists: VND 23.8 billion (USD 1.68 million) in comparison with VND 21.6 billion (USD 1.52 million), respectively. The surplus gained per person is double than that of domestic tourists, implying that foreign tourists glean greater enjoyment from the Hon Mun Islands than their local counterparts, meaning international tourists value the natural resources of Hon Mun more than domestic tourists do. Survey results showed that foreign tourists are also more active and participate in most of the recreational activities on the islands, while Vietnamese tourists mainly just enjoy the scenery.

4.6 Conclusions and Policy Implications

4.6.1 Conclusions

In the year 2000, the recreational value of the Hon Mun Islands from domestic visitors was estimated at VND 57.3 billion, VND 21.6 billion of which went to recreational benefit (consumer surplus). Similarly, a demand curve for Hon Mun foreign visitors was plotted in linear form. The recreational value from foreign visitors in the year 2000 was VND 202.4 billion, of which the consumer surplus was VND 23.8 billion. Therefore, the recreational value of the Hon Mun Islands is estimated to be VND 259.8 billion, annually, of which Hon Mun's consumer surplus was estimated to be VND 45.4 billion, based on statistics from the year 2000.

Clearly, the Hon Mun Islands represent a valuable environmental resource, and, even though people do not presently pay admission, there is a large consumer surplus of welfare gained from the islands. In the future, as the number of visits to the islands increases, it is expected that the islands will also become more valuable. Although the estimated recreational value is only one aspect of the total value of the islands, it shows that, with proper conservation and management, tourism can be a significant source of benefit.

4.6.2 Policy Implications

Sustainable Tourism The number of visitors going to the Hon Mun Islands averages 290,000/year yielding a total recreational value of VND 259.8 billion (USD 18.3 million). However, only a small part of this amount (VND 48.9 billion, i.e., USD 3.4 million) is earned by the local community as tourism service revenue. Hence, it is fair to create funds for the proper management and conservation of the islands.

Establishment of a fund based on donations from visitors would be feasible because estimations from the ZTCM show that consumer surpluses derived from the site for both models are quite large (18 % of the total recreational value). However, although the TCM-derived estimates of consumer surplus show that there is a considerable potential revenue to support a fund, nearly half of the respondents surveyed reported unwillingness to contribute, due to skepticism that a fund would be well managed. This skepticism is consistent with findings from similar studies elsewhere.

This suggests that, while the revenue potential exists, it can only be realized if tourists feel that their payment will translate into improved management. This suggests that the fund should:

1. Be available to local resource managers.
2. Be managed by an accountable entity with transparent transactions.
3. Yield meaningful visible results within a short time period.

The visible benefits need not be direct conservation benefits (e.g., healthier corals). They could be of more support for infrastructure (such as mooring buoys to prevent boats from dragging anchors, thus avoiding damage to coral) or improvements that enhance tourist appreciation of the sites (e.g., signboards). If tourists notice visible improvements to infrastructure, it will signal to them that funds are indeed being used for local benefit.

Adjustment of the Port Expansion Plan Although the recreational value is only one part of the total value of the islands, the results show that tourism can generate significant revenue (some VND 259.8 billion/year or USD 18.3 mil./year). At the time of the study, there was a port upgrading plan that aimed to extend the pier. According to the Nha Trang Port Upgrading Feasibility Study (Ministry of Transportation 1997), the estimated revenue of the new port is about VND 45.8 billion per year (USD 3.2 million). Since the new port would not cause a total loss of the recreational benefit of the Hon Mun Islands,[6] a direct comparison of the revenue of the new port and the recreational value of the Hon Mun Islands is not appropriate. A full cost-benefit analysis (CBA) of the port expansion versus recreational development would be the best basis for comparison but is, however, not feasible within the scope of this study.

This research estimates the maximum recreational value that would be lost (in other words, the value at risk) and compares it with the benefits projected from the port expansion. The large recreational value of the islands is a strong indicator of the potential of the islands' tourism industry. The new port is expected to take on 1.8 times more goods and 3 times the number of passengers, which risks increased air, water, and noise pollution in the surrounding areas, including the coral islands. If the islands' tourism activities (VND 259.8 billion, i.e., USD 18.3 million) were to be reduced by 20 % due to increased pollution created by the new port, the resulting decrease in the recreational value of the islands would be more than the projected annual revenue of the port (VND 45.8 billion; USD 3.2 million). There is no national substitute for the Hon Mun Islands in terms of coral-related tourism and research. Hence, the proposed port expansion plan needs to be seriously reconsidered.

Coral Reef Management Coral reefs are the most important and unique characteristic that makes Hon Mun Islands stand out among all marine recreational sites in Vietnam, but they have not been marketed appropriately to attract tourists. In 2000, about 4000 tourists have scuba dived to look at the coral reefs of Hon Mun. Compared to the 290,000 odd visitors to the islands, this figure is small. There are three reasons for this:

First of all, the coral around the islands has been seriously damaged. The opportunity to view good coral reefs decreases by the day. It is obvious that consumers always choose the best goods; as the quality of the coral around the islands deteriorates, fewer tourists would want to pay for a diving trip to look at it.

The second problem lies in disproportionate pricing. The price of a scuba diving tour is considered expensive, even for foreigners. The average price is VND 426,000 or USD 30/h whereas the price for a day's tour around the islands, complete with lunch and tourist guide, is only VND 99,400 or USD 7.

The third reason centers on marketing and advertising. Tourists are not provided with enough interesting information about the natural properties of the islands. Many visitors to the Hon Mun Islands are not even aware of the existence of coral reefs there, so are content to just look at the scenery and swim; the survey data confirms about 80 % of visitors to the islands participate in these two activities. The demand for the Hon Mun Islands would increase, and their tourism value would rise if their coral reefs were conserved, and their inherent beauty and worth were marketed appropriately.

[6] Except for accidents like oil spills.

Appendix: Questionnaire

A Marine Protected Area (MPA) in Nha Trang Bay around The Hon Mun Islands and the other islands now is in plan. The Bay now is being damaged by over-exploitation, including too much fishing, harmful fishing methods, careless use by tourists, and pollution. The purposes of the MPA are to maintain biodiversity, protect coral reefs, improve fisheries, control pollution, manage tourism, and create new jobs for local people who will be hired to manage the MPA. This survey is about your use of the area. Please tick the appropriate boxes to indicate your choice. Your answers to these questions will be used to help plan and manage the MPA. Keep in mind there are no right or wrong answers to these questions. Your best opinions are fine. Thank you for your cooperation.

Name of interviewer: ________________________
Date:____________________________________
Reviewed by:_____________________________

1. What country and city are you from?
   Country ________________
   City __________________

2. How many times have you visited these islands, including this trip? _______ times

3. How many people are in the group you are traveling with in NhaTrang? _______

4. How many nights is your visit to NhaTrang? _______ nights

5. Why are you visiting NhaTrang? (Please tick)
   ☐ Vacation or holiday
   ☐ Work
   ☐ Study and research

   ☐ Other reason ____

6. How did you get to NhaTrang from your home? (Please tick one or more)
   ☐ Airplane
   ☐ Train
   ☐ Tour bus
   ☐ Hired car

7. In Vietnam, which places did you visit or are you going to visit, apart from NhaTrang?
   (Please specify the name of the places) ____________________________

8. What activities have you participated in at the islands? (Please tick all that apply)
   ☐ Using beaches/Sunbathing
   ☐ Swimming
   ☐ Snorkeling
   ☐ Scuba-diving
   ☐ Boating/Sailing/Jet-skiing
   ☐ Just visiting/Relaxing/Looking at scenery
   ☐ Eating seafood
   ☐ Visiting fishing villages

9. Please indicate your expenditure (estimate thereof) in the islands
   - Return trip ticket ____________ USD/person
   - Food & drinks ____________ USD/person
   - Souvenirs ____________ USD/person
   - Scuba-diving ____________ USD/person
   - Others ____________ USD/person

10. Please rank the places you have visited in NhaTrang in the order of their satisfaction to you

Place	Rank
The islands	______
NhaTrang beach	______
Ponaga tower	______
Hon Chong rocks	______
Long Son pagoda	______

Willingness to Pay for the Marine Protected Area

Experts and people on the islands believe that creating the Marine Protected Area is the best approach to preserving the environment around the islands, but they are not sure if the MPA will be successful. New source of funds will be needed to pay programs and for jobs for people who no longer will be able to earn their living from fishing.

The next questions concern your willingness to pay new fees to visit the islands and use the Marine Protected Area.

11. Would you be willing to pay more fee each time you visit and use the islands to help fund new programs to manage the Marine Protected Area?
    - □ Yes → go to question 12
    - □ No → go to question 13

12. If you answered Yes to question 11, what is the **highest user fee** that you would be willing to **pay more** (not including paying the return trip ticket to travel agency) for new programs to manage the Marine Protected Area?
    - □ 0.5 USD / 7,000 VND
    - □ 1.0 USD / 14,000 VND
    - □ 1.5 USD / 21,000 VND
    - □ 2.0 USD / 28,000 VND
    - □ 2.5 USD / 35,000 VND
    - □ 3.0 USD / 42,000 VND
    - □ 3.5 USD / 49,000 VND
    - □ 4.0 USD / 56,000 VND
    - □ 4.5 USD / 63,000 VND
    - □ 5.0 USD / 70,000 VND
    - □ 6.0 USD / 84,000 VND
    - □ 7.0 USD / 98,000 VND

13. If you answered No to question 11, what is the main reason that you said no:
    - ☐ I do not care about the Marine Protected Area
    - ☐ The Marine Protected Area is not needed
    - ☐ It costs too much already to visit the islands
    - ☐ The money would be wasted
    - ☐ Other people and business that pollute should pay
    - ☐ Not enough information

Questions About You

14. Are you male or female?
    - ☐ Male
    - ☐ Female

15. How old are you? _________ years

16. What is the highest grade you completed in school?
    - ☐ Primary school
    - ☐ Secondary school
    - ☐ High school
    - ☐ College/University
    - ☐ Masters or other graduate degree

17. Are you married?
    ☐ Yes ☐ No

18. What is your approximate net MONTHLY income?

For foreigners:
- ☐ 0 - 1,000 USD
- ☐ 1,001 - 2,000 USD
- ☐ 2,001 - 3,000 USD
- ☐ 3,001 - 4,000 USD
- ☐ 4,001 - 5,000 USD
- ☐ 5,001 - 6,000 USD
- ☐ 6,001 - 7,000 USD
- ☐ 7,001 - 8,000 USD
- ☐ 8,001 - 9,000 USD
- ☐ 9,001 - 10,000 USD
- ☐ More than 10,000 USD

For Vietnamese:
- ☐ 0 - 400.000 VND
- ☐ 400.000 - 600.000 VND
- ☐ 600.000 - 800.000 VND
- ☐ 800.000 - 1.000.000 VND
- ☐ 1.000.000 - 1.200.000 VND
- ☐ 1.200.000 - 1.500.000 VND
- ☐ 1.500.000 - 2.000.000 VND
- ☐ 2.000.000 - 3.000.000 VND
- ☐ More than 3.000.000 VND

Thank you very much!

References

DeShazo JR (1997) Using the single-site travel cost model to value recreation: an application to Khao Yai National Park. EEPSEA research report. Economy and Environment Program for Southeast Asia, Singapore

DOSTE (Department of Science, Technology and Environment) (1996) Investment project – establishment of Hon Mun Marine protected area. Unpublished paper, Nha Trang, Khanh Hoa, Vietnam

Garrod GD, Willis KG (1999) Economic valuation of the environment: methods and case studies. Edward Elgar, Cheltenham/Northampton

Georgiou S, Whittington D, Pearce D, Moran D (1997) Economic value and the environment in the developing world. Edward Elgar Publishing Ltd, London

GSO (General Statistical Office) (1998) Socio-economic statistical data of 61 provinces and cities in Vietnam. Statistical Publishing House, Hanoi

Institute of Oceanography (1998) Preliminary assessment of suitability of Hon Mun as a site for establishment and management of a marine protected area. Unpublished paper. Nha Trang, Vietnam

Ministry of Transportation (1997) Feasibility of Nha Trang port upgrading. Ministry of Transportation, Hanoi

Mwebaze P, MacLeod A (2013) Valuing marine parks in a small island developing state: a travel cost analysis in Seychelles. Environ Dev Econ 18(04):405–426

Scheaffer RL, Mendenhall W III, Ott RL (1996) Elementary survey sampling, 5th edn. Duxbury Press, North Scituate

So Du Lich Khanh Hoa (1997) Bao Cao Tong Ket 5 Nam Ve Du Lich 1993–1997. Report of the 5-year period on tourism 1993–1997. Khanh Hoa Tourism Department, Nha Trang, Vietnam

So Du Lich Khanh Hoa (1998) Bao Cao Tinh Hinh Thuc Hien Thang 12. Report of December 1998. Khanh Hoa Tourism Department, Nha Trang, Vietnam

So Du Lich Khanh Hoa (1999) Bao Cao Tinh Hinh Thuc Hien Thang 12. Report of December 1999. Khanh Hoa Tourism Department, Nha Trang, Vietnam

So Du Lich Khanh Hoa (2000) Bao Cao Tinh Hinh Thuc Hien Thang 12. Report of December 2000. Khanh Hoa Tourism Department, Nha Trang, Vietnam

Tran VHS (1998) Environmental economic assessment activities aimed at developing a proposal for establishing and managing a marine protected area. Unpublished paper. Nha Trang, Vietnam

Vo ST (1998) The establishment of marine protected areas based on coral reefs in Vietnam. Institute of Oceanography. Unpublished paper. Nha Trang, Vietnam

Ward FA, Beal D (2000) Valuing nature with travel cost models. Edward Elgar, Cheltenham/Northampton

Chapter 5
An Economic Valuation of Coastal Ecosystems in Phang Nga Bay, Thailand

Udomsak Seenprachawong

Abstract Conserving mangroves and coral reefs has become an increasingly important topic of public debate in developing countries like Thailand. This is particularly important to the mangroves and coral reefs in areas such as Phang Nga Bay, a bay along the coast of the provinces of Phuket, Phang Nga, and Krabi, Thailand. This study estimated the economic value of the changes to the quality of the mangroves and coral reefs ecosystems in Phang Nga Bay by using a choice experiment to value the economic impacts of the changes to the quality of coastal ecosystems in Phang Nga Bay. The welfare estimate indicates a willingness to pay (WTP) of THB 1133 (USD 28) per year for improved diversity of flora and fauna. Local livelihood (e.g., income from fishery) and ecological function (e.g., flood protection benefits) are tied as the second most important attributes of Phang Nga Bay. The rare and endangered species attribute is held as the least important attribute of Phang Nga Bay. The aggregate benefits were computed by multiplying the number of people in the beneficiary groups with the welfare estimate of improving Phang Nga Bay ecosystem—THB 2263 (USD 57) per person—yielding THB 5784 million (USD 144.6 million) per year.

Keywords Economic valuation • Coastal ecosystem • Choice experiment • Entrance fee • Thailand

5.1 Introduction

It is widely accepted that mangroves and coral reefs are very valuable because they possess unique, rare, or endangered plant or animal species. They play multifunctional roles like nutrient purification, provision of habitat for fish and migratory birds, and erosion control. Therefore, conserving mangroves and coral reefs has become an increasingly important topic of public debate in developing countries like Thailand. This is of particular concern for mangroves and coral reefs

---

U. Seenprachawong (✉)
School of Development Economics, National Institute of Development Administration, Bangkok, Thailand
e-mail: udomsak.s@nida.ac.th

N. Olewiler et al. (eds.), *Marine and Coastal Ecosystem Valuation, Institutions, and Policy in Southeast Asia*, DOI 10.1007/978-981-10-0141-3_5

in areas such as Phang Nga Bay, along the coast of the provinces of Phuket, Phang Nga, and Krabi, Thailand.

Coastal ecosystems are complex systems that provide numerous goods and services and perform critical ecological functions that arise out of the interdependence between their different components. In particular, these goods and services are classified into:

(a) Ecological function: The existence of biodiversity helps to keep ecosystems stable and functioning. Mangroves and coral reefs perform many important ecological functions for man. They are valuable in providing protection against coastal erosion and coastal storms.
(b) Consumption: Plants and animals in the mangroves and coral reefs provide many goods which satisfy human needs. Mangroves and coral reefs provide natural nurseries for large numbers of commercially important species. Many coastal people have lived, fished, and hunted within mangroves, deriving valuable commodities from them such as timber, fuel, medicine, and food.
(c) Aesthetic value: Coastal tourism appears to be on the rise throughout coastal cities around the world. Mangroves and coral reefs' potential for tourism are increasingly being explored in recent years.
(d) Medicinal value: The importance of conservation of the mangrove and coral reef ecosystems in general takes on an added dimension as scientists are increasingly turning to the biodiversity of the sea in their search for medical cures and unique compounds.
(e) Existence: The diverse plant and animal species in the mangrove and coral reef ecosystems need to exist, regardless of their use to people. Some people wish to see them preserved, even though they do not currently make use of them.

Each of these functions has an economic value, and changes in their ecosystems bring about corresponding changes in the flow of benefits derived from these ecosystems. The general objective of this study is to estimate the economic value of changes to the quality of the mangrove and coral reef ecosystems in Phang Nga Bay. The specific objectives are as follows:

(a) To estimate the economic value of changes to the quality of the flora and fauna of coastal ecosystems in Phang Nga Bay
(b) To estimate the economic value of changes to the quality of the local livelihood of rural community living in Phang Nga Bay
(c) To estimate the economic value of changes to the quality of the ecological function of coastal ecosystems in Phang Nga Bay
(d) To estimate the economic value of changes to the quality of rare and endangered species in Phang Nga Bay

5.2 Description of the Site

Phang Nga Bay is a large bay in the Andaman Sea, located on the west sea border with a coastline along the coast of Phuket, Phang Nga, and Krabi provinces. The bay is wide and has many archipelagic islands. Mangroves, coral reefs, and sea-grass beds are dominant ecosystems along the coast of Phang Nga Bay. The existing mangrove area, totaling 60,227.60 ha, is located along the coast of Phang Nga Bay (Charuppat and Charuppat 1997 cited in Aksornkoae 1998). One of the most important marine habitats in the bay area is the coral reef, which is distributed along the coastline and offshore areas in relatively deep water. The abundant coral reef is distributed mostly along the nearshore islands of Phang Nga and Krabi. Sea-grass bed areas of 1700 ha exist along the coast of the bay area at shallow water depths of about 5 m. The major beds are located around the islands of Ao Nang and Ao Tha Len of Krabi province and the Yao Yai Island of Phang Nga province. The stakeholders in the bay area include:

- The local rural population, which derives a use value from the bay, mainly through the collection of fuel wood and fish
- The local urban population, which benefits from incomes generated by the hospitality and tourism industries
- The management, which is responsible for the maintenance of the ecological health of the bay such that the economic benefits continue to accrue
- The community of scientists and ecologists who put a premium on existence value of the coastal and marine ecosystem
- People from the other parts of Thailand who understand the importance of the use and nonuse value of coastal ecosystems in Phang Nga Bay

Phang Nga Bay's principal attraction is Ao[1] Phang Nga National Park. The Park was declared a national park on April 29, 1981. It is comprised of a large, shallow marine component (34,700 ha) in Phang Nga Bay and a small terrestrial area (5300 ha) where over 100 limestone islands rise steeply from almost perpetually calm waters. There are a few areas of coral in the west, and some islands have beaches (UNEP/IUCN 1987). Ao Phang Nga has been featured in several international movies. Many islands are riddled with aquatic grottoes and are best explored aboard sea canoes, which are able to enter inner chambers.

5.3 Research Methodology

Valuing the characteristics of biodiversity within coastal ecosystems is a difficult task, particularly as many of the benefits arising from such have no market value. This research uses a choice experiment to value the economic impacts of changes to

[1] The word "Ao" in Thai means bay.

the quality of coastal ecosystems in Phang Nga Bay. In the choice experiment, individuals were given a hypothetical setting and asked to choose their preferred alternative from among several. Each alternative is described by a number of attributes or characteristics. A monetary value is included as one of the attributes, along with other attributes of importance. Thus, when individuals make their choices, they implicitly make trade-offs between the levels of the attributes and monetary value in the different alternatives presented in a choice set.

5.3.1 Theoretical Framework

The random utility model is the theoretical framework used to analyze the data from the choice experiment exercise. The choice of an alternative (one of three scenarios in the choice experiment) represents a discrete choice from a set of alternatives.

According to this framework, each alternative is represented with the indirect utility function that contains two parts: a deterministic element (V_i) and a stochastic element (ε_i), which represents unobservable influences on individual choice. The overall utility of alternative i is shown in Eq. 5.1:

$$U_i = V_i + \varepsilon_i \tag{5.1}$$

An individual will choose alternative i if $U_i > U_j$ for all $j \neq i$. However, since the utilities include a stochastic component, one can only describe the probability of choosing alternative i as

$$prob(i\ chosen) = prob\left(V_i + \varepsilon_i > V_j + \varepsilon_j; \forall j \in C\right) \tag{5.2}$$

where C is the set of all possible alternatives. In the choice experiment, the V_i contains attributes of the situation and there are three alternatives (status quo, plan A, and plan B). McFadden (1974) showed that if the error terms in Eq. 5.2 are independently and identically distributed (IID) with a type I extreme value distribution (a Gumbel distribution), then the probability of choosing alternative i has the following closed-form representation:

$$prob(i) = \frac{e^{\lambda V_i}}{\sum_{j \in C} e^{\lambda V_j}} \tag{5.3}$$

An estimated linear-in-parameters utility function for alternative i often takes the form

$$V_i = \alpha_i + \sum_{j=1}^{n} \beta_j X_j + \sum_{k=1}^{m} \gamma_k \alpha_i Z_k \tag{5.4}$$

where α_i is an alternative specific constant; X_j represents the ecosystem attributes associated with the alternative; Z_k is a vector representing individual characteristics; and α_i, β_j, and γ_k are parameters. Individual characteristics can be included in the model by introducing them to the alternative specific constants (as shown in Eq. 5.4) and/or the attributes (not shown). All ecosystem attributes are entered into the model using effect codes, which indicate that the utility of the average quality level is the negative sum of the utilities of the good and excellent quality levels.

Welfare estimates are obtained in choice experiment studies, using the general formula described by Hanemann (1984):

$$CV = \frac{1}{\mu}\left[\ln \sum_{i \in C} e^{V_{i1}} - \ln \sum_{i \in C} e^{V_{i0}}\right] \tag{5.5}$$

where μ is the marginal utility of income, V_{i0} and V_{i1} represent the indirect observable utility before and after the change under consideration, and C is the choice set. When the choice set includes a single before and after policy option, Eq. 5.5 reduces to

$$CV = \frac{1}{\mu}\left[\ln e^{V_{i1}} - \ln e^{V_{i0}} l\right] = \frac{1}{\mu}[V_{i1} - V_{i0}] \tag{5.6}$$

From Eq. 5.6 it is easily seen that, for a linear utility function, the marginal rate of substitution between two attributes is simply the ratio of their coefficients (Hensher and Johnson 1981) and that the marginal willingness to pay for a change in attribute is given by

$$MWTP_j = -\beta_j / \mu \tag{5.7}$$

5.3.2 *Survey Design*

This research estimates the value that resource users place on marine and coastal ecosystem quality changes. It is assumed that the current quality of ecosystems in Phang Nga Bay is at its average level (status quo). Respondents are presented with two new management plans (plan A and plan B) to protect endangered birds and other marine species, maintain fish stocks, and improve recreational and educational facilities for residents and tourists. The plans will ensure that coastal ecosystems will be improved to higher quality levels (good and excellent).

Each plan is defined using four ecosystem attributes: living coral cover (a proxy for recreational use), income from fishery (a proxy for consumptive use), flood occurrence (a proxy for indirect use), and area protected (a proxy for nonuse value).

The increase in income tax is included as a willingness-to-pay (WTP) measure attribute, which will provide the link between the parameter weights of the ecosystem attributes (recreational use, consumptive use, indirect use, and existence value) and money.

5.3.2.1 Questionnaire Design

The questionnaire and the attributes used in the choice experiment were developed in cooperation with researchers specialized in marine ecosystems from Marine Science Institute of University of the Philippines. Several focus group discussions and a pilot study were conducted in the process. The questionnaire in the survey contains three sections. The main sections of the questionnaire addressed attitude toward the environment, current use of the mangroves and coral reefs, the choice experiment, and socioeconomic characteristics.

In the introduction of the choice experiment, the purpose of Phang Nga Bay ecosystem valuation was briefly explained. The respondents were then informed about the particular coastal area in Phang Nga Bay that was about to be improved and that the researchers were interested in their views on the best possible management plan. Next, the attributes used in the choice experiment (see Table 5.1) were explained; the respondents were provided with a separate fact sheet describing them. Maps, text, graphics, and concepts such as biodiversity were used to communicate information on the mangroves and coral reefs of Phang Nga Bay. The current uses of the mangroves and coral reefs of Phang Nga Bay, its global importance as a habitat for rare and endangered birds, and the current threats the area is facing were presented to respondents.

The four ecosystem attributes were offered at three different levels (average, good, and excellent) and varied to reflect consistent ecological linkages. The cost variable varies between THB 200 (USD5) and THB 1500 (USD37.5). In the choice experiment, each respondent responded to four choice sets. The upper and lower bounds of the cost variable were identified in the piloting stages of the questionnaires; nonlinear spacing of the fee levels ensured that choice experiment questions incorporated the maximum possible number of implicit price for each ecosystem quality levels. For each choice set, respondents were asked to choose among three alternatives (see Table 5.2). The first alternative was always the base alternative, in which there would be no improvements to the Bay, at no cost. The two other alternatives implied a number of improvements to the Bay.

Ecosystem attributes were arranged into three hypothetical options for the respondents to elicit their most preferred choice. This information together with their most preferred choice indicates the relative importance among these ecosystem attributes and money. Socioeconomic characteristics of the respondents were included in the estimation and entered in the estimating equation interactively. The conditional logit model was used to analyze the choice experiment data, and the unknown parameters were estimated by maximizing the likelihood function. This information was then used to calculate the value of each ecosystem attribute, which

Table 5.1 Attributes and attribute levels

Attributes	Level
Increased living coral cover	Average (no change), good (25 %), excellent (65 %)
Increased income from fishery	Average (no change), good (35 %), excellent (60%)
Flood occurrence	Average (every year), good (every 2 years), excellent (every 4 years)
Increased area protected	Average (no change), good (20 %), excellent (50 %)
Increased income tax in 2002 (THB)	0, 200, 700, 1000, 1500

THB 40 = USD 1

Table 5.2 Example of a choice set from Phang Nga Bay questionnaire

Your choice ⇒			
	Status quo	Plan A	Plan B
Increased living coral cover	No change	25 %	No change
Increased income from fishery	No change	No change	60 %
Flood occurrence	Every year	Every 2 years	Every 4 years
Increased area protected	No change	20 %	50 %
Your increased income tax in 2002 (THB)	0	200	700

THB 40 = USD 1

is essentially the marginal rate of substitution between ecosystem attribute and money.

5.3.2.2 Main Survey

Most of Phang Nga Bay benefits accrue at the local level (local livelihood) and national level (recreational values), and only to some degree at the global level (existence values). Therefore, foreigners were excluded from the survey. The population sample was from Thais between the ages of 18 and 75 years, living in and traveling to Phang Nga Bay area; out of a total population of 2,555,703, a sample of 300 individuals was randomly selected.

The main survey was conducted from November 2001 to January 2002. Four teams of surveyors conducted face-to-face interviews; each team consisted of two persons. A 2-day training course was given to minimize potential biases caused by interviewers misunderstanding the questions. The training course consisted of a careful explanation of all questions, simulation of interviews between the surveyors, and a pilot interview on an individual in order to check that the questions would be clear to the respondents and that the time required to complete the interview would not exhaust the respondents. Each individual answered four choice sets.

5.4 Empirical Results

The welfare estimates of improving Phang Nga Bay ecosystem, as shown in Table 5.3, indicate that diversity of flora and fauna, which provide recreational and tourism benefits, is regards as the most important attribute of Phang Nga Bay. The welfare estimate indicates a WTP of THB 1133 (USD 28) per year for improved diversity of flora and fauna. Local livelihood (e.g., income from fishery) and ecological function (e.g., flood protection benefits) are tied as the second most important attributes of Phang Nga Bay. The rare and endangered species attribute is held as the least important attribute of Phang Nga Bay. The aggregate benefits were computed by multiplying numbers of people in the beneficiary groups[2] with welfare estimate of improving Phang Nga Bay ecosystem—THB 2263 (USD 57) per person—yielding 5784 million Baht (USD 144.6 million) per year.

5.5 Conclusion and Policy Recommendations

The rapid growth of tourism makes many demands on the environment, such as pressure on the beaches, use of precious resources for craft items, use of wetlands for waste disposal, removal of sea grass for swimming beaches, and blocking of visual and public access to the coast. These impacts could reduce the desirability of Phang Nga as a tourist destination, as they have at the most developed tourist beach, Patong (Bunapong and Ausavajitanond 1991, cited in Wong 1995). The natural tree cover was removed, wetlands were filled in or converted into open sewers, and the sea and beach were polluted by poorly treated wastewater. Since 1979, researchers of the Phuket Marine Biological Center have studied the condition of reefs around the southeast tip of Phuket. This study shows that corals were experiencing significant changes both on a daily and seasonal basis. In addition to natural changes in the environment, the construction of resorts, hotels, and other structures and increased pollution have greatly affected the corals around Phuket.

The recommendations for managing tourism impact are supported by findings from these studies and presented below.

5.5.1 *New Hotel Development*

The projected growth in tourist accommodation far exceeds the development goal of Phuket province. For example, hotel development in Phuket is more than 1000 rooms/km of beach, far exceeding the 200 rooms/km of beach for sound

[2] The number of people in beneficiary groups was taken from the National Statistical Office, 1999. The groups included 348,369 adjacent dwellers, 133,767 tourism operators, and 2,037,567 tourists.

Table 5.3 Welfare estimates of moving from status quo to non-status quo

Attributes/welfare estimates	WTP to improve Phang Nga Bay THB (USD)/person/year	Percentage (%)
Flora and fauna (direct use values, e.g., recreation and tourism)	1133[a] (28)	50
Local livelihood (direct use values, e.g., fishery income)	514[b] (13)	22
Ecological function (indirect use values, e.g., flood protection)	504[c] (13)	22
Rare and endangered species (nonuse values)	112[d] (3)	6
Total	2263 (57)	100

[a]Welfare estimate of moving from average to excellent was selected for flora and fauna
[b]Welfare estimate of moving from average to good was selected for local livelihood
[c]Welfare estimate of moving from average to excellent was selected for ecological function
[d]Welfare estimate of moving from average to excellent was selected for rare and endangered species

environmental conditions recommended by the Japan International Cooperation Agency (Bunapong and Ausavajitanond 1991). The results of this study suggest that the government impose an official moratorium on all new hotels and related tourist accommodation developments, including condominiums and time-sharing projects, for an initial period of 7 years during which no new permits for such developments would be issued. In addition, all existing permits for new lodging developments and extensions to existing properties should be reviewed and, where appropriate, renegotiated so as to ensure that the total number of lodging units is capped at 200 rooms/km of beach.

Such a policy would result in a shift away from emphasis on attracting large new projects, toward facilitating the upgrading of existing tourism-related products and stimulating small business growth. Investments in other viable tourism-related facilities (both local and foreign investors) would be encouraged, provided such projects conform to the guidelines set out in development policy of Phuket Province.

5.5.2 *Protection of the Environment*

Every major development proposal must be accompanied by a thorough environmental impact study conducted by an independent expert. The government can submit a developer's plan to an outside independent appraisal funded by the developer. A permit should be required for the erection of advertisement signs and billboards, which must conform to certain specified standards. The operation of jet skis and similar high-powered marine crafts should be totally prohibited in the environmentally sensitive areas. Examples of specific guidelines on tourism development of Phang Nga Bay might be:

- Exclude development from geologically unstable areas (e.g., high erosion areas, flood-prone areas) and provide adequate buffer zones between development and existing shoreline
- Define zones suitable for development (e.g., along specific beaches)
- Ensure that construction materials and methods are not harmful to the reef and environment
- Control effluent disposal to prevent disturbances to the reef ecology

Further research should be undertaken by the relevant government agencies to determine the carrying capacity of the parks, whether in terms of the maximum number of users or the intensity of use by various categories of users. Establishment of an ongoing program for monitoring changes in the levels of coral covers and in the species composition of corals, fishes, and other marine organisms would be a first step toward determining the carrying capacity of Phang Nga Bay. Gathering information on user attitudes to congestion and apparent degradation within the marine parks is also recommended.

This research emphasizes the need to consider the impact of ecotourism on the conditions of the ecosystems that support such industry. Coastal tourism development must be accompanied by measures to reduce environmental degradation by way of putting in place an effective management program to mitigate or prevent negative impacts by ecotourism. Decisions to undertake reclamation and modification of coastal resources in Phang Nga Bay should consider the loss in ecosystem services such actions would entail. There is a need to look into how these development projects would impact coastal resources, and their values should then be estimated using values such as those derived in this research.

In addition, the ecosystem values generated in this research could be used in designing conservation financing mechanisms for Phang Nga Bay as discussed in the following section.

5.5.3 Rent Capture to Fund Management

The environment is the greatest economic asset Phang Nga Bay has—it is what the area is marketing to the world—but environmental resources are fragile and require investments to maintain and manage them. The bottom line is the environmental resources in Phang Nga Bay generate economic rents that could be used to pay for improved management of Phang Nga Bay's marine parks. This study focused on assessing the willingness of the Thai people to pay for improved environmental quality in Phang Nga Bay and on their attitudes toward environmental protection.

The annual welfare gain of improving recreational and tourism attributes in Phang Nga Bay is estimated to be THB 1133 (USD 28) per person. By capturing a portion of this welfare gain, the park can finance management activities to protect and restore its mangrove forests and coral reefs, as well as fulfill the broader social objective of supporting scientific research and education. To get an optimal entrance fee to the park would need the underlying theory for the optimal pricing of protected areas used in recreational activities, from the perspectives of a park authority interested in welfare maximization. Hence, the welfare estimates derived through this research are insufficient for setting the optimal entrance fee for the park.

Based on the existing benefit capture mechanisms used to finance marine protected area management elsewhere (Gustavson 2000), it is recommended that the Thai government introduce a two-tiered basic entrance fee for marine parks in Phang Nga Bay with different rates for residents and foreigners.[3] Thais would pay an entrance fee of THB 40 (USD 1)[4], while foreigners would pay THB 400 (USD 10).[5] This practice both captures sizeable amounts of rent to help pay for the management of the park system and recognizes the income differences between residents and foreigners. Supplementary user fees should also be levied when visitors receive additional services from the variety of recreational sites on offer in Phang Nga Bay. The park could impose charges for tourists visiting certain special and ecologically sensitive areas. For example, after having charged a basic entrance fee, the park could impose a diving fee of THB 200 (USD 5) per person per visit if the visitor chooses to dive in the Coral Island. This user charge would help raise additional revenue for the park by transferring economic rents from high-end consumers to gains while ensuring that the low-income visitors are unaffected.

The use of a hotel room tax could also be implemented. This is preferred since the tax is proportional to resource use. Given the inelastic nature of the demand, the impact on the bottom line for hotel operators is likely to be modest. It is recommended that a room tax of THB 40 (USD1)/bed/night be imposed; this tax should be reduced or waived for Thais. Finally, it is a question of marketing. Brochures could be provided to newly arrived tourists, explaining the sensitive nature of the Phang Nga Bay that they will be enjoying and the need to fund the care and preservation of this beautiful bay. Instead of a "lodging tax," a "resource conservation fee" on the hotel bill might be more acceptable to the tourist. The key point is that the tax should not be perceived as something designed to discriminate against foreigners, but rather as a user fee for the enjoyment of the environment. The rents collected can be shared between earmarked activities (environmental resource management, trash collection, coastal patrolling, etc.) and general revenues for the national treasury (e.g., a statement that 60 % of room tax goes to resource management or pollution control). In cases where a

[3] There is certain logic to this. Foreigners do not pay income tax or business tax to the local government. They take the extra benefit from the use of the resource home with them when they leave. As such, failing to charge higher user fees on foreign visitors implies that the country is subsidizing an increase in the social welfare of the visitors from richer countries.

[4] Entry fees for marine parks in Phang Nga Bay have traditionally been very low. At present, a basic entrance fee of THB 20 (USD 0.50) is charged per person.

[5] Earlier studies on Phi Phi Island (Seenprachawong 2001) indicate an annual consumer surplus, as estimated by the travel cost method, on the order of 10 to 1 (USD 2010 for each foreign visitor and USD 184 for each Thai visitor).

division of the revenues is not possible, it may be necessary to create a separate entity to collect and manage funds, and in some cases to help provide management services.

5.5.4 Fund-Raising at the Site Level

The results on environmental attitude surveys show that Thais place a very high value on the environment. A high percentage of respondents (80 %) believe that natural resources are of value because of the benefits they will provide to future generations. In terms of existence values respondents place on rare bird and marine lives, 88 % agreed with the statement, "Mangroves and coral reefs should be protected because rare birds and marine lives depend on them"; 63 % placed option value on mangroves and coral reefs. These findings show that the park authority can potentially raise money from the general public. Basically there are three steps to successful solicitation from individual donors (Phillips 2000):

- Inform and educate them about the conservation program and what needs to be done.
- Inspire and help them to develop personal views of the difference their contributions would make.
- Ask them to help make that difference.

There are many opportunities for raising money at the site level. One option is to develop "site membership" and "friend" schemes. Contrary to the "pay-per-visit" concept of user fees, membership programs provide a vehicle for voluntary support by a constituency that may or may not actually visit the marine parks. The park manager at Mai Khao beach could implement a "Friends of the Sea Turtle" program. Such programs provide excellent opportunities to channel individual contributions directly to marine fauna protection. Staff could collect donations on site or capture visitor information (names and addresses) for future fund-raising campaigns. Members can make other contributions as well: volunteer work, word-of-mouth publicity, providing information, buying products and tickets to benefit events, and identifying potential donors.

Alternatively, an "adoption program" could be used to generate revenue for a specific site. For example, the park manager can raise money for sea turtle protection by selling "deeds" for THB 100 (USD 2.50)/turtle protected. The donor would receive a certificate acknowledging adoption of the turtle. The work would involve producing certificates and mailing them, writing thank-you letters, and answering correspondence; having a group of volunteers to help would be an advantage as it would be time-consuming. The park manager could also obtain revenue by charging a "publicity fee" to corporations using the site as a location or backdrop for advertisements, posters, and other uses. Sales of locally made crafts could be excellent ways of bringing financial benefits to local communities near marine parks. Even if the direct financial returns from these sales were small, the support of the local people would be an advantage.

Tourism, if carefully planned, managed, and promoted, could become the mainstay of the economy of Phang Nga Bay. The hospitality industry affects and embraces people from all walks of life, and the increasing economic benefits it could bring will benefit everyone. However, it is necessary to strike a balance between the economic benefits that it could bring and its possible negative impacts on the natural and cultural environment of Phang Nga Bay.

Appendix 1: Questionnaire on Mangroves and Coral Reefs in Phang Nga Bay

- INSTRUCTION TO INTERVIEWERS ARE IN CAPITALS
- IN THE CASE OF A REFUSAL TO RESPOND NOTE THIS WITH A CAPITAL 'R'. DO NOT MERELY LEAVE A BLANK
- NEED RESPONDENTS WHO ARE PAYING INCOME TAX
- INTRODUCE YOURSELF AS FOLLOWS:

Good morning /afternoon, sir/madam. My name is … I am involved in a study being conducted by Sukhothai Thammathirat Open University on the mangroves and coral reefs in Phang Nga Bay. We are carrying out a survey to find out how much people value the mangroves and coral reefs in Phang Nga Bay, and would like to ask you a series of questions. All answers are confidential and there are no right or wrong answers. Your opinion and the information provided will be used to improve the quality of mangroves and coral reefs in Phang Nga Bay. Therefore your honest response is essential for the success of this research project.

Name of Interviewer: …………………… Date: …/… /2001 Serial No. ……..
Time interview starts: ………….. Time interview ends: ……………
Area: 1. Phuket 2. Phang Nga 3. Krabi 4. Others, please specify …………

A. GENERAL ATTITUDES AND BEHAVIOR

A1. Suppose that the Thai government is going to invest money to help with one of the problems listed below. Which of these problems do you consider to be the most important one to solve in Thailand?

☐ 1. Increasing agricultural productivity
☐ 2. Inflation
☐ 3. Reducing water pollution
☐ 4. Protecting natural habitats and wildlife
☐ 5. Improving quality of education
☐ 6. Others, please specify ______________________________

A2. What in general do you think are the three most important problems related to nature and human impact on the natural environment in Thailand, which you personally find worrying?

☐ 1. Water pollution
☐ 2. Deforestation
☐ 3. Air pollution
☐ 4. Degraded mangroves and coral reefs
☐ 5. Floods
☐ 6. Others, please specify _______________________

A3. Please indicate your opinion on a scale of 'strongly agree' to 'strongly disagree'. There is no right or wrong answer.

	Strongly Agree	*Agree*	*No Opinion*	*Disagree*	*Strongly Disagree*
(1) We have a duty to protect the environment from development regardless of the cost	1	2	3	4	5
(2) We should minimize the damage to the environment now so that out grandchildren may benefit from it	1	2	3	4	5
(3) Thailand needs to develop her forests, sea, and land to increase jobs and incomes, regardless of the environmental damage	1	2	3	4	5
(4) Mangroves and coral reefs should be protected because rare birds and marine lives depend on them	1	2	3	4	5
(5) I should pay for the protection of parks and nature reserves even if I do not visit them	1	2	3	4	5
(6) Even if I do not use the mangroves and coral reefs now, I am prepared to pay now to protect them in case I want to use them in the future	1	2	3	4	5
(7) It is worth spending money to protect mangroves because they help to protect agricultural productivity in the area	1	2	3	4	5
(8) We have more important things to think about than the loss of the mangroves and coral reefs	1	2	3	4	5

B. USE OF MANGROVES AND CORAL REEFS IN PHANG NGA BAY

SHOW FIGURE 1: This map shows the major mangroves and coral reefs in Thailand. You can see that the mangroves and coral reefs of Phang Nga bay are one of these areas.

SHOW FIGURE 2: MANGROVES AND CORAL REEFS OF PHANG NGA BAY

B1. Have you ever visited mangroves and coral reefs of Phang Nga Bay?

☐ 1. Yes ☐ 0. No

B2. What benefits do you currently get from using marine and coastal resources of the Phang Nga Bay ?

☐ 0. No benefit
☐ 1. Swimming
☐ 2. Bird watching
☐ 3. Boating/sailing
☐ 4. Diving/Snorkeling
☐ 5. Research/Education
☐ 6. Income from fishery
☐ 7. Tourist related income
☐ 8. Mangrove forest products
☐ 9. Enjoy the scenery
☐10.Savour seafood from the Bay in the past 5 years

☐11.Others, please specify ___________________

B3. Are you likely to visit mangroves and coral reefs of Phang Nga Bay in the next 5 years?

☐ 1. Yes/Likely ☐ 0. No/Unlikely

B4. Have you ever heard of the concept of biodiversity before?

☐ 1. Yes ☐ 0. No

SHOW CARD A: READ INFORMATION ON MANGROVES AND CORAL REEFS BIODIVERSITY

SHOW CARD B: READ BACKGROUND INFORMATION ON THE MANGROVES AND CORAL REEFS OF PHANG NGA BAY

PRESENT CARD C: READ SCENARIO A AND SCENARIO B

C. WILLINGNESS TO PAY

* * BIODIVERSITY FUND * *

As described, management and protection of mangroves and coral reefs in Phang Nga Bay is necessary to protect the area's rare bird species and to enhance the quality of life of the local communities by providing a continuous source of seafood and by protecting agricultural land. Obviously, the implementation of this protection project would cost money and people would have to pay their share of the costs on a continuing basis if they want to enjoy the benefits protection of mangroves and coral reefs offer.

As such, in order to protect the mangroves and coral reefs, you would be asked to pay a fee to a PHANG NGA BAY BIODIVERSITY FUND, which will be established and managed by the local government unit to help protect the mangroves and coral reefs of Phang Nga Bay. Possible projects to increase biodiversity in Phang Nga Bay are

- planting mangroves and coastal plants to reduce impacts of run-off
- treating of sewage to a high standard
- building new drainage system for storm waters
- monitoring of fish, plant life and mangroves
- promoting environmental sensitive tourism activities
- establishing visitor centers
- encouraging proper disposal of garbage and other waste to reduce pollution

Suppose a proposal to establish a PHANG NGA BAY BIODIVERSITY FUND was on the ballot in the next nationwide election. How would you vote on this proposal? Remember that by law, the funds could only be used to protect the mangroves and coral reefs in Phang Nga Bay. If the PHANG NGA BAY BIODIVERSITY FUND was the only issue on the next ballot, how would you rank the desirability of each of the proposed management plans for Phang Nga Bay with one (1) being most desirable and three (3) being least desirable?

Please keep in mind:
The issues discussed here are only a few among many other environmental problems Thailand faces.

This interview is on the mangroves and coral reefs of Phang Nga Bay and not on other environmental issues or other mangroves and coral reefs around the country.

Your own personal income is limited and has important alternative uses. There is no right or wrong answers and you should answer for yourself. Please choose one of these three options according to your preference

YOUR CHOICE ⇒			
	Status quo	*Plan A*	*Plan B*
Increased living coral cover	No change	No change	65 %
Increased income from fishery	No change	35 %	60 %
Flood occurrence	Every year	Every 2 years	Every 4 years
Increased area protected	No change	20 %	No change
Your increased income tax in 2002 (THB)	0	200	700

Note: 40 THB = 1 USD

D. SOCIO ECONOMIC BACKGROUND

The following are a few questions on your background that will only be used for statistical purposes

D1. Gender of the respondent: ☐ 1.Male ☐ 0.Female

D2. Age: _________years

D3. What is the highest level of education you have obtained?

☐ 1. No formal education
☐ 2. Primary
☐ 3. Secondary
☐ 4. Technical Diploma
☐ 5. Bachelor Degree
☐ 6. Masters Degree
☐ 7. Others, please specify __________

D4. What is your occupation?

☐ 1. Civil servant
☐ 2. Own business
☐ 3. Private employee
☐ 4. Laborer
☐ 5. Student
☐ 6. Retired
☐ 7. Others, please specify ____________

D5. Number of members in your household: ___________persons

D6. Your monthly income

0-2,500 THB	1
2,501-5,000 THB	2
5,001-7,500 THB	3
7,501-10,000 THB	4
10,001-15,000 THB	5
15,001-20,000 THB	6
20,001-25,000 THB	7
25,001-50,000 THB	8
50,001 THB and above	9

Note: 40 THB = 1 USD

D7. Your household income

0-2,500 THB	1
2,501-5,000 THB	2
5,001-7,500 THB	3
7,501-10,000 THB	4
10,001-15,000 THB	5
15,001-20,000 THB	6
20,001-25,000 THB	7
25,001-50,000 THB	8
50,001 THB and above	9

D8. Present address : City __________ Province ____________

D9. Last of all, what do you think of this questionnaire?

	Yes	No
1. Interesting	1	0
2. Too long	1	0
3. Difficult to understand	1	0
4. Educational	1	0
5. Unrealistic / not credible	1	0
6. Others, please specify ____________	1	0

END INTERVIEW, THANK RESPONDENT

Appendix 2: Cards

CARD A

BACKGROUND INFORMATION ON MANGROVE AND CORAL REEF BIODIVERSITY

Biodiversity is defined as the totality of genes, species and ecosystems in a region. Genetic diversity refers to the variation of genes within species. Species diversity refers to the variety of species within a region. Ecosystem diversity refers to the variety of systems, of living things and their environment, within a region.

Mangrove biodiversity refers to the total number and variety of plants, animals and fish species found in the mangroves. These mangrove plants, animals and fishes live and interact within different types of mangrove environments.

Coral reef biodiversity refers to the different habitats for fish, coral, mollusks, shellfish and other sea animals, and also vegetation, fungi and bacteria. The kind and number of such habitats depend upon the total number of coral species, dominant species in an area, and the complex patterns that occur in coral reefs over time and space.

To protect the individual plant and animal species diversity, it is necessary to protect mangrove and coral reef environment

CARD B

BACKGROUND INFORMATION ON MANGROVES AND CORAL REEFS IN PHANG NGA BAY

Mangroves and coral reefs are dominant coastal ecosystems in Phang Nga Bay. Mangrove forests of Phang Nga bay serve as nursery and feeding grounds for marine organisms like fish, shellfish, and crabs. They also help prevent coastal soil erosion and provide protection from storms. Coral reefs in Phang Nga Bay are found around Ko Hong, Ko Dam Hok, and Ko Yao Noi. They provide habitats for marine organisms like fish and shellfish. They also provide recreational opportunities and natural beauty.

Phang Nga Bay has 60,791 ha of rich mangrove resources all around its coastline. However, only 20% of these are protected. Currently, mangrove habitats in Phang Nga Bay are diminishing due to urban development, and conversion to fishponds and agricultural land. If mangrove areas are further reduced, some fishermen dependent for their livelihood on coastal fisheries will be affected. As a result, these people will observe a decline in the fishery yields. In addition, a plan to develop a southern seaboard project in Phang Nga has been proposed. This project coupled with the expanding tourism-related business will result in the destruction and degradation of the coastal ecosystem, starting from the mangrove forests to coral reefs.

A coastal management plan prepared for Phang Nga Bay by the Ministry of Agriculture and Cooperatives recommend that:

"The mangrove forests and coral reefs in Phang Nga Bay should be designated as a coastal protected area"

That is it should be protected. The benefits of this would include:

- conservation of important flora and fauna
- maintenance of a substantial offshore fishery
- provision of a sustainable harvest of mangrove products
- protection of coral reefs, marine turtle nesting beaches, and recreation areas.

CARD C
MANAGEMENT SCENARIOS

SCENARIO "A": PRESENT STATE OF AFFAIRS – NO PROTECTION

- mangroves in Phang Nga Bay vulnerable to illegal encroachment and deforestation
- loss of mangrove areas to urban development
- potential loss of globally important bird species
- deterioration of recreational facilities (sandy beaches, coral reefs) and aesthetic beauty

SCENARIO "B": PROPOSED MANAGEMENT PLAN: PROTECTION OF MANGROVES AND CORAL REEFS IN PHANG NGA BAY AS A COASTAL PROTECTED AREA

- protection of globally significant birds and other wildlife and habitat currently under threat
- maintenance of fish stocks and shell fish which is of benefit to local communities
- improved recreational and educational facilities for residents and tourists
- reduced pollution
- protection from illegal activities

References

Aksornkoae S (1998) Mangroves: ecology and management. Kasetsart University Press, Bangkok

Bunpapong S, Ausavajitanond S (1991) Saving what's left of tourism development at Patong Beach, Phuket, Thailand. In: Coastal zone'91, 1685–1697. American Society of Civil Engineers, New York

Gustavson K (2000) Capturing coral reef benefit values- financing marine environment conservation in Montego Bay, Jamaica. The World Bank, Washington, DC

Hanemann M (1984) Welfare evaluations in contingent valuation experiments with discrete responses. Am J Agric Econ 66:332–341

Hensher DA, Johnson LW (1981) Applied discrete choice modeling. Wiley, New York

McFadden D (1974) Conditional logit analysis of qualitative choice behavior. In: Zarembka P (ed) Frontiers in econometrics. Academic, New York, pp 105–142

Phillips, A. 2000. Financing protected areas. Financing protected areas task force of the World Commission on Protected Areas (WCPA) of IUCN, in collaboration with the Economics Unit of IUCN. IUCN – The World Conservation Union, Gland/Cambridge, UK. viii + 58pp

Seenprachawong U (2001) An economic analysis of coral reefs in the Andaman sea of Thailand. EEPSEA research report No. 2001-RR7. Economy and Environment Program for Southeast Asia, Singapore

UNEP/IUCN (1987) Directory of coral reefs of international importance, vol 2, Indian Ocean, Red Sea and Gulf. UNEP regional sea directories and bibliographies. FAO, Rome

Wong PP (1995) Coastal and marine tourism in the Asian and pacific region: experience and impacts in coastal and marine environmental management: proceedings of a workshop bangkok, Thailand. Asian Development Bank, Bangkok

Chapter 6
Willingness to Pay for Whale Shark Conservation in Sorsogon, Philippines

Anabeth L. Indab

Abstract This study estimated the willingness to pay (WTP) of Sorsogon residents to preserve the continuous presence of whale sharks by using the contingent valuation methodology (CVM). CVM is a method for estimating the monetary value that a person places on a good or service that is not normally sold in markets. The approach directly asks people if they are willing to pay for the good rather than inferring this from observed behavior. Using the drop-off survey method, this study achieved a response rate of 98 %; out of the 600 questionnaires distributed, 588 usable questionnaires were retrieved. The results showed that Sorsogon residents had low positive WTP, which ranged from PHP 17 (USD 0.34) to PHP 35 (USD 0.70) per household per month. Furthermore, more than half of the respondents (60.71 %) strongly agreed that everyone is responsible for ensuring that plants and animals—as they are known today—will exist for mankind in the future. This suggests that the respondents hold a perceived bequest value. Overall, the survey results show that people are aware and concerned about environmental issues, including the precarious condition of whale sharks in Sorsogon. They are not, however, able or willing to pay for the implementation of a conservation program due to poverty, employment, and other economic concerns taking precedence over environmental issues.

Keywords Willingness to pay (WTP) • Philippines • Whale shark • Bequest value • Contingent valuation methodology (CVM)

6.1 Introduction

Societies continually trade-off between choices that affect their welfare. Trade-offs are especially difficult to make in the context of a developing country where even the basic needs are sometimes hard to finance. It then becomes an empirical question whether people from a developing country perceive any benefit from

A.L. Indab (✉)
Resources, Environment and Economic Center for Studies, San Juan City, Philippines
e-mail: bates_89@yahoo.com

N. Olewiler et al. (eds.), *Marine and Coastal Ecosystem Valuation, Institutions, and Policy in Southeast Asia*, DOI 10.1007/978-981-10-0141-3_6

what can be considered as a nonessential good like the preservation of certain endangered species and would therefore be willing to sacrifice certain things for it. This study explores this particular issue in the case of whale sharks in Sorsogon, Philippines, using contingent valuation methodology (CVM) to capture the value to people of preserving the species.

One of the most publicly known endangered species found in Philippine waters is the whale shark (*Rhincodon typus)*, which is known as the world's largest fish. It has a migratory nature, making it vulnerable to a wide range of threats, hence requiring collaboration of various nations for its protection. The International Union for Conservation of Nature (IUCN) has declared this species as vulnerable.

Donsol, Sorsogon, is located on the southernmost island of Luzon, Philippines. The November 2004 issue of *TIME* magazine identified the whale shark interaction program in Donsol waters as the *Best Place for an Animal Encounter*. Despite the precarious condition of this species and the privileged position of Donsol as "whale shark capital" of the world, certain existing factors threaten the continuous migration and presence of whale sharks in the area. The local municipal government is already investing in the protection of whale sharks in the area, but its resources are insufficient. Toward this end, it is imperative to look at the viability of implementing a more comprehensive conservation program for whale sharks in Sorsogon.

This paper aims to estimate the willingness to pay (WTP) by Sorsogon residents to preserve the continuous presence of whale sharks using contingent valuation methodology (CVM). CVM is a method of estimating the monetary value that a person places on a good or service that is not normally sold in markets. The approach directly asks people if they are willing to pay for the good rather than inferring this from observed behavior.

The study also extracts some rough estimates of certain use-values of whale shark that might be relevant considerations in making an informed policy decision on whale shark conservation in Sorsogon.

6.2 CVM Design

This CVM survey presents a hypothetical scenario about a proposed comprehensive whale shark conservation program in Sorsogon. The different aspects and components of the program, including the institutions that would pay and the payment vehicle, are presented in the contingent or hypothetical scenario. On the basis of the given scenario, the respondents were asked to decide whether they are willing to pay, or in this case to vote, for or against a public program at a particular price provision.

The key element in a CVM study is a properly designed questionnaire. A great deal of the study was spent crafting this hypothetical scenario and generating the appropriate survey instrument that could best assess and capture the existence of whale shark preservation benefits for the people of Sorsogon.

6.2.1 *Procedures Observed*

Qualitative and quantitative information was obtained from three main sources: personal interviews and focus group discussions, pilot surveys of households, and the main survey of households.

The project's initial activities focused on gathering scientific information and proper conservation measures for whale sharks, which were then presented in terms of a hypothetical conservation program. The program was framed around consultations with local whale shark experts and existing scientific literature.

6.2.1.1 Personal Interviews and Focus Group Discussions (FGDs)

The personal interviews gathered respondent insights into institutions that could be involved, payment vehicles, timing of payments, the technical and political feasibility of such a program, and other information. In the course of the series of FGDs held, information was presented, refined, revised, and finalized for inclusion in the survey pretest. Several FGDs were conducted for young professionals, government bureaucrats, and middle-income and low-income household heads in Sorsogon.

6.2.1.2 Pretests

Both the personal interview and self-administered survey format were pretested. Pretesting the questionnaire on a small sample of respondents helps to identify and correct potential problems (Bateman et al. 2002). The survey pretest also provided the indicators that were used to plan the required overall sample size. Based on the pretest results, the questionnaire was revised and back-translated by language experts to ensure that the researcher was able to communicate the questions based on the desired information that the research meant to convey. Pretest observations suggested the surveys should be dropped off and self-administered survey to avoid researcher biases.

6.2.2 *Structure*

The questionnaire provided a description of the commodity to be valued, the mechanism by which the good would be delivered, and the method of payment. To determine whether the scenario had been communicated successfully, each respondent was debriefed at the end of the questionnaire. For example, some questions served to assess whether a respondent who says she/he would pay nothing for whale shark conservation is stating his/her true valuation or just making a protest bid (Table 6.1).

6.2.2.1 Structure of the Whale Shark Questionnaire

Table 6.1 Structure of the questionnaire

Part 1	To get people to start thinking about social or public priorities. This section includes questions on what public issues they think are currently being prioritized and what are being neglected
Part 2	To establish attitudinal behavior about species conservation and to assess the respondent's familiarity with the existence, condition, threats, and importance that they place on the presence of whale sharks
Part 3	To describe the conservation plan to protect the existence of whale sharks
	This section contains the referendum or WTP question
Part 4	Debriefing questions, including those designed to bring out the problems in the choice of payment vehicle, are included in the questionnaire
Part 5	To identify the socioeconomic characteristics of the respondent

6.2.3 Elicitation Method

The study adopted the referendum method recommended by the NOAA panel because of its incentive properties. The single-bounded dichotomous choice with an open-ended follow-up question was used in the survey. Each respondent was asked whether she/he would vote (i.e., pay) in favor of the proposition that, if approved by the majority, would provide for the conservation of whale sharks in Sorsogon or be willing to donate to the program. The elicitation question was presented in a take-it-or-leave-it format that imitates the choice facing an individual in a real market.

6.2.4 Payment Design

The electricity bill was used as the vehicle for payment because of recommendations from the series of FGDs conducted and the literature, which proposes that the payment vehicle must appear familiar and believable to respondents. The questionnaire asked the survey respondents about their willingness to pay on a monthly basis for a 5-year time frame. Two split sample conditions were used in the payment design: one as a surcharge to the electricity bill and the other as a voluntary contribution, also through the electricity bill.

Cheap talk was incorporated in the hypothetical script to address certain potential protest issues, like the absence of direct connection between electricity bill and whale shark conservation and why it is still the most efficient vehicle to use, concerns about the poor, and others. Cheap talk is an element in the hypothetical script which provides explicit explanations or warnings about certain aspects of the hypothetical scenario that remind and encourage the people to respond in the valuation question as if they are in a real situation.

These aspects were intentionally added because of the pretest results to make sure that the respondents understand and accept the plausibility of scenario.

6.2.5 *Price Bid Determination*

For bid prices, three sets of prices were used in the pretest given the following targets: 90 % of the respondents would be willing to pay (WTP) given Price 1 (low price), 50 % WTP given Price 2 (middle price), and 10 % WTP given Price 3 (high price). For each price bid, there were approximately 40 samples. Price 1 was tested in a low-income neighborhood, Price 2 in an average neighborhood, and Price 3 in an upper-middle-income neighborhood. A mixed neighborhood survey site with all income levels was selected so that the pretest interviews could be applied in just one site. Two follow-up pretest surveys were done to recalibrate the price bids. In all, three waves of pretests were conducted.

From the pretest results, a set of the final bids was obtained and used in the actual survey. It was comprised of five bid prices from PHP 1, 5, 10, 50, to 100 (USD 0.02, 0.1, 0.2, 1, and 2).

6.2.6 *Valuation Question*

The exact wording of the WTP question is as follows:

> "Would you vote in favor of ***Whale Shark Conservation Program?*** The **Whale Shark** Conservation Program would mean that you and the other households in Sorsogon would pay a fixed monthly surcharge of PhP___ through the electricity bill. Remember that the surcharge of PhP___ will be collected every month for 5 (or 10) years. All the money that will be raised will go to the **Whale Shark Trust Fund to finance the Conservation Program** described above."

6.2.7 Dealing with Protest Response

Various possible reasons (based on a series of FGDs) as to why respondents would not be willing to pay were included in the questionnaire as a debriefing question. Table 6.2 classifies the possible reasons as a valid "No" or as a protest response. Two of the reasons—"Majority of the poor will be affected" and "Only those from higher income groups should pay for this"—could be either a valid "No" or a protest response. They were classified under the latter category if the respondent was considered poor (according to the asset list she/he reported in the survey). Respondents were allowed to choose a maximum of three reasons as to why they answered "No" to the referendum question. If the sample chose at least one of the "protest No" reasons, she/he was considered a scenario rejecter under the condition that she/he also answered "No" to the follow-up WTP open-ended question. Otherwise, if she/he indicated willingness to pay an amount other than the bid given in the follow-up WTP question, his/her response is considered a valid "No."

Giving the respondents the opportunity to choose three reasons, however, created a problem of interpreting scenario rejecters, as it is possible that a sample would be rejected or censored for being a low-ranked choice.

Table 6.2 Reasons for unwillingness to pay for a whale shark conservation program

Reason
Valid "No" response
I do not think conservation of whale sharks is worth doing
I cannot afford to pay for the program
I think that other species are more important than whale sharks
I prefer giving money to humanitarian cause instead
Only people who will directly benefit from whale shark conservation should pay for this
Only those from higher income groups should pay for this[a]
Majority of the poor will be affected[a]
Protest response (scenario rejecters)
I do not believe that the money I will pay will actually be used for whale shark conservation
I do not like adding the amount to my electricity bill. Find other alternative means
I do not like adding the amount to my electricity bill. Find other alternative means except electricity charge
Only those from higher income groups should pay for this[a]. Majority of the poor will be affected[a]

[a]Protest bid or valid "no" if respondent is actually poor

6.3 Survey Administration

6.3.1 Sampling Procedure

The household survey was administered to Sorsogon residents who are considered to be household heads, e.g., husband, wife, or adult child who is already earning an income. The household head is defined as an adult member of the family who can decide on financial matters for the family. A multistage stratified sampling was conducted to obtain a representative sample.

Sample Size With an initial assumption of approximately 70 % response rate, the survey targeted about 600 households in Sorsogon.

Sample Selection Procedure This procedure involves constructing a sampling frame and selecting from the sampling frame such that the probability each person would be selected was known. Four municipalities (a political unit, next to provincial level), out of the 15 municipalities of Sorsogon, were selected. To distribute the 600 questionnaires/samples proportionally to the 4 municipalities, a simple ratio and proportion step was taken.

The four municipalities were then further subdivided into *barangay* clusters. A *barangay* is the basic political unit in the Philippines consisting of about 100 or more households clustered in a certain locality. For each of the municipalities, *barangays* were classified into four groups by ranking their respective populations in ascending order from the least populated to the most populated. Finally, for the random selection of respondents within the randomly selected *barangays*, a systematic sampling with a random start was followed.

Ten sets of questionnaires were divided between the voluntary and mandatory payment modes, each with five price bid levels (Table 6.3). Given 600 target samples, the questionnaires were randomly distributed to randomly selected respondents. The respondents (samples) were divided in the following test groups:

Table 6.3 Number of samples for the whale shark survey

Split sample	Mandatory payment	Voluntary contribution
No. of actual samples	60 samples/bid	60 samples/bid
No. of bid prices	5	5
Total	300	300
Target (usable questionnaires)	200	200
Pilot testing	40 samples/bid × 3 bids = 120	

6.3.2 Field Protocol

The household survey was implemented over 3 weeks, beginning 10 September 2005.

Drop-Off Protocols The CVM survey was dropped off to randomly selected households and collected after 1–3 days, depending on the commitment made by the respondent. If after two follow-ups the household head had not yet filled out the questionnaire, the enumerator retrieved the unanswered questionnaire and returned it to the survey supervisor for redistribution to other potential respondents.

Before leaving the house of a respondent, the enumerator had to assess the questionnaire to make sure that the respondent did not miss any questions. The enumerator was also instructed to conduct a follow-up personal interview for items in the questionnaire that were not answered. After checking the questionnaire, a debriefing statement was given to the respondent, thanking him/her for his/her participation and informing him/her that there were split samples, so that she/he would not be surprised if she/he found out that a neighbor was asked a different price bid.

6.3.3 Model Estimation Procedure

Parametric Mean and median WTPs in the dichotomous choice format are traditionally calculated from the coefficients relating the WTP responses against a constant and the bid (*BID*). An additional vector of socioeconomic variables (*X*) can also be factored into the model. The coefficients are typically estimated based on a logistic function:

$$P_i\,(Yes) = F_\varepsilon(\Delta V) = \frac{1}{1+\exp(-\Delta\, V)} = \frac{1}{1+\exp\left[-\left(\alpha+\beta_1 BID + X'\beta_2\right)\right]} \tag{6.1}$$

where α and β are coefficients to be estimated and *BID* is the amount the respondent was asked to pay. The logit model above is then estimated using the maximum likelihood estimation method, which is the most common technique (Lee 1997). The log-likelihood function is

$$Log\ L = \sum\nolimits_{k=1}^{N} I_k \ln\ F_\varepsilon(\Delta V_k) + (1-I_k)\ln\,(1-F_\varepsilon\,(\Delta V_k)) \tag{6.2}$$

where I_k is an indicator variable for observation k. If the answer is *yes*, $I_k = 1$; otherwise, $I_k = 0$.

Finally, the mean WTP can be determined using the formula

$$Mean\ WTP = -\frac{\alpha}{\beta} \tag{6.3}$$

where α is the covariate parameters and β is the parameter of *BID*. In this case, median WTP will be equal to mean WTP.

Nonparametric Given the single-bounded, dichotomous choice question with a series of positive price bids, denoted as B_j, $j = 1,2,\ldots,j$, respondents were presented with one of these bid prices, and they indicated whether or not they were willing to pay such an amount. It is assumed that there exists a bid level at a value of zero, which, if presented to a sample, all of the respondents would be willing to pay this amount. Point estimates of the survivor function (S) can only be calculated at each of the J bid levels. The following notations are used for estimating the point estimates:

- If the number of households in the sample is denoted as N, then the subsample facing bid level B_j can be denoted N_j.
- The households replying "Yes" WTP are those respondents that have a higher WTP than the B_j amount. This amount is denoted as n_j.
- Considering the assumption that everyone is willing to pay a nonnegative amount the nonmarket environmental good, it can therefore be set to $S\wedge(B_0) = 1$.

From the equation, a point estimate of the survivor function at each bid level can be generated by dividing the number of respondents who answered "Yes" to that bid level by the number of respondents in that subsample:

$$\hat{S}(B_j) = \frac{n_j}{N} j = 0\ to\ J \tag{6.4}$$

In cases wherein the survivor function is not a nonincreasing function, it will not generate a valid survivor function. To correct such a problem, a technique known as the pooled adjacent violators algorithm can be employed. The technique includes pooling data for two adjacent bid levels for which the estimate of the survivor function for the higher bid level is greater than that for the lower bid:

$$\hat{S}(B_j) = \hat{S}(B_{j+1}) = \frac{n_j + n_{j+1}}{N_j + N_{j+1}} \tag{6.5}$$

Mean WTP is then calculated using (Bateman et al. 2002):

$$MeanWTP = \sum_{j=1}^{J} \hat{S}(B_j)\left[B_j - B_{j-1}\right] \tag{6.6}$$

6.4 Empirical Results

6.4.1 Data

According to the 2000 Family Income and Expenditures Survey (FIES) of the National Statistics Office (NSO 2002), Sorsogon, had an average monthly household income of PHP6,300 (USD)—approximately the same as that determined through the survey—and an average monthly household expenditure of PHP5,800 (USD 116) (Table 6.4). Also, according to the same NSO records, food accounted for the bulk (53 %) of the total family expenditures.

While whale shark conservation is not a household essential, it is a charismatic good that could potentially be subject to a warm glow response and even "yea saying." As such, aside from using drop-off as the survey method and including budget reminder in the cheap talk, the study only targeted respondents who were considered household financial decision-makers, banking on the assumption that they would be more objective with regards to their respective households' financial standings. This explains why the majority of respondents are married, approximately half of them are male, all of them are adults, and most are currently employed (Table 6.4).

Table 6.4 Demographic information (Sorsogon physical framework plan)

Characteristics	Value
Number of households	
Sorsogon province	125,191
Sample observations	588
Mandatory payment samples	293
Voluntary payment samples	295
Sample observation characteristics	
Age (average)	42
Gender: male (% to total)	43.2
Civil status: married (% to total)	77.72
Educational attainment (no. of years)	10.16
Ave. monthly household income PHP (USD)	6,052 (121)
Sorsogon province household economic characteristics	
Ave. monthly household income in PHP (USD)	6,300 (126)
Ave. monthly household expenditure in PHP (USD)	5,800 (116)
Ave. monthly household savings in PHP (USD)	500 (10)

"How certain are you that you would vote "YES"/"NO" in Question Number 11 if such an election whether to implement Whale Shark Conservation Program were to really take place?

Completely sure/Definitely vote YES/NO	
Somewhat sure	
Not very sure	
Not sure at all	

Using the drop-off survey method, a response rate of 98 % was achieved; out of the 600 questionnaires distributed, 588 usable questionnaires were retrieved. For further consideration of potential biases, the WTP responses of the 588 observations were subjected to certainty consideration. Based on the study by Champ and Bishop (2001), when less certain WTP respondents were treated as if they had responded negatively, a statistically significant hypothetical bias disappeared. Uncertainty about the answer to a valuation question implies uncertainty about the trade-off between the amenity in question and the monetary good and therefore needs to be addressed prior to welfare estimation.

Incorporating uncertainty responses in the actual analysis requires that the positive WTP responses by those who answered "Not very sure" and "Not sure at all" in response to the WTP certainty question were changed to negative responses to the referendum question.

Results show that the uncertainty is not considerable, the highest change being in the initial bid price (PHP 1; USD0.02), with 4 % of positive WTP converted to negative WTP (i.e., from 50 % to 46 %; Table 6.5). All of the other bids registered a 1–2 % decline in "Yes" WTP answers. The minimal certainty adjustment signifies the respondents understood the hypothetical scenario.

The 588 observations were divided into two groups, with approximately half of them responding to the WTP question under a mandatory type of payment and the other half to a voluntary type of payment. Factoring in the certainty consideration, results of the WTP question show that if there were to be a referendum on

Table 6.5 Distribution of WTP responses

		Yes response (%)	
Bid price PHP (USD)	Total number of observations	Uncensored	Adjusted to certainty
1 (0.02)	118	50	46
5 (0.10)	119	39	38
10 (0.20)	121	33	32
50 (1.00)	115	17	15
100 (2.00)	115	16	15

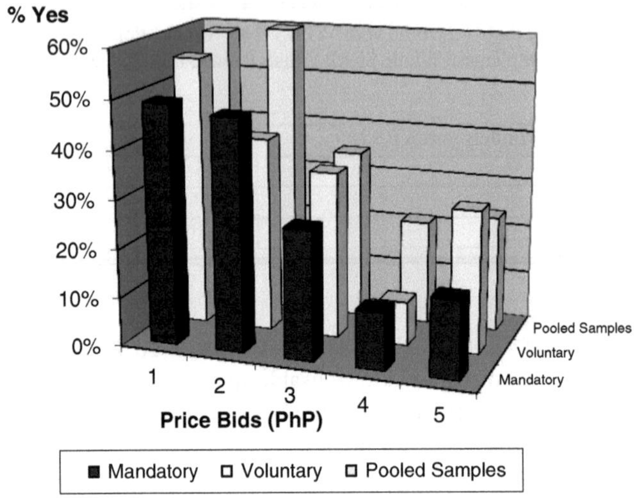

Fig. 6.1 Percentage of respondents

mandatory payment for the implementation of whale shark conservation program, the vote would probably not pass in Sorsogon. As Fig. 6.1 shows, the lowest bid used in the survey was already PHP1 (USD 0.02), and yet only a little over 40 % of the observations in the mandatory split sample survey expressed a positive WTP on this bid level. To use a bid lower than this level would no longer be taken seriously by the respondents as it would appear ridiculously low or without any real economic value. Likewise, only a borderline majority would agree, even to a voluntary payment for conservation.

6.4.2 Logistic Regression

Models and Hypothesized Signs Table 6.6 outlines the hypothetical direction of effects of explanatory variables on WTP, based on theoretical expectations and empirical regularities. The variables are categorized as attitude and awareness, program design, and socioeconomic characteristics. Familiarity with whale sharks, particularly the experience of seeing a live one (*SEEN*), and knowledge of whale sharks (*SCORE1*) are expected to affect WTP positively. The perceived threat to whale sharks (*THREAT*) and people's trust and belief regarding the current situation (*BELIEF*) are also expected to affect WTP positively. On the other hand, the bid price (*WTPAMT*) would negatively affect WTP, implying that as prices increase, the demand for the good—or in this case willingness to buy—decreases. Various socioeconomic indicators are also predicted to influence WTP in different

Table 6.6 Hypothetical direction of effects of explanatory variables on WTP

Significant variable	Description	Expected direction/sign
Attitude and awareness		
Seen	Has seen a live whale shark	+
Score1	Knows about whale sharks (knowledge score)	+
Threat	Threat to whale sharks is known	+
Believe	Belief in the description of the situation described	+
Program design		
Payveh	Payment design (mandatory and voluntary)	+/–
Bid	Bid price	–
Elctrcbill	Liked the proposal to collect people's contribution as surcharge on their electricity	+
Sociodemographics		
Age	Age of the respondent	+/–
Sex	Gender of the respondent	+/–
Status	Civil status of the respondent	+/–
Educattain	Educational attainment in years	+
Tothh	Total number of household members	+/–
Envtorg	Membership in environmental organization	+
Revinc	Total monthly income of the household	+

directions. Age (*AGE*), gender (*SEX*), and civil status (*STATUS*) have no predetermined sign because both younger and older residents, male or female, married or single, are equally likely to prefer such an environmental good. Educational attainment (*EDUCATTAIN*), membership in environmental organizations (*ENVTORG*), and monthly household income (*REVINC*) are expected to affect WTP positively also.

An initial multivariate logit regression analysis was done using Stata to identify the factors associated with responses to the WTP question. Preliminary analysis revealed some of the independent variables were either not significant or highly correlated with other variables. Hence, these variables were excluded in the succeeding logit regression analyses.

Several data sets were generated: one for the raw (uncensored) data/sample responses, another for protest response-adjusted dataset, and another for certainty-adjusted WTP responses. Overall, three models were considered in the logit/econometric analysis using selected covariate specifications related to socioeconomic characteristics.

The three models considered two approaches to dealing with protest votes: (a) consider the "protest No" responses as legitimate zero valuation (Model 1)

Table 6.7 Statistics of important variables included in the preliminary logit analysis

	(Model 1) uncensored	(Model 2) adjusted for protest	(Model 3) adjusted to certainty
Bid	−0.17***	−0.02***	−0.17***
Age	(0.00)	(0.00)	(0.00)
	0.00	0.01	0.01
Sex	(0.00)	(0.01)	(0.01)
	0.06	−0.06	0.11
Status	(0.19)	(0.26)	(0.19)
	−0.38*	−0.24	−0.39*
Educattain	(0.23)	(0.32)	(0.23)
	−0.05	0.02	−0.02
Tothh	(0.03)	(0.05)	(0.03)
	0.04	0.10**	0.04
Revinc	(0.04)	(0.05)	(0.04)
	0.00***	0.00***	0.00***
Envtorg	(0.00)	(0.00)	(0.00)
	0.07	0.18	0.15
Constant	(0.26)	(0.41)	(0.27)
	−0.32	−1.21	−0.92
Pseudo R^2	(0.58)	(0.82)	(0.59)
	0.07	0.22	0.07

Standard error values in between parenthesis
***At 1% level of significance
**At 5% level of significance
*At 10% level of significance

and (b) censor the protest votes (Model 2). Supporting the former approach, Halstead et al. (1992) criticized that the censoring of protest bids may aggregate WTP estimates in a manner that is not easily predicted. On the other hand, Lindsey (1994) argued that when the CVM survey is designed as a referendum, it is less concerned with whether a zero response represents a true valuation and more focused on whether it reflects intended behavior. The referendum model is less stringent in terms of the classification of protest because protest reasons are mostly legitimate influences on actual behavior. A parallel argument was provided by the study of Mc Guirk et al. (1984), which states that protest bids "should be considered legitimate WTP bids as respondents are essentially valuing a proposed policy, not just a commodity."

Model 3 deals with certainty responses wherein the answers of respondents who were not certain about their positive WTP responses were converted to negative answers to the valuation question. In their study of the Mexican spotted owl,

Loomis and Ekstrand (1998) recognized that survey respondents may be uncertain about their values for threatened and endangered species. They used a follow-up question asking how certain respondents were about their "Yes" and "No" answers. The study reported that incorporating uncertainty tends to lower WTP estimates but that such adjustments do not improve the estimate efficiency.

Estimates of the Models Overall, the behavior or direction and significance of the variable in the various models did not register any significant change, although there was an improvement of the overall performance of the model or the Pseudo R2 in the Adjusted to Protest model (Model 2) is summarized in Table 6.7.

The dependent WTP variable (yes = 1 and no = 0) is regressed against a *Constant* term, price bids (*Bid*), income (*Revinc*), household size (*Tothh*), years of schooling (*Educatt*), age (*Age*), gender (*Sex*), civil status (*Status*), and membership to environmental organization (*Envorg*). The majority of variables had positive coefficients; however, only Revinc and Bid variables turned out to be highly significant.

Consistent with a priori economic expectations, the income variable (*REVINC*) is significant to 0.01 and positively associated with a higher WTP for whale shark conservation. The Bid variable also turned out highly significant in all models (as expected) with a negative influence on the probability of the respondent saying "Yes" to the WTP question. This means that the larger the bid value presented to the respondent, the less willing the respondent would be to pay for whale shark conservation.

Effect of Payment Design The survey samples were disaggregated in terms of payment vehicle. Of the total 600 respondents, 588 were deemed valid and included in the analysis. A total of 293 respondents were asked about their WTP a mandatory payment for the conservation of whale sharks, while 295 respondents were asked about voluntary payment vehicle. The study applied hypothesis testing using logistic regression to assess payment design effect (mandatory vs. voluntary). If the P-value for the dummy variable of payment vehicle (*Payveh*) is less than the conventional 0.05, it will result in acceptance of the H_o (null hypothesis) value. It would then imply that there is no difference between pooling the data or treating the data by payment design:

$$\text{Ho} : \text{E}(\text{WTPmandatory}) = \text{E}(\text{WTPvoluntary})$$
$$\text{Ha} : \text{E}(\text{WTPmandatory}) \neq \text{E}(\text{WTPvoluntary})$$

To start off, the socioeconomic characteristics (Tothh, Age, Educattain, and Revinc variables) of the respondents were compared. The *T*-test results show no statistical difference in the characteristics of the respondents for the payment design group comparison. The regression results in Table 6.8 show that some important variables such as Bid and Revinc are highly significant and have the expected signs, but the payment vehicle is not a significant factor affecting respondent WTP

Table 6.8 Logit regression results for payment design effect

Variable	Coefficient	Std error
Bid	−0.02***	0.00
Age	0.01	0.01
Sex	0.11	0.19
Status	−0.39*	0.23
Educattain	−0.02	0.03
Tothh	0.04	0.04
Revinc	0.00***	0.00
Envtorg	0.15	0.27
Payveh	0.03	0.19
Constant	−0.94	0.60
Pseudo R2		0.08

***At 1% level of significance
**At 5% level of significance
*At 10% level of significance

decision. The respondent indifference to the payment design justifies pooling samples in the regression analyses shown earlier.

6.4.3 WTP Estimates

Parametric (univariate and multivariate models) and nonparametric welfare estimation results were compared to assess the consistency and behavior of survey data and how these might affect the WTP estimates. Similar to the multivariate models, logit equations were generated for the univariate models. The parametric results for the logit univariate models are consistent with each other in that the bid coefficients are negative and significant.

Table 6.9 shows that the median and mean WTP have nearly zero values for both univariate and multivariate estimates, except in Model 2, the Adjusted to Protest model, since a lot of negative WTP samples were omitted. The nonparametric estimate generated a positive but relatively low mean WTP value, ranging from PHP 17 (USD 0.34) to PHP 35 (USD 0.70) per household, per month.

The results are not implausible as they are consistent with the socioeconomic characteristics of the samples, particularly the income factor. The estimate is consistent with the survey results of a very low positive WTP response even to the lowest bid, signifying a likely failure of a referendum for implementing a whale shark conservation program. Also, only a borderline majority is agreeable to a voluntary type of payment.

Table 6.9 Parametric and nonparametric mean WTP estimates in PHP and (USD)/HH per month

	(Model 1) uncensored	(Model 2) adjusted for protest	(Model 3) adjusted to certainty
Multivariate model	0	46 (0.92)	0
Univariate model	0	38 (0.76)	0
Nonparametric estimate	18 (0.36)	35 (0.70)	17 (0.34)

Table 6.10 Respondents' perceptions on environment and natural resources issue

Issue	% of observations (588 in total)
National priorities	
Economic problems	43.71
Poverty	29.25
Environment	0.68
Environment not cared for	81.63
Three most important environmental problems	
Solid wastes	31.29
Deforestation	26.87
Water pollution	11.73

6.4.4 *Motivation to the Payment Question*

Perception of Environmental Issues and Familiarity with Whale Sharks The majority of respondents (82 %) believe the environment is not being properly cared for. The three main environment-related concerns are solid waste, deforestation, and water pollution. Although respondents generally believe the environment is neglected, still its environment does not rank as a top priority among the other national concerns (Table 6.10).

Although 70 % of the respondents had not seen a live whale shark, the majority of them had a good level of knowledge of them. As Table 6.11 shows, most respondents scored highly in the four general questions pertaining to whale sharks; almost half earned a perfect score.

Conservation Program's Perceived Effectiveness and Acceptability The majority (81.12 %) of respondents recognize there are real threats to the existence of whale sharks (Table 6.12), and respondents who believe in the effectiveness of the conservation program are more inclined to give a positive WTP response.

On the proposal to collect contributions through surcharges on electricity bills, only a few respondents (32.62 %) agreed to such a scheme (Table 6.11). Most respondents who gave a negative WTP response consistently answered that they do not like the electricity bill surcharge. At first glance, this alone appears to be grounds for considering negative WTP responses on the basis of the payment

Table 6.11 Assessment of importance of species and knowledge of whale sharks

Indicator	% of observations
Seen a live whale shark	25.34
Knowledge of whale sharks	
Whale shark is the largest fish in the world	74.32

Table 6.12 Perceptions on the proposed conservation program

	% of "believes/accepts" observations				
Group	Threats	Status	Program effectiveness	Electricity bill surcharge	Total respondents
All observations	81.12	90.65	67.69	32.65	588
Respondents with WTP	94.05	97.30	92.43	80.00	185
Respondents without WTP	75.19	87.59	56.33	10.92	403

vehicle as a clear case of protest responses. Closer probing of the responses to the debriefing questions reveals that the primary reason for nonacceptance of the electricity bill surcharge proposal is due to respondent fear of electricity bills increasing. This information, reinforced by the economic status of households in Sorsogon, bolsters the assumption that these respondents are merely expressing that they could not afford further strain on their finances.

Willingness to Support the Conservation Program Having established the respondents' attitudes toward environmental issues and familiarity with whale sharks, other motivation questions were asked to assess potential commitment to support the conservation program. More than half of the respondents (60.71 %) strongly agree that everyone is responsible for ensuring that plants and animals, as they are known today, will exist for mankind in the future. This suggests that the respondents hold a perceived bequest value. Likewise, some strongly agree to their existence value, saying that the endangered species are important even if they do not get to see the whales. Though they agree to the importance of preserving endangered species in principle, when it comes to actual cash outlay to finance the cause, only 9.69 % strongly agree that cash donation is the best payment vehicle and fewer still (5.27 %) agree on increasing taxes to pay for endangered species conservation makes the most sense (Table 6.13).

6.5 Implications on Whale Shark Conservation

The survey results show that people are aware and concerned about environmental issues, including the precarious condition of whale sharks in Sorsogon. They are not, however, able or willing to pay for the implementation of a conservation

Table 6.13 Respondents' opinions on endangered species conservation

Issue	% of observations
It is everyone's duty to ensure that plants and animals as we know them today will exist for mankind in the future	60.71
Endangered species are important even if I don't get to see them	29.08
Citizens should contribute to endangered species conservation by making cash donations to this cause	9.69
Government should raise taxes to pay for more endangered species protection	5.27
Total respondents	588

Table 6.14 Approximate benefit and cost values related to whale shark conservation in Sorsogon

Benefit and cost	Value in '000 PHP ('000 USD)
Potential revenue of the conservation program	0
Use (nonconsumptive) value of whale sharks	35,000 (700) per year[a]
Program cost	7,375 (148)[b]

[a]Manila Times, August 2006
[b]Excluding yearly operation and maintenance cost and the potential foregone use value

program because issues of poverty, employment, and other economic concerns take precedence over environmental issues. This result confirms the ubiquitous notion about developing countries and environmental improvements that Solow (1993) summarized in the following way: "The poor countries seem to have a choice between cooperating in the degradation of their own environment or acquiescing in their own poverty." Choe et al. (1996) pointed out that the policy message for developing countries seems to be to, "Wait until your economy has grown more before spending much effort trying to take care of the environment."

The cost of conservation is not cheap. Based on the experience of other conservation programs, roughly PHP 7.38 million (USD 148,000) is needed to enhance the existing conservation project of Donsol's local government and implement a more comprehensive whale shark conservation program in Sorsogon (details provided in Annex). This estimate does not yet include the yearly operation and maintenance to keep the proposed comprehensive program going. What the study results say is that the Philippines cannot afford to pay for whale shark conservation (Table 6.14).

Wildlife conservation efforts that depend on public funds often face major systemic barriers that hamper effective conservation efforts. There is chronic underfunding of wildlife conservation in developing countries. Whale sharks are migratory by nature, and the scale and urgency of the response toward their conservation require a significant amount of resources and international collaborative partnerships. The international community must act to protect failing whale shark populations.

Aside from the inherent ecological/biological value of whale sharks, the Philippines would gain significant economic benefits if whale sharks are protected.

Ecotourism through whale shark watching is continually gaining ground as a strong economic force in the region. In 1998, Sorsogon had only about 900 visitors; in recent years, the area has seen an average of 8,000 tourists every whale shark season—a third of whom are foreigners. This industry creates about 1,000 seasonal jobs and raises revenues in the food, transportation, and hotel sectors. In 2005, the World Wildlife Fund (WWF) reported that whale shark watching contributed at least PHP 35 million to the country's economy during the whale shark season from February to May—such income is vital to Sorsogon and nearby areas.

Aside from gaining support by partnering with the international community through bilateral and multilateral cooperation, the Philippine government should explore other conservation financing mechanisms. A WWF-funded study by Koteen (2004) identified multilateral agencies that provide species conservation funding: the Global Environment Facility (GEF), the United Nations Development Programme (UNDP), United Nations Environment Programme (UNEP), and the World Bank. Some multilateral agencies, such as regional development banks, provide loans to governments rather than grants or donations. Regional development banks are also beginning to collaborate with the GEF to leverage additional resources for species conservation. Koteen's study also provides a menu of schemes that governments can use to stabilize and even increase allocations for species conservation such as earmarking revenues collected through various fiscal instruments (e.g., government bonds, lotteries, premium-priced motor vehicle license plates, wildlife stamps, and tax incentives) (Table 6.15).

Table 6.15 Menu of options for financing conservation

Financing mechanism	Source of revenue
Government revenue allocations	
Taxes and bonds earmarked for conservation	Taxpayers, investors
Real estate and development taxes	Property owners, property developers
Lottery revenues	Gamblers
Premium-priced motor vehicle license plates	Vehicle owners
Wildlife stamps	Postal customers, hunters, fishers
Economic instruments to stimulate environmental investment	Investors
Debt relief	Donors, governments, NGOs
Grants, donations, and loans	
Bilateral and multilateral agencies	Donor agencies
Foundations	Individuals, corporations
Conservation trust funds	Multisource
Nongovernmental organizations	NGO members and supporters
Private sector	Private companies

Source: Koteen (2004)

It is interesting to note that after an elaborate preparation and implementation of the survey and analysis of the CV data, results show zero (or close to zero) welfare values, and that is what the exercise was intended to do: to illustrate that with meager resources, people in a developing country like the Philippines are not willing (or able) to pay for the conservation of wildlife. The government, as it stands now, has limited a budget to provide for basic services for its citizens; hence, funding for species conservation is low (as expected). Financing of conservation programs, therefore, necessitates assistance from the international community. Be that as it may, there is a lot at stake for the Philippines to ensure the continuous existence of whale sharks, and the country would do well to proactively explore alternative financing options for conservation.

Annex: English Version of the Questionnaire

Conservation Value of Whale Sharks: Survey of Sorsogon Residents

Introduction:

Good day! This is a survey sponsored by the Economy Environment Program for Southeast Asia (EEPSEA). The purpose of this survey is to find out how people in SORSOGON think about a conservation program to protect whale sharks in Donsol, Sorsogon. A similar survey is being done in three other countries: Thailand, Vietnam and China, but each of this country looks into a different endangered species. Here in the Philippines, we are focusing our study on whale sharks in Sorsogon.

Your household was randomly chosen to be part of the study. It will probably take you about 30 minutes to complete the questionnaire. I would like to assure you that whatever information you will reveal in the survey form will only be used for this research. Please take note that the information presented in the questionnaire below regarding the Conservation Program for whale shark is not yet in existence. The Program is presented, for the purpose of this survey, so that we can get your opinion on this matter.

If anything is unclear, please make note of any questions you might have. We will try to answer your questions when we come back to pick up the questionnaire.

There is no right or wrong answer to the questions. We only want to find out your honest opinion. We would like to request that only the <male/female> *(to be identified by the enumerator)* household head (husband/wife/ employed sons or daughters) should answer this questionnaire. However, you may consult with other members of your household when answering the questionnaire if you wish. We also request that you NOT discuss the questions with your neighbors or other people outside your immediate household before you provide your answers.

If this questionnaire is fully accomplished, you will qualify for a raffle, which will give you a chance to win any **ONE** of the following prizes: (Please check the prize of your choice – check only ONE)

______ two-year subscription of National Geographic magazine
______ All expenses paid tour for Donsol whale shark watching for two (2)

______ Grocery coupon worth ₱ 10,000

This raffle will be drawn on December 15 at Sorsogon Provinicial Capitol. We will call you in case you will win.

Name of the respondent (optional)	
Complete Address	
Phone Number	
Date form was left	
Date form was picked up	
Interviewer/Facilitator	

SECTION 1: Problems Facing the Country

1. In your opinion what are the **THREE BIGGEST PROBLEMS** facing our country today? On the right column, place **1** if you think it is the biggest problem, **2** if it is the second biggest, **3** if it is the third biggest problem.

Problem	Rank (Choose which is 1, 2 & 3)
a. Economic Problems (e.g., inflation)	
b. Poverty	
c. Education	
d. Health	
e. Crime, violence, inequality	
f. Government and Governance	
g. Infrastructure (e.g. roads, water)	
h. Environment	
i. Terrorism	
j. Relations with other countries	
k. Others: ______________________	

2. Do you think our environment and natural resources are properly taken care of? Please tick your choice.

☐ Yes ☐ No

3. What do you think are the **THREE (3) MOST** important issues related to nature and human impact on the natural environment? On the right column, place **1** if you think it is the most important problem, **2** if it is the second important, **3** for the third most important.

(Choose only 3)

Natural Resource & Environmental Problem	**Rank (Choose which is 1, 2 & 3)**
Air pollution	
Water pollution	
Solid waste	
Increasing number of endangered species	
Deforestation	
Traffic problems	
Flooding due to soil erosion	
Enhanced greenhouse effect (reason why heat cannot exit earth)	
Destruction of coral reefs (this is the spawning and feeding area for fish)	
Others, pls. specify ________________	

SECTION 2: Attitude Toward Conservation & Knowledge of Whale Sharks

4. In terms of endangered species protection, which species do you believe is more deserving of protection? On the right column, place **1** if you think it is the **most important species**, 2 if it is the second important, 3 for the third most important species.

	Species	**Rank**		**Species**	**Rank**
A	 **Dugong**	________	**D**	 **Whale Shark**	________

B	 **Rhino**	______	E	 **Spoonbill**	______
C	 **Marine Turtle**	______	F	 **Eagle**	______

5. Please read the following statements and tell us your opinion by choosing a grade between Strongly agree/Agree/ Indifferent/Disagree and Strongly disagree for all the issues enumerated below. Please remember that there is no right or wrong answer to these questions. Please check (√) the column to enter your answer.

    Note: Endangered species are plants and animals considered to be facing a high risk of extinction.

STATEMENT TO AGREE/DISAGREE ON...	Strongly Agree	Agre e	Indifferent/ Neutral	Disagree	Strongly Disagree
The government should raise more funds to deal with environmental programs in this country.					
There are more important environmental concerns than endangered species conservation.					
Poaching of wildlife species should be punishable by law.					
It is everyone's duty to ensure that plants and animals as we know them today will exist for mankind in the future.					
Citizens should contribute to endangered species conservation by making cash donations to this cause.					
Endangered species are important even if I don't get to see them.					

The government should raise taxes to pay for more endangered species protection.					
The government should first invest in helping people before it spends money on endangered species protection.					
Households who earn more income should pay higher taxes in order to pay for endangered species conservation.					
Endangered species conservation should not be a high priority concern of the government.					

The following questions are about whale sharks.

6. Have you ever seen a live whale shark?

Yes	
No	

Please tell me whether the following statements are "True" or "False" – or just indicate "Don't Know" if you are unsure of the answer.

7. The whale shark is the largest fish in the world.

True	
False	
Don't Know	

8. Whale sharks do not stay in one place. They migrate to different

parts of the ocean.

True	
False	
Don't Know	

9. The congregation of whale shark in Donsol is claimed to be the

largest in the world.

True	
False	
Don't Know	

10. The whale sharks have slow reproductive rates and the continuing presence of one or a combination of these threats could easily wipe out the entire population.

True	
False	
Don't Know	

Important: Please read the brochure attached to this survey form where you can find the details about the Whale Shark Conservation Program before you proceed with this survey. The succeeding part of the survey would not make sense to you and it would be very difficult to answer the following survey questions without reading the brochure first. ***You can tear off the brochure from this survey form and keep it for your own use after you've read it thoroughly.***

SECTION 3: A Conservation Plan for the Protection of Whale Shark

Implementing the many good ideas in the Conservation Program would require a lot of money. So far, there is no clear agenda where to get funds to carry out the activities in the Conservation Program.

The local government of Donsol regularly earmarks a portion of its budget for whale shark protection. However, this is not enough to cover the expenses for the implementation of the Conservation Program. One possible source of fund is to set up a **Whale Shark Trust Fund** wherein residents of Sorsogon will give an amount to the Fund by way of the **Whale Shark Conservation Fee**. However, setting up a Whale Shark Conservation Fund is not easy. It takes considerable effort.

We are undertaking this survey to find out if enough people from Sorsogon would be willing to pay a share to the fund to make it worthwhile. One proposal for fund collection is that people would give to the Trust Fund by paying a **surcharge to their SORECO electricity bill which would serve as their part for setting up Whale Shark Trust Fund**. The amount of the surcharge would be fixed for all households (meaning it would not vary by the level of household's electricity consumption and it would not vary in time). We know that there is not such a close connection between whale sharks and electricity bill. But after a series of consultations, we found out that it is only by this way would it be possible to collect payments from almost all households so that the responsibility for protecting whale sharks in Donsol will not just fall on select people or sectors. Given this situation, we are assuming that SORECO would be willing to collect the surcharge and **ALL** the money would be transferred to the Trust Fund and will only be used for Butanding Conservation Program. Naturally, the very poor households in our society who do not have electricity connection will not be required to pay anything for the conservation.

Let us suppose that there would first be *voting*. The purpose of this referendum would be to see how many people in Sorsogon would be willing to support a plan *to impose a monthly fixed surcharge on everyone's electricity bill*. Suppose that the SORECO bill surcharge would be adopted to support the *Whale Shark Conservation Program* only if greater than 50% of the people would vote positively to this plan.

Assume that this fixed surcharge would be added to your SORECO ***bill every month for the next five (5) years*** . **All funds** raised through the SORECO surcharge **would go to the Whale Shark Trust Fund** that will be **managed and administered** by the ***Whale Shark Conservation Management Board,*** a multi-sectoral entity. In other words, SORECO will not have any role in managing the funds, it will only collect and will automatically transmit the money to the Trust Fund every month.

The reason that the surcharge would end after five (5) years is that it is expected that this would raise enough money to establish a trust fund. The income from this trust fund could be used by the *Whale Shark Conservation Program* far into the future.

PLEASE REMEMBER: The **survey** you are participating in today **is only to find out your opinion** about this matter. We are simply interested in finding out how you would vote **IF** an actual referendum would take place.

Past studies have found that many people say that they agree with a program like this when they are asked their opinion in a survey, but they actually vote "No" when the actual voting takes place. In other words, respondents have a tendency to say things in the survey that they do not really mean to do. Researchers are not sure why people do this. It may be because it is feels good to say "Yes" in a survey when people do not actually have to pay. Or it may be to please the person dropping off the survey.

We are requesting that you would answer based on how you would really vote if there will be an actual poll. Please vote positively in this survey only if you are really willing to pay a surcharge in your electricity bill for Whale Shark Conservation Program.

11. Would you vote in favor of ***Whale Shark Conservation Program?***

The **Whale Shark** Conservation Program would mean that you and the other households in Sorsogon would pay a fixed monthly surcharge of PHP___ through the electricity bill. Remember that the surcharge of PHP ___ will be collected every month for 5 (or 10) years. All the money that will be raised will go to the **Whale Shark Trust Fund to finance the Conservation Program** described above.

Yes (*Go to Number 15*) No (*Go to Number 12*)

12. If you voted "No", what are the reasons why you did not vote for the program? Please tick the most appropriate answers (limit your answers to 3).

- ☐ I do not think the conservation of Whale Sharks is worth doing.
- ☐ I do not believe that the money I will pay will actually be used for Whale Shark Conservation.
- ☐ I think that other species are more important than Whale Sharks.
- ☐ I do not like adding the amount to my electricity bill. Find other alternative means except electricity surcharge.
- ☐ A lot of poor people will be affected.
- ☐ I prefer giving money to humanitarian cause instead.
- ☐ I do not earn enough money that can pay for the Program.
- ☐ Only people who directly benefit from Whale Shark conservation should pay for this.

☐ Only those from higher income groups should pay for this.
☐ Others (pls. specify) ______________________________

13. IF you voted "No" to the proposal of setting up the Whale Shark Conservation Program given that this will cost your household a monthly payment of <bid> over 5 years, is there any amount that you would be willing to give to support the Whale Shark Conservation Program?

Yes *(Go to Number 14)* No *(Go to Number 16)*

14. If YES, what amount would this be? <Currency>___________/month--*(Go to Number 15)*

15. If you voted "Yes", what was it about the Conservation program that convinced you to vote positively for it? Check the most appropriate answers (limit your answers to 3).

☐ The whale shark is a special animal and should be protected.
☐ It is high time that people in Sorsogon do something concrete about protecting the whale shark in Donsol.
☐ This initiative can lead to more protection efforts for other endangered species in the country.
☐ Other, please specify____________________________________

3. Before you began answering this questionnaire, did you think there were real threats to whale sharks as described? Please check your answer.

☐ Yes ☐ No

17. Please rank the following according to how effective you think they would be in encouraging people to contribute to a Whale Shark Conservation Fund. **Write number 1 in the box** for the method you think is **most effective**, and then **2, 3, 4, etc. for the methods you think are less effective**.

	Rank (1, 2 and 3)
Provide more information to the public about the problems of Whale Shark	
Provide more information about the charitable organizations and their activities so people know the channels how to help	
Create more transparency and accountability about how to help	

Rely on celebrities to promote and disseminate information about the importance of protecting whale sharks	
Make it convenient for people to donate	
Get organizations to publicize their activities	
Others, specify ___________________________________	

You're almost finished answering this survey. At this point, we'd like to give you a chance to think back and further evaluate your answer to the referendum question I asked you earlier regarding the Conservation Plan to protect whale sharks.

18. How certain are you that you would vote "Yes"/"No" in Question Number 11 if such an election whether to implement Whale Shark Conservation Program were to really take place?

Completely sure/Definitely vote YES/NO	
Somewhat sure	
Not very sure	
Not sure at all	

0.

19. When you decided on your vote, did you believe the information provided in this survey with regards to the status of the whale sharks? Please check your answer.

☐ Yes ☐ No

20. If no, why not? Please limit your answers to 3.

21. When you decided on your vote, did you believe that the Whale Shark Conservation Program will actually be effective in saving the Whale Sharks? Please check your answer.

☐ Yes (*Go to Number 23*) ☐ No (*Go to Number 22*)

22. If no, why not? Please check the most appropriate answers (limit your answers to three)?

☐ 1. The funds may not be used to support the program activities due to graft and corruption.

☐	2. The funds may not be used to support the program activities since the government may channel it to other uses.
☐	3. The funds may not be remitted on time by the collecting agency
☐	4. Others, pls. specify________________________________

23. When you decided on your vote, did you like the proposal to collect people's contribution as a surcharge on their electricity bill?

☐ Yes(*Go to Number 26*) ☐ No (*Go to Number 24*)

24. If No, why not? Please check your answer/s. (Limit your answer to three).

	a. The electricity bill is always increasing. I'm afraid that paying for the Conservation program will further cause the increase of my bill.
	b. I can't see any connection between electricity and whale shark conservation – it doesn't have any basis.
	c. Not everyone is connected to SORECO – how can you collect from those not connected?
	d. The collection should not be mandatory every month - why can't we just pay when we want to?
	e. Monthly collection is too complicated - why not make this an annual payment?
	4. SORECO is not bound by law to do this – I don't know what will make them agree to do this task for Whale Shark Conservation Program.
	g. Others, pls. specify : ______________________________

25. If you don't agree to the electricity bill surcharge, do you have other suggestion as means of collecting payment from the people of Sorsogon for Butanding Conservation Program?

We are almost finished but before we wrap up, we'd like to ask some general information about you and your household that we will use as basis for analyzing your answers.

SECTION 4: Household Data

26. What is the main source of drinking and cooking water in the house? Please check all applicable.

Private or shared water connection	
Own shallow well/hand pump	
Bottled water	
Water vendor	
Public (communal) tap	
Public (communal) well or hand pump	
River/Spring	
Rain water	
Other (*please specify*)	

27. How much was your household's own water bill last month? Or if you share the bill, how much was your share?

Monthly water bill	PHP

Electricity

28. Does your household have electricity?

☐ Yes ☐ No *(Go to Number 31 & following)*

29. Does your household pay the entire electricity bill, or share the bill with anyone outside your household?

Household paid entire bill	
Shared bill with someone else	

30. How much was your household's own electricity bill last month? Or, if you share the bill, how much was your share?

PHP_____________ - last month

31. What is the main fuel used for cooking in the house? Please check your answer.

Electricity	
Bottled Gas (LPG)	
Kerosene	
Charcoal	
Firewood	
Others	

House Characteristics

32. Does your household own this house/apartment? Please check

One. Own	
Rent	
Provided by Employer	
Others, pls. Specify	

33. How many rooms does your house have (excluding the kitchen and bathrooms)

# of rooms_____________

Socioeconomic profile of the HOUSEHOLD HEAD-respondent (information should pertain to source of information).

34. . How old are you?

Age

35. Gender

☐ Male ☐ Female

36. Civil Status

☐ Male ☐ Female ☐ Others, pls. specify ____________

37. Religion

Roman Catholic	
Protestant	
Islam (Muslim)	
Buddhism	
Hindu	
Confucianism	
No religion	
Other (please specify)____________	

38. Are you able to read newspaper? Please check one

	Able to Read
	Yes, but with some difficulty in understanding
	No, I can't read it

39. Highest Educational Attainment. (PLEASE GIVE the highest level attended on the left column. For example, college – 4th year)

No formal schooling	
Elementary	
High school	
Vocational	
College	
Higher than Master's degree	
Master's	

40. Please list number of household members per age group

Children (<12 yrs)	
Teens (13-17 yrs)	
Adults (above 18 yrs)	

41. How many in your family, including yourself, are earning cash income?

☐ Male ☐ Female

42. Please check the average monthly HOUSEHOLD income bracket where your household belongs (include the cash earnings of all family members who are working or gainfully employed, including yourself). Please check one

Below PHP 5000	
PHP 5,000-10,000	
PHP 10,001-15,000	
PHP 15,001-20,000	
PHP 20,001-25,000	
PHP 25,001-30,000	
PHP 30,001-35,000	
PHP 35,001-40,000	
PHP 40,001-45,000	
PHP 45,001-50,000	
PHP 50,000 & above	
Below PHP 5000	
PHP 5,000-10,000	

43. Please indicate **how many** items your household owns for each of the following

Fan	
Radio	
Television	
Refrigerator	
Washing Machine	
Motorcycle	
Air conditioner	
Computer	
Car	
Fan	
Radio	

Economic Status and Access to Credit

44. How would you classify the economic status of your household relative to others in this country? Please check one.

Much better than most people (rich)	
Better than most people (relatively well off)	
About average	
Below average	
Much worse than average	
Don't know	

45. How easy would it be for your household to borrow PHP 5,000 from a bank or someone who was not a relative?

Very Easy	
Somewhat easy	
Somewhat difficult	
Very Difficult	
Impossible	
Don't know/not sure	

46. In the past year, did your household make donations to any charitable causes?

☐ Yes ☐ No

47. Are you a member of any environmental organizations?

☐ Yes ☐ No

48. Did you discuss the questions in this survey with other household members before you answered them?

☐ Yes *(Go to Number 49)* ☐ No *(Go to Number 51)*

49. If Yes, which of the following best describes how your household answered the questions in this survey?

I discussed some of the questions with others, but the answers I gave represented my own opinions	
We discussed together how to answer the questions together and gave our household's best judgment	
Somewhat difficult	

50. How much time do you think you spent discussing the questions with other members of your household?

<5 minutes	
6-15 minutes	
16-30 minutes	
31-60 minutes	
> 60 minutes	

51. Do you think your answers to the questions would have been different if you had been able to discuss them with other members of your household?

☐ Yes ☐ No ☐ Don't know

References

Bateman I, Carson RT, Day B, Hanemann M, Hanley N, Hett T, Jones-Lee M, Loomes G, Mourato S, Ozdemiroglu E, Pearce D, Sugden R, Swanson J (2002) Economic valuation with stated preference techniques: a manual. Edward Elgar Publishing, Cheltenham

Champ P, Bishop R (2001) Donation payment mechanisms and contingent valuation: an empirical study of hypothetical bias. Environ Resour Econ 19:383–402

Choe K, Whittington D, Lauria D (1996) The economic benefits of surface water quality improvement in developing countries: a case study of Davao, Philippines. Land Econ 72:519–527

Halstead JM, Luloff AE, Stevens TH (1992) Protest bidders in contingent valuation. Northeast J Agric Resour Econ 21(2):160–169

Koteen S (2004) Financing conservation: a menu of options. World Wide Fund-Philippines, Makati City

Lee Choong-ki (1997) Valuation of nature-based tourism resources using dichotomous choice contingent valuation method. Tour Manag 18(8):579–591

Lindsey G (1994) Market models, protest bids, and outliers in contingent valuation. J Water Res Plan Manag 120:121–129

Loomis JB, Ekstrand E (1998) Alternative approaches for incorporating uncertainty when estimating willingness to pay: the case of Mexican Spotted Owl. Ecol Econ 27:29–40

Mc Guirk AM, Stephenson K, Taylor DB (1984) The use of Tobit analysis in the valuation of non-market resources. Department of Agricultural Economics SP89-25, Polytechnic Institute and State University, Blacksbury

NSO (National Statistics Office) (2002) Family income and expenditure survey. National Statistics Office, Manila

Solow R (1993) An almost practical step towards sustainability. Resour Pol 19(3):162–172

Chapter 7
Mobilizing Resources for Marine Turtle Conservation in Asia: A Cross-Country Perspective

Orapan Nabangchang-Srisawalak, Jin Jianjun, Anabeth L. Indab, Truong Dang Thuy, Dieldre Harder, and Rodelio F. Subade

Abstract This study reports the results of a comparative study conducted in China, the Philippines, Thailand, and Vietnam that assessed households' willingness to pay for marine turtle conservation and the potential to mobilize funds. Results suggest that many people place a low priority on marine turtle conservation as compared to other public policy issues. When asked whether they would vote "for" or "against" a policy that would impose a monthly surcharge on residential electricity bill, majority of the respondents answered that they would only vote to support this

---

This article has been expanded from an earlier version under the same title published in the ASEAN Economic Bulletin. Vol. 25 No. 1 April 2008.

O. Nabangchang-Srisawalak (✉)
School of Economics, Sukhothai Thammatirat Open University, Bangkok, Thailand
e-mail: orapann@gmail.com

J. Jianjun
College of Resources Science & Technology, Beijing Normal University, Beijing, China

Resources, Environmental & Economic Center for Studies, Beijing, China
e-mail: jjjin@bnu.edu.cn

A.L. Indab
Resources, Environment and Economic Center for Studies, San Juan City, Philippines
e-mail: bates_89@yahoo.com

T.D. Thuy
Faculty of Development Economics, Environmental & Natural Resources Unit, University of Economics, Ho Chi Minh City, Vietnam
e-mail: thuy.truong@ueh.edu.vn

D. Harder
Department of Economics, College of Economics & Management, University of the Philippines Los Baños, Laguna, Philippines
e-mail: dieldre@yahoo.com

R.F. Subade
Division of Social Sciences, University of the Philippines Visayas, Iloilo, Philippines
e-mail: redabus@yahoo.com

N. Olewiler et al. (eds.), *Marine and Coastal Ecosystem Valuation, Institutions, and Policy in Southeast Asia*, DOI 10.1007/978-981-10-0141-3_7

policy if the surcharge is only USD 0.02 per month. If the poor were to be exempted, this modest surcharge would only generate a sum of USD 1.52 million per year (less than 8 % of the total global expense for marine turtle conservation). Nevertheless, there is some potential for voluntary contributions. Based on the percentages of respondents who would voluntarily pay USD 1 per month, the potential revenue could reach USD 50 million per year, but mobilizing these also presents problems. For example, the voluntary payment was explored by asking the respondents to *check off* the option to contribute to a marine turtle conservation program on their monthly electricity bills. While that might work once, it is unlikely that this can be repeated for other endangered species and environmental causes. The traditional prescription of *raising awareness* is unlikely to generate support, as urban Asians are already well informed about the existence and plight of marine turtles. Efforts to develop conservation-financing mechanisms should therefore be directed to a different, albeit more difficult, direction, that is, to improve people's trust in the government tax collection and expenditure systems. Charities could explore the potential for voluntary contributions from the relatively small population segment willing to contribute voluntarily and develop cost-effective ways of collecting payments. Finally, until Asia develops higher per capita incomes and trustworthy payment vehicles, the international community will need to take on a significant role in financing conservation in the region.

Keywords Willingness to pay • Turtle conservation • Contingent valuation method (CVM) • Voluntary payment • Southeast Asia

7.1 Introduction

Marine turtles are important, not only for their economic and intrinsic value, but because the presence of an adequate population of marine turtles is often an indicator of healthy marine ecosystem (Perrine 2003). Of the seven species of marine turtles, four are classified by the International Union for the Conservation of Nature and Natural Resources (IUCN) as critically endangered, while two are in the next highest risk category (IUCN 2002). The marine turtle status in Asia is of interest for two reasons. First, human activity in the region presents a wide variety of threats, including excessive and illegal harvesting for meat, shells, skin, and eggs; habitat loss from development of beaches; destructive fishing methods, such as dynamite fishing and use of drift nets; and pollution from shipping and tourism. Many of these threats are increasing rapidly with economic growth (IUCN 2002; Safina 2006). Second, marine turtles are a migratory species; their habitat is spread throughout a large number of countries such as China, the Philippines, Cambodia, Thailand, Malaysia, Vietnam, and Indonesia. Thus, international coordinated policies to conserve marine turtles are more likely to be effective than those pursued by countries on their own. There is evidence of willingness of countries in East and

Southeast Asia to collaborate, but so far the measures taken have not been adequate to the challenge.[1]

This paper reports the results of a comparative research project carried out in China, Thailand, the Philippines, and Vietnam. The study explored the prospects for increased regional or national efforts to conserve marine turtles in Asia, whether Asians value turtles more for their use as food, shells, etc., than for nonuse values; whether Asians are aware of marine turtles and their plight; and whether there is sufficient local willingness to pay to support larger conservation efforts. Using a common survey instrument, the contingent valuation method (CVM) was applied to assess the willingness of local populations to pay for the conservation of marine turtles. This study also explores how various payment vehicles affect people's decisions to support national and regional conservation plans. The survey instrument included an extensive set of attitudinal questions that allowed assessment of the relationship between respondent attitude, socioeconomic characteristics, and willingness to pay. The surveys were dropped off at residences and administered using similar procedures and protocols in each study country. Altogether, 3680 respondents participated in the survey; recipients were randomly selected from across all administrative districts in Beijing, Ho Chi Minh City, Hanoi, Bangkok, and Davao City. The most populous of these cities is Beijing with over 15 million people; the smallest is Davao City in the Philippines, with about 1 million. Average annual per capita income ranges from USD 540 for Ho Chi Minh City/Hanoi to USD 2490 for Bangkok. There are also variations in culture and familiarity with the uses of marine turtles.

7.2 Attitudes Toward Environment, Wildlife, and Marine Turtles

Respondents ranked ten public policy issues: economic problems, poverty, education, health, crime/violence/inequality, government/good governance, infrastructure, environment, terrorism, and relations with other countries. The survey revealed that people in all four countries accord relatively low priority to environmental protection—only in Beijing does it appear among people's top 3 concerns (see Table 7.1). While environmental concerns do not feature as priority concerns, over 70 % of the respondents in all four countries agree that environmental problems are not properly addressed.

Respondents were asked to rank nine issues: air pollution, water pollution, solid waste, loss of endangered species, deforestation, traffic congestion, soil erosion, global warming, and destruction of coral reefs. Among the top 3, there were few surprises. It appears people are primarily concerned with environmental problems

[1] An example is the Turtle Islands Heritage Protected Area (THIPA), which is a transboundary protected area initiated in 1997 and has been jointly implemented by the Philippines and Malaysia.

Table 7.1 Social and economic priorities

Rank	Beijing $N=600$	Davao city $N=847$	Bangkok $N=789$	Ho Chi Minh city/Hanoi $N=1444$
1	Economic problems	Economic problems	Economic problems	Economic problems
2	Good governance	Poverty	Poverty	Good governance
3	Environment	Good governance	Good governance	Education

N = the total number of respondents surveyed in each city

Table 7.2 Environmental priorities

Ranking	Beijing $N=600$	Davao city $N=847$	Bangkok $N=789$	Ho Chi Minh city/Hanoi $N=1444$
1	Air pollution	Deforestation	Traffic congestion	Air pollution
2	Water pollution	Solid waste	Deforestation	Water pollution
3	Traffic congestion	Air pollution	Air pollution	Deforestation

N = the total number of respondents surveyed

that affect their daily lives. Davao City was the only city where a nonurban issue (deforestation) ranked first, probably because it is located close to natural forests and has faced increasing episodes of flooding. In none of the cities did wildlife conservation appear among the top 3 environmental concerns (see Table 7.2).

Studies have highlighted that conservation efforts for lesser-known species have less public appeal than for those species that are more charismatic, cute, or familiar (Tisdell and Wilson 2006). If this is the case, marine turtles may have fairly strong appeal.

First, as Table 7.3 shows, the marine turtle is not an obscure species. Asians are familiar with them, even though they may have neither seen live turtles nor consumed their meat or eggs. These are, after all, urban populations with access to television and, as the survey results suggest, this may be a more important source of information about environmental issues than formal education. Second, respondents in the five cities surveyed showed common preferences for marine turtles.

The survey asked respondents to prioritize six endangered species for priority of conservation resources; the species were marine turtles, dugongs, whale sharks, Philippine eagles, black-faced spoonbills, and Javan rhinos. In all cases, marine turtles were ranked either first or second (Table 7.4).[2] An individual may be

[2] The survey was conducted by dropping off a questionnaire and information packet at each household and collecting it later. It is possible that some respondents read the entire survey instrument—including the willingness to pay question about turtle conservation—before filling out the attitudinal questions. If so, this may partially explain the high ranking for marine turtles, but it does not explain the relatively low ranking for environmental issues and for wildlife conservation among environmental issues. So, there is some confidence that the high priority accorded to marine turtles is not an artifact of the survey instrument.

Table 7.3 Familiarity with and knowledge about marine turtles (% of total respondents)

	Beijing	Davao city	Bangkok	Ho Chi Minh city/ Hanoi
	$N = 600$	$N = 847$	$N = 789$	$N = 1444$
Seen a live marine turtle?	53 %	67 %	57 %	24 %
Have eaten marine turtle eggs or meat?	3 %	9 %	7 %	2 %
Have purchased or owned a product made from the shell of a marine turtle?	6 %	4 %	10 %	5 %
Watch Discovery Channel, National Geographic, or any television show or video document about marine turtles or other animals?	81 %	79 %	98 %	78 %
Know that marine turtles have cultural value in some societies in Indonesia, Thailand, China, and Malaysia (temple ceremonies, "release of life" ritual, etc.)	44 %	27 %	51 %	45 %
Marine turtles lay their eggs on land[a]	80 %	77 %	85 %	84 %

[a]Figures reflect the proportion that gave the correct answer that this statement is "TRUE"

Table 7.4 Ranking for priority in resource allocation

	Beijing	Davao city	Bangkok	Ho Chi Minh city/Hanoi
Rank	$N = 600$	$N = 847$	$N = 789$	$N = 1444$
1	Marine turtles	Marine turtles	Dugongs	White rhino
2	Dugongs	Philippine eagles	Marine turtle	Marine turtles
3	Whale sharks	Dugongs	Whale shark	Dugongs

N = the total number of respondents surveyed

concerned about animals in general or specific species, without taking action to prevent their extinction; she/he may assume that someone else will or should solve the problem. There is some evidence of this in the survey results. When 57–65 % of respondents strongly agree that "it is everyone's duty to ensure that plants and animals as we know them today will exist for mankind in the future," one might conclude that most people have *pro-environmental* attitudes. When faced with making personal trade-offs, however, there is a notable drop in supporting opinions. Only 4 % of the respondents in Davao strongly agreed that "Governments should raise taxes for more endangered species protection"; the percentage of "strongly agree" in the other cities was similarly low.

7.3 The Potential for Private Contributions for Marine Turtle Conservation

To assess WTP, a hypothetical marine turtle conservation program was constructed. Respondents were provided information about the importance of marine turtles to coastal and ocean ecosystems and the threats and risks of extinction. Once the hypothetical marine conservation program was introduced, respondents were asked whether they would be willing to contribute to the program by paying a monthly surcharge on their electricity bills or a period of 5 years.[3]

Two variants of the surcharge were tested: a mandatory charge and a voluntary charge. In the mandatory payment scheme, respondents were presented with a hypothetical referendum and asked to vote for or against a specified monthly surcharge on all household electricity bills for a period of 5 years. Respondents were told to assume that if more than 50 % voted in favor of the referendum, the surcharge would be imposed on all households, regardless of how they voted. In the voluntary payment scheme, respondents were asked whether they would voluntarily make a private contribution, with no assurance that anyone else would pay it. Those who are willing to voluntarily donate would check off in their monthly water bills that they are willing to contribute the specified sum as contribution for marine turtle conservation.

Separate groups were asked their WTP for one of three marine turtle conservation packages: (1) a region-wide program financed through a mandatory charge; (2) a region-wide program financed through voluntary contributions; and (3) a national program financed through a mandatory charge. Region-wide programs would involve the collaboration of many countries, including the four surveyed. In principle, the likelihood of success of the regional program would be greater than that of a single-country program, given the species' transboundary habitat. For the region-wide program to be financed by the mandatory charge system, over 50 % of respondents in each of the four countries would have to vote in favor for the program to be implemented. In other words, respondents were told to assume that if the referendum did not pass in one of the countries, the international effort would not go ahead.

All together, 1249 respondents were randomly selected to vote on the region-wide program with mandatory payment, 1220 on the region-wide program with voluntary payment, and 1211 on the single-country program with mandatory payment. Each set of respondents was divided into five groups, each of which was asked to give a yes-or-no response to one of the five amounts, or bid levels. In the research design, the common bids used in all cities surveyed are USD 0.02, USD 1, and USD 5. The other two bids are set by the researchers in each country. The Table 7.5 shows all the bids used in each country. For example, the five bids for Beijing are USD 0.02, USD 0.5, USD 1, USD 5, and USD 7.5. The five bids for Bangkok are USD 0.02, USD 0.25, USD 1, USD 2.5, and USD 5.

[3] The decision to use a monthly surcharge on electricity bills is based on the results of Focus Group Discussions in the each of the countries. The participants in the FGDs were asked to consider several options which include income tax surcharge and a monthly surcharge of water bills. The majority of the participants settled with the monthly electricity surcharge. During the surveys, the respondents were informed that the electricity authority in their countries would simply collect and transfer the money to the Regional Marine Turtle Conservation Trust Fund and would not be involved in any way in the administration of the Fund.

Table 7.5 Number of respondents saying "Yes" to each bid under the Regional Mandatory Program

Bid (USD/month/hh)	Beijing	Davao	Bangkok	HCMC&Hanoi
(.02USD)	30 (75)	32 (56)	39 (67)	78 (81)
0.1		30 (56)		
0.25			37 (66)	
0.5	19 (48)			68 (69)
(1USD)	16 (40)	17 (28)	20 (36)	44 (44)
2USD		17 (31)		
2.5 USD			18 (33)	
(5 USD)	8 (20)	11 (20)	7 (13)	17 (17)
7.5USD	2 (5)			20 (21)

[a]The shaded rows are the three common bid values used in all four countries
[b]Figures in parentheses are percentages of "Yes" responses from the number of responses in the split sample

The findings show that, while respondents are familiar with marine turtles and believe them to be important, they are not concerned to a degree that would lead them to make personal trade-offs by making private contributions. Table 7.5 shows that only the lowest surcharge (USD 0.02) would pass a referendum in all four countries. For Davao City, Bangkok, Hanoi, and Ho Chi Minh City, the referendum would also pass at the next lowest bid prices which were USD 0.1, US 0.25, and USD 0.5, respectively. The second lowest bid for Beijing was USD 0.5 and only 48 % of the respondents voted to pass the referendum. At bids of USD 1 and above, the referendum would not pass in any of the countries.

For each city surveyed, respondents indicate a WTP only a small payment. In the Regional Mandatory Conservation Program, the mean WTPs (MWTP) for Davao City was found to be USD 0.17/household/month and for Ho Chi Minh City and Hanoi, USD 0.83/household/month.[4] MWTP for the two higher income cities were found to be USD 1.16 and USD 1.41/household/month for Beijing and Bangkok, respectively. These values are comparable to the WTP values from other studies on

[4] For details of the methodology of this study and the parametrics and nonparametrics analysis, see Indab et al. (2006). The results are from parametric estimates assuming normal distribution.

Table 7.6 Number of respondents saying "Yes" to each bid under the Regional Voluntary Program

Bid (USD/month/hh)	Beijing	Davao	Bangkok	HCMC&Hanoi
(USD.02)	29(73)	36(60)	42(81)	60(65)
0.1		38(69)		
0.25			31(61)	
0.5	23(58)			60(61)
(USD1)	20(50)	18(33)	19(35)	27(29)
2		17(30)		
2.5			20(38)	
(USD5)	7(18)	11(19)	9(17)	20(20)
USD7.5	3(8)			11(12)

The shaded rows are the three common bid values used in all four countries
Figures in parenthesis are percentages of "Yes" responses from the number of respondents to each scenario

species conservation in Asia.[5] The pattern of response to the Regional Voluntary Program in Table 7.6 indicates that, similar to the Mandatory Program, the percentages of the respondents who would be willing to contribute decreases significantly as the bids rise.

No significant difference was found in the level of support for variations in the conservation programs we offered. With the exception of Ho Chi Minh City/Hanoi, where results suggest that respondents would be willing to pay significantly higher for the mandatory program, respondents in Beijing, Davao City, and Bangkok appear to be indifferent. Also, there were no significant differences in the MWTP between international and national scale of efforts. Again, Ho Chi Minh City/Hanoi is the exception with where the MWTP for the conservation efforts is higher at the international scale.[6] One encouraging finding is that a significant percentage of respondents would voluntarily pay, regardless of whether people paid. While the estimated MWTP values are lower than WTP values for conserving endangered species in developed countries, given the large difference between the gross

[5] Mean willingness to pay (MWTP) for the black-faced spoonbill in Macao was estimated at 9.51 MOP (USD 1.19)/household/month (Jinjuan 2006); for Philippines' whale sharks, USD 0.5/household/month (Indab 2006); and for the Javan Rhino of Vietnam, USD 0.21/household/month according to a CVM study (Truong 2007). In contrast, the estimated lump-sum contribution to conservation measures for the Philippines eagle was USD 63/household (Harder et al. 2006).

[6] We asked separate groups of respondents their WTP for a national and a (larger) international program, partly with the intention of using it as a "scope" test to see if respondents were making rational choices. However, there is reason to believe that such "external" scope tests are unrealistic; valuation is an inherently comparative process (M. Hanemann, personal communication 2008). For this reason, the apparent scope insensitivity of respondents does not invalidate the findings. See Jianjun Jin et al. (2010).

Table 7.7 Reasons why respondents decided to contribute money for the establishment of a marine turtle conservation fund (% of respondents who said "Yes" to the WTP question)

Reasons	Beijing	Davao city	Bangkok	HCM/ Hanoi
The marine turtle is an important animal and should be protected	48	41	39	34
I believe that marine turtles can only be protected through the collaboration of several countries	19	6	10	29
It is high time that the people in East Asia do something concrete about protecting the marine turtles since this is the center of illegal trade in the world	4	16	9	9
This initiative can lead to more protection efforts for other endangered species in the region	22	15	21	9

national income of the United States and the four countries surveyed (16 times greater than that of Thailand in 2005), this difference in MWTP was not unexpected. For example, the average US citizen reported a WTP of USD 7.5/hh/month (Loomis et al. 1996) for conservation of the spotted owl and USD 2.77/hh/month for the gray-blue whale conservation (Bulte and Van Kooten 1999).[7]

Even if those who agreed to make voluntary contributions are not in the majority, and they are willing to make only small contributions, they do constitute a potential source of finance for conservation.

Commonalities were observed in the reasons as to why respondents decided to contribute money for the establishment of a marine turtle conservation fund. There appeared to be consensus that the most important reason was because respondents think the marine turtle is an important animal and should be protected. Table 7.7 shows that Vietnamese respondents placed more importance on the "collaborative efforts" of countries, whereas Thai respondents saw the merit of this initiative for marine turtle conservation as paving the way for other endangered species conservation efforts. This was also the sentiment of those in Beijing, although this respondent group also attached importance to international collaborative efforts.

As to why respondents decided not to support the proposed marine turtle conservation fund, with the exception of Thailand, the most common response was that they could not afford the amount. Affordability was less of an issue for Bangkok respondents than was attitude toward the government; at the time of the survey period in Bangkok, perceptions of the government tended to be negative, and the most common reason for deciding not to support marine conservation was because they thought it should be the responsibility of the government which, after all, already had their tax revenue (Table 7.8).

[7] From World Bank's country statistics, GNI for the United States in 2005 was USD 43,740/capita.

Table 7.8 Reasons why respondents voted "against" the referendum (% of respondents who said "No" to the WTP question)

Reasons	Beijing	Davao city	Bangkok	HCM/ Hanoi
I cannot afford that amount	39	36	13	27
I do not trust the institutions that will handle the money for this conservation work	20	5	14	19
I do not believe paying will solve the problem	18	5	4	11
It should be the government's responsibility since it already has money from my tax revenues	7	27	41	10

Table 7.9 Variables included in the multivariate analysis and definitions

Variable	Definition
Bid	Bid used in WTP question (USD)
HHsize	Number of household members living together
Age	Age of respondents
Gender	1= male, 0 = female
Education	Education of respondents (years of schooling)
Income	Total household income (1000 USD/month)
Marriage	1= married, 0 = others
Member	1= for environmental organization member and 0= otherwise
Cn	1= Beijing, 0 others
Ph	1= Davao, 0 = others
Th	1= Bangkok, 0 = others

7.4 Factors Influencing Decision Making

To analyze factors that influenced respondent decisions, the data for all the three scenarios (Regional Mandatory, Regional Voluntary, and National Mandatory) were pooled together. The variables included in the multivariate analysis and definitions are presented in Table 7.9.

Results in Table 7.10 conformed to a priori expectations for all of the models where the *Bid* variable was significant at a level of 0.01 with a negative sign, indicating that as the Bid price increases, respondents are less likely to answer "Yes" to the WTP question. In the pooled sample, *education* and *income* were also significant to the 0.01 level, both with positive coefficients, indicating that the higher the level of education and income, the higher the probability those respondents would vote in favor of a mandatory payment. Respondents who are already members of environmental organizations were also more likely to vote "Yes" and in the pooled sample, this variable was significant to 0.01.

Apart from *Bid* and *Income*, which were statistically significant for all country models, the influence of other variables varied. *Education*, for instance, was

Table 7.10 The logit regression

	China		Philippines		Thailand		Vietnam		Pooled	
	Coeff	Std err	Coeff	Std err	Coeff	Std err	Coeff	Std err	Coeff	Std err
	−0.365	0.046	−0.422	0.052	−0.519	0.061	−0.503	0.053	−0.438	0.026
Bid	***		***		***		***		***	
Hhsize	0.085	0.077	−0.008	0.032	0.018	0.048	−0.065	0.032	−0.007	0.019
							**			
Age	0.001	0.009	−0.005	0.006	0.001	0.009	0.008	0.005	−0.002	0.003
Gender	0.036	0.204	0.129	0.160	−0.040	0.168	0.132	0.122	0.109	0.077
Education	0.070	0.032	−0.013	0.028	0.038	0.028	0.036	0.015	0.034	0.011
	**						**		***	
Income	0.001	0.000	0.002	0.001	0.0003	0.000	0.001	0.000	0.001	0.000
	***		***		*		***		***	
Marriage	0.385	0.315	0.044	0.199	0.203	0.187	−0.408	0.155	−0.0002	0.087
							***			
Member	0.894	0.355	0.808	0.235	0.432	0.417	0.215	0.345	0.658	0.156
	**		***						***	
China									0.049	0.136
Phil.									0.440	0.102
									***	
Thai									−0.159	0.112
Cons	−2.077	0.706	−0.003	0.501	−1.027	0.615	−0.713	0.309	−0.821	0.225
	***				*		**		***	
Log-likelihood	−302.2		−502.5		−433.6		−810.1		−2069.6	
Pseudo R2	0.17		0.09		0.12		0.10		0.10	
N	598		839		789		1430		3656	

***Significant at 1 %
**Significant at 5 %
*Significant at and 10 %

significant only for China and Vietnam, at a 0.05 significance level. The *Member* variable was significant only for China and the Philippines. The size of the household and whether the respondent was married were significant only in the Vietnam model; both had negative coefficient signs, suggesting that the larger the household size, the lower the probability they would be willing to pay. This is consistent with the expectations that larger families would have larger expenditures; similarly, married people would have larger expenditures than respondents who were single.

7.5 Summary and Conclusions

This study shows that people in several cities in Southeast Asia are already exposed to an abundance of information; they are at least generally aware, if not well informed, about marine turtles. They believe that conservation is important, but at the time of the study, their priorities understandably lie with other public policy issues, such as improving governance and reducing poverty. A mandatory surcharge on electricity bills to support marine turtle conservation would only pass at the lowest bid of USD 0.02 in all cities surveyed.

The results provide some support for the proposition that voluntary contributions could provide considerable sums for marine turtle conservation; based on the percentages of respondents in the cities who would voluntarily pay USD 1 per month, the potential revenue would be around USD 50 million/year.[8] This is much less than what could be mobilized were the mandatory payment referenda passed in all four countries surveyed (USD 135 million). However, it is more than the current global expenditures on marine turtle conservation of all 162 conservation organizations combined, which is estimated at some USD 20 million per year (Treung and Drews 2004). Having said that, the harsh reality is that actually mobilizing these contributions would be difficult. The voluntary payment vehicle explored was a "checkoff" for a marine turtle conservation program on a household's monthly electricity bill. While this might work once, it is not feasible to put checkoff boxes on utility bills for every species or environmental cause. In the long run, these

[8] For Beijing, for example, the estimation is based on the assumption that 50 per cent of 5.12 million households would pay (50 per cent being the percentage of respondents in our survey whi agreed to voluntary contribution of US 1 per month). A similar approach was used for Davao City and Bangkok. For the other cities, the voluntary contributions were stimated using the percentages of respondents agreeing to voluntarily 1 per month, that is, 33 per cent of the 0.23 million households in Davao City, 35 per cent of the 2.091 million housedholds in Ho Chi Minh City and Hanoi.

Estimated by using the MWTP for each city and the estimated number of households in each city at the time of this study.

efforts must be financed primarily out of general government revenue or "user-pay" schemes like environmental service payments.

An important implication of the findings is that the traditional prescription of "raising awareness" is unlikely to yield results: people in Asia are already well informed about the existence and plight of marine turtles. Efforts to develop conservation finance mechanisms should therefore be directed in a different and more difficult direction: improving the trustworthiness of government tax collection and expenditure systems. Conservation agencies could play a role by working with governments to set up trust funds in which public funds could be deposited with confidence. Charities could also explore the potential for voluntary contributions suggested by this study; their efforts should go primarily into identifying the relatively small segment of the population that is willing to contribute and to developing cost-effective ways of collecting payments. Eventually, as incomes rise and governance improves, Asia's ability to pay for conservation will increase. In the meantime, contributions from the international community will continue to be important in conserving what is, after all, a global resource.

Acknowledgments We would like to thank the Economy and Environment Program for Southeast Asia (EEPSEA) for providing the financial support to undertake this study. We thank Dr David Glover, Founding Director of EEPSEA, and Dr Herminia Francisco, present Director of EEPSEA since January 2007, for their technical support, comments, and opportunities for series of meetings which facilitated organization and coordination of cross-country efforts. We would also like to thank Dr Dale Whittington (University of North Carolina at Chapel Hill, USA) and Dr Wiktor (Vic) Adamowicz (University of Alberta, Canada) for their valuable suggestions and support which helped us carry through this study.

References

Bulte EH, Cornelius Van Kooten G (1999) Marginal valuation of charismatic species: implications for conservation. Environ Resour Econ 14(1):119–130

Harder D, Labao R, Santos FI (2006) Willingness to pay for conservation of endangered species in the Philippines: the Philippines eagle. EEPSEA technical report. Economy and Environment Program for Southeast Asia, Singapore

Indab A. "Conservation Value of Endangered Species in the Philippines: A CVM Exercise". EEPSEA Research Report, 2006

Indab A, Jin Jianjun, Truong Dang Thuy, Orapan Nabangchang, Harder D, Subade R (2006) Valuing the marine turtle conservation using the contingent valuation method: a cross-country study in Asia, 2007. Paper presented at the third world congress of environmental and natural resource economists, Kyoto, 3–7 Jul 2006

IUCN (International Union for the Conservation of Nature and Natural Resources) (2002) Red List. www.iucnredlist.org

Jianjun J, Indab A, Nabangchang O, Truong DTT, Harder D, Subalde RF (2010) Valuing marine turtle conservation: a cross-country study in Asian cities. Ecol Econ 69:2020–2026

Jinjuan J (2006) Economic valuation of the black-faced spoonbill conservation in Macao. EEPSEA technical paper. Economy and Environment Program for Southeast Asia. Singapore

Loomis JB, Gonzalez-Caban A, Gregory R (1996) A contingent valuation study of the value of reducing fire hazards to old growth forests in the Pacific Northwest. USDA Forest Service Research Paper PSW-RP-229-Web. United States Department of Agriculture, Washington, DC

Perrine D (2003) Sea turtles of the world. Voyageur Press, Stillwater

Safina C (2006) Voyage of the turtle: in pursuit of the Earth's last dinosaur. Henry Holt and Company, New York

Tisdell C, Wilson C (2006) Information, wildlife valuation, conservation: experiment. Contemp Econ Pol 24(1):144–159

Treung S. and C Drews. "Money Talks: Economic Aspects of Marine Turtle Use and Conservation". WWF-INternational-Gland, Switzerland, 2004. www.panda.org

Truong DT (2007) Willingness to pay for conservation of endangered species in Vietnam: rhino. EEPSEA technical report. Economy and Environment Program for Southeast Asia. Singapore

Part II
Pollution and Land Degradation

Nancy Olewiler

Overview

The maritime countries of Southeast Asia are home to approximately one third of the world's population with 70 % of its residents living in coastal areas, and large numbers are heavily dependent on coastal resources for their livelihood and survival, many of whom are in the lowest-income groups. Economic activity combined with natural events and changing climate is putting increasing pressure on coastal marine resources. The focus of this section is on threats to marine resources emanating from pollution and land degradation. Pollution comes from both land- and sea-based sources and includes untreated sewage, industrial effluents, runoff of fertilizer nutrients and pesticides/herbicides, toxic compounds from industry, and oil and chemical spills from shipping. Coastal land degradation is the result of aquaculture practices, dredging, land reclamation, mining, and coastal development.[1] Degradation and destruction of coral reefs, fish stocks, and buffers, such as mangroves and sea grass that help protect against storm surge and sea level rise, are

[1] See UNEP (2006) for information about the threats to marine ecosystems created by human activity. They report the following: the amount of untreated wastewater discharged in the Indo-Pacific region can be as high as 80–90 %. In Southeast Asia, more than 600,000 t of nitrogen are discharged annually from the major rivers. These numbers may become further exacerbated as coastal populations are depicted to increase from 77 people per km^2 to 115 people per km^2 in 2025. Wetlands and mangroves are also declining rapidly, typically by 50–90 % in most regions in the past four decades. Also see Brown et al. (2008), Appendix 3, for a detailed analysis of the relationships between changing levels of ecosystem services from marine and coastal areas and the impact on people living in the region dependent on these services for their livelihood and well-being.

N. Olewiler (✉)
School of Public Policy, Simon Fraser University, Vancouver, BC, Canada
e-mail: olewiler@sfu.ca

just a few of the damages to the marine ecosystem that result from activities that are classical externalities—actions that create deleterious spillovers affecting the natural environment and those dependent on it. Prices of goods do not reflect the damages their production and consumption incur, thus necessitating the introduction of government policies to regulate or price the external effects.

Case Studies

Four case studies comprise this section: (1) water pollution created by economic activity that reduces fishery harvests in Thailand, Pornpinatepong et al. (2010); (2) overstocking at fish farms in the Philippines that can lead to massive fish kills both within the aquaculture and to wild species harvested by local fisherfolk, Sumalde et al. (2003); (3) barriers to the restoration of mangrove forests that have been destroyed by fish farms in the Philippines, Ferrer et al. (2011); and (4) potential land reclamation projects in the Philippines that destroy coral reefs and fishing grounds, Montenegro et al. (2005). Each paper is reviewed with highlights on the key points of the study and a summary of lessons learned that are transferable across Southeast Asia. The first two papers illustrate a classical externality problem: the impact of pollution on the natural environment from economic activities. The next two papers examine the relationships between land degradation and the health of coastal marine resources.

Methodologies

The papers in this section use multiple methods to analyze and quantify the damages to marine coastal resources and, in turn, impacts on the people dependent on these resources and society, more generally. They explore means of reducing damages through policies that price pollution and regulate the economic activities generating that pollution with the goal of restoring nature. The primary methodology used in the papers is to estimate the loss of economic activity using production functions and market valuation of damages. The framework first establishes that the losses due to pollution are tangible and of significant magnitude to warrant policy intervention. The next step is to determine the best package of policy reforms to reduce emissions and help sustain the coastal resources. Two of the papers, Pornpinatepong et al. and Ferrer et al., also make use of multiple attribute analysis to assess policy options.

Pollution Control and Sustainable Fisheries Management in Southern Songkhla Lake, Thailand

Pornpinatepong et al. examine the impact of pollution from human activity and industrial development on fish harvests in Songkhla Lake, Thailand, a tidal lake. The growth in economic activities surrounding a tidal lake rich in aquatic resources leads to release of pollutants that threaten the harvests and long-term sustainability of the capture shrimp fisheries operating in the lake. Potential policy solutions extend the existing command and control regulations to include price incentives in the form of emission taxes and tradable emission permits.

The water pollution comes from land and sea sources: domestic wastewater, pesticides, fertilizers, nutrients from shrimp farming, and livestock waste, as well as oil and grease from shipping. The impact of the pollution is felt strongly in the most profitable economic activity in the lake, its shrimp fishery. Local fisherfolk are dependent on the fishery for their livelihoods and their harvests decreased by over 70 % from 1996 to 2003. The goals of the study are to develop policy recommendations for a sustainable fishery in a portion of the lake called Southern Lake as a pilot study and then to extend those results to the entire lake.

The authors use secondary data to estimate the impact of water pollution on shrimp harvests, controlling for fishing effort and the salinity of the lake. They design a water quality index to capture the relative impact of the key contaminants that reduce shrimp productivity. The results show a loss of productivity as water quality deteriorates, ranging from 3 % to 10 % for each unit reduction in the index, depending on the region of the lake. Current policies addressing water pollution are assessed and found insufficient to quell the flows of pollution for the usual problems of insufficient monitoring and enforcement, low levels of penalties for noncompliance, regulations covering concentrations of pollutants but not emission volumes, and communities and manufacturers being unaware of their contribution to the problems or what they can do to reduce their negative impact. New policy directions are thus indicated to address these problems. Their candidates include incentive-based policies in the form of pollution charges and tradable discharge permits, as well technological improvements in waste treatment. Using available literature, they undertake a cost effectiveness study of the potential technologies. They find that the effectiveness of a treatment technology is sensitive to the size of the plant and the influent load. It is less effective for businesses with a lower influent load or a smaller-size operation to invest in individual treatment plants, while it is more cost-effective for small-scale polluters to share a common treatment plant. Large-scale plants are less effective because they tend not to be run at capacity. They thus recommend smaller-scale treatment facilities.

Incentive-based policies compared to command and control policies are assessed using a set of criteria that include effectiveness, legal feasibility, implementation complexity, public acceptance, capital and operating costs, compliance costs, equity, and co-benefits or costs. While each policy involves trade-offs, the authors feel that a tradable discharge permit system applied to the larger firms would be the most likely to address the problems identified due to its flexibility in allowing firms to meet targets using a market mechanism; the price of pollution is determined by the market, not set in the tax rate; and there is less potential opposition from stakeholders, once they understand the mechanism and how it could work. The latter is a barrier that could be addressed with a substantial information and education campaign. There are many facets to setting up a tradable discharge permit system so these design factors would require considerable study and consultation, as recommended by the authors.

Pollution-Induced Fish Kills in Bolinao: Effects of Excessive Aquaculture Structures and Overstocking

Sumalde, Francisco, and Peñales examine the adverse impact that aquaculture can have on water quality. The pollution studied is created by the aquaculture fishery's practices of overstocking their fish pens. The waste products from the fish and their feed reduce water quality in the marine environment. Damages accrue not only to those in the vicinity, but the aquaculture owners themselves can suffer catastrophic loss of fish if dissolved oxygen levels fall to levels that threaten survival of fish (among other aquatic species).

Aquaculture is the cultivation of edible fish and other marine species in an artificial system in both freshwater and marine environments. Aquaculture has changed over the centuries from traditional methods that use natural or manufactured ponds and natural feeds to produce a typically small but sustainable harvest over time. Modern aquaculture is quite different. The "ponds" can be large containment areas that rear thousands of kilograms of fish, using commercial feeds, antibiotics, and pesticides to sustain very intensive cultivation. To warrant high capital and/or operating costs, operators may stock large numbers of species, requiring large amounts of feed and antibiotics to control disease. The result can be water pollution and degradation and destruction of aquatic ecosystems. This paper focuses on the marine ecosystem in a region of the Philippines. The research question is motivated by a major fish kill that occurred in February 2002, at that time the largest such kill in the Philippine history. Thousands of fish were wiped out due to a combination of low levels of oxygen in the water and the bloom of toxic plankton. The two factors are interdependent, and a major source of the problem can be traced back to the large number of fish farms in the region, far more operations than allowed by government policy and permitting.

The authors work on all facets of the problem to come up with strong evidence that the overstocking of fish in the farms contributed to the low oxygen levels due to biological oxygen demand created by the fish waste products and decomposition of the feed. They trace the explanatory variables by doing water quality analysis, estimating the profitability of the fish farms, and investigating how water quality affects fish productivity and why the regulatory system did not work to control the number of operators in the region. They find that the number of fish per farm is well above what they estimate is an optimal level of stocking, that there were too many farms in operation, and that new policies are needed to regulate the industry. While factors in addition to the fish farms contributed to the low levels of oxygen in the water and a strictly causal relationship cannot be established, the correlation is strong between fish farming and detrimental environmental impacts. The irony of this story is that the very people who contributed to the problem—the fish farms—are the ones most adversely affected by the low water quality. They had lost not only large numbers of the fish they had bought, fed, and hoped to sell but were also subjected to a moratorium imposed on farming after the fish kill event. Those dependent on the ocean for capture of wild species also suffered a loss, and the overall marine environment in the region was at least temporarily badly degraded.

The specific methodologies highlighted in our abridged version of their paper include the estimation of a production function and use of survey data to compute average profitability of the fish farms. While the average farm in their sample, covering one quarter of the farms in the Bolinao region, falls within recommended guidelines for stocking, some operators overstock their milkfish in a contained area by almost 75 %. They also do not know how representative their sample is, and they may have bias with those who are likely to be more compliant with permits responding. Visual investigation shows that there may be more than twice as many operations in the region than allowed. Thus, although stocking for many of the operations may fall within the norms for good management, the sum of all the operations likely pushes total stocking well over the levels able to sustain good water quality in the ocean. This is an illustration of the problem of cumulative effects where the sum of a large number of small operators, each of whom may be within regulatory norms, exceeds the carrying capacity of the ecosystem. In this case, the excess operations are also illegal, and there doesn't seem to be the capacity or will to enforce the regulatory framework. This calls for two main measures: first, there need to be more and better information flows to fish farm operators that overstocking is detrimental to the environment and may well destroy their own operations if a major fish kill arises; second, a stronger, enforceable regulatory environment is required. The authors do not go into detail about how the regulations and policies should change but offer the obvious observations that more monitoring and enforcement are needed along with stiffer penalties for noncompliance.

Reverting Disused Fishpond Lease Agreement Areas to Mangrove Forests in Region VI, Philippines

Ferrer et al. investigate the loss of mangrove forests in the Philippines and the potential to reclaim and restore them through acquisition of abandoned, underutilized, and undeveloped fish farms. In this case, the damage to the mangrove resource has been done, so the issue is what policy framework is needed to accelerate the restoration of the resource. The Philippines has lost around half of all its mangroves to aquaculture along its coasts, despite widespread recognition that mangroves provide fish habitat, improve water quality, and protect against storm surges during extreme weather events. The need to protect and restore mangroves is a priority in Southeast Asia, even more so with the growing threat of sea level rise and severe typhoons associated with climate change. Ferrer's team explores the barriers to restoring mangrove forests in the region of the country with the greatest loss of mangroves (95 % have been converted to fish farms) and largest number of fish farms and thus has large potential for restoration. The focus is on the aquaculture operations that are abandoned, underdeveloped, and undeveloped (AUU) fishponds.

The study employs a two-stage, five-step process in evaluating the existing leasing arrangements for fish farms (the Fishpond Lease Agreement, FLA). Stage 1 examines the process by which a FLA reverts from jurisdiction by the agriculture ministry to the environment ministry; the former governs the fish farms and the latter mangrove restoration. The team assesses in detail the FLA cancellation process under the authority of the agriculture ministry and finds it deficient in many regards. Far too many idle fish farms exist and fail to be processed. Reasons for this situation include a lack of approved guidelines covering FLA cancellation, the power of connections whereby lessees have more than ample opportunity to use their political connections to appeal a cancellation order, poor monitoring and recording, and lack of personnel to handle the work load. Step two assesses the process of returning jurisdiction over disused FLA areas to the environment ministry. The process in place simply does not work due to an absence of approved guidelines for reversion of disused FLA areas, lack of cooperation among agencies, no definition of what underutilized or abandoned means in the legislation, power politics from having more lands under one's jurisdiction, and finally concerns about agency competence. Ferrer et al. thus find a myriad of institutional, economic, and political problems that have prevented action. Progress on identification of AUU lands has been incredibly slow, and changes are needed all the way through the system from monitoring and enforcement of existing FLA holders to identifying and seizing AUU sites and addressing jurisdictional conflict among agencies. One of their key recommendations is to formulate and enforce guidelines on cancellation of FLAs and on better protocols for turning over the AUU lands to the environment ministry for restoration from the agriculture ministry department—way too few lands have made their way from identification as an AUU to action on restoration.

The second stage examines the process by which an AUU site can be restored to a mangrove forest, once identified and under the jurisdiction of the environment ministry. This stage encompasses three sub-steps: assessment of disused FLA areas for reversion to mangrove forest, determination of the net benefits (benefits minus costs) of restoration of AUU lands, and analysis of potential policies to finance mangrove reforestation. The study uses the data from FLA areas officially declared canceled or identified as abandoned and undeveloped. The disused fishponds were assessed in terms of suitability for reforestation. They find that only 23 % of the total area (1240 ha) of disused FLA areas in the region is suitable for restoration. The reasons for less than one in four sites being viable include conversion to residential use; inundation by sea—or more often lake—water making replanting impossible; illegal use as a fishpond despite a cancellation order; unsuitable terrain; and inhabitation by squatters leading to community opposition to conversion. The small percentage of suitable sites is both good and bad news. The good news is that the cost of restoration is a less daunting barrier than it could be. The bad news is that so little can be done to restore the mangrove ecosystem.

For the sites found suitable, the costs and benefits of reforestation were calculated. There is only good news here. Under a range of plausible scenarios, all the present values of net benefits are positive, indicating that, in aggregate, it is in society's interest to restore these sites. The authors look at the relative merits of restoration that allow subsequent harvesting of mangrove timber once the site is restored versus those that keep the mangroves intact. Their estimation shows that the net benefits are virtually the same when one does not factor in the ecosystem benefits from sustaining mangroves without timber harvests. Non-timber benefits are not explicitly included because the authors had no reliable method of quantifying them. If those values are positive (and they could be quite large), the net present value calculation would favor a ban on timber harvests.

Upon establishing that viable sites warrant restoration, their evaluation concludes with an analysis of seven potential policy options for financing reforestation. They have a fulsome and creative list of options divided into policies that place the burden of financing restoration on the lessees versus those by which the costs are borne by society through general taxation or voluntary donations by international organizations or companies. Payment by lessees falls into two categories, those that are imposed before any damage from fish farming occurs and those that are assigned after damage is measured. In the first category, a performance bond is posted at the time the fish farm begins operation. The notion here is that the party responsible for the environmental degradation should be liable for its cleanup. The bond serves as insurance that upon termination of fish farming, the site is restored to its previous state. There is considerable interest in this concept for all sorts of activities that degrade the environment, including mining, oil and gas extraction, and pipeline operation, and more it is a "polluter pays" concept. Another pre-damage charge would be a lease fee that reflects the economic rent, which could be based on the value of the land in alternative uses, where such uses reflect the ecosystem goods and services provided by mangroves. In the post-damage category is a green tax on fish farm operations, which is also a polluter pays approach, but this would be an ongoing charge based on some estimation of the incremental damages incurred for each year that the farm operates. A beneficial use tax is the second post-damage fee that charges for the use of the land. Examples of third party sources for funds include grants from the environment ministry that would need to come from base budgets (funded by citizens' tax revenues) or foreign or domestic donors.

It was beyond the scope of Ferrer's analysis to go into detailed examination of each policy's characteristics. The analysis proceeds by evaluating each policy using the following criteria: feasibility, ease of implementation, amount of funds raised and their sustainability, reduction in inequality, production efficiency, and improvement in compliance with existing laws and regulations. They are looking for the policies that have the following characteristics: high feasibility, easy implementation, high positive impacts on other desirable objectives (equity, compliance, etc.), and ability to generate funds on a sustainable basis. No single option meets this condition—there are trade-offs, a common outcome with complex problems. Policies that are capable of raising substantial funds are also those that are more complex to implement and incur opposition from some key stakeholders. Weighing all of the trade-offs, the authors find beneficial taxes and lease rates based on economic rent promising solutions. That said, while funding from donors is likely to be smaller than bonds and taxes, efforts to elicit donations could begin immediately as there would be little to no opposition to this source.

They recommend the following additional reforms and actions to improve the regulatory process and interagency cooperation: (1) reduce the incentives to continue fish farming by increasing rental rates, shortening lease periods, and reducing the size of land holdings, (2) implement a better screening process for renewal and transfer of fishpond leases, and (3) survey all FLA areas in the region to determine the scale of potential mangrove restoration. Their research supports the creation of a nationwide program for mangrove reforestation—the sooner the better.

Environmental Trade-Offs from Coastal Reclamation: The Case of Cebu, Philippines

Montenegro et al. look at the potential damages that would be created by a massive land reclamation project planned for the Philippines. The analysis shows the value of undertaking damage assessment prior to making decisions about major infrastructure projects that damage vital ecosystems. The research takes as its starting point a proposed project that was slated to infill 2700 ha of the inshore in the coastal City of Cordova, in Metropolitan Cebu. The goal of the study is to develop a clear conception of the types and magnitudes of the environmental damages that land reclamation can create, so as to identify the trade-offs between environmental protection and economic development and guide decision-making and policy implementation. The history of land reclamation in other jurisdictions is typically identification of damages after the projects are completed, rather than before their initiation.

The approach uses standard damage assessment techniques to measure the direct impacts of reclamation activities on the "use value" of the marine ecosystem—the benefits lost due to reclamation to those whose livelihoods and utility come from the marine coastal environment. Because of limited time and scope, indirect benefits—"nonuse" values such as option, bequest, and existence values—are not estimated.[2] Damages are measured in the form of loss of income to those fishing in the coastal waters, loss of recreational benefits from snorkeling and scuba diving, and damage to the environment from dredging activities that would be used to provide the fill materials for the project. These represent a lower bound of loss due to the exclusion of nonuse benefits and because Montenegro et al. estimate their magnitude over a fixed period of 30 years.[3]

Damages to the intertidal reef flats and the marine environment above them are the main focus of the study. These reef areas are often the target for reclamation because they are easy to fill and provide a cost-effective means of increasing the land base of the municipality. Intertidal reef flats are also highly productive ecosystems. Montenegro et al. list the many ecosystem goods and services they provide, including shoreline protection and storm damage buffer zones, fisheries and fish nurseries, nutrient cycling, storage and pollution control, and homes to a diverse set of species that are a food source for humans and other species. The first set of damages measured are for those directly benefitting from the on-site fishery, reef flat gleaning, and recreational use of the affected coral area.

To measure the impact on the on-site fishery and reef flat gleaning, a stratified and random sample of 377 Cordova households was conducted to gather data on socioeconomic variables and dependence on the marine environment (e.g., share of income by household members, amount and type of fish species harvested). Using the market prices of the fish species harvested and data on harvests, Montenegro et al. compute the value of lost fishing output; an additional loss in income was the increased cost of fishing at more remote sites outside the reclamation area. She estimates that the losses in net incomes to fisherfolk and reef gleaners would be with the project. The present value of the loss is estimated at $5 million US dollars (using a discount rate of 8 % in all calculations) for fisherfolk and $3 million for reef gleaners.

The big losses come from the destruction of the coral reefs and loss of recreational activity. The coral reefs in the region, while not pristine, are still of sufficient beauty and diversity to attract upward of close to 3400 visitors per hectare per year. If the project does not go ahead, the number of visitors would grow to 3500–4000 visitors per hectare per year or 2.2–2.5 million visitors annually. If the project occurs, approximately 640 ha of reef would be destroyed, thus leading to a total loss in recreational use. Using a benefit transfer approach, Montenegro et al. apply a willingness to pay estimate to derive the present value of lost recreational benefits. Based on a study of diving and snorkeling at three different dive sites in this region of the Philippines, she estimates that the present value of loss would be close to US $32 billion if the reclamation project were to go forward. There would also be losses to the municipality in the form of foregone licensing fees from tour operators and user fees from tourists.

---

[2] See Part I for case studies that measure indirect benefits derived from coastal marine resources.

[3] When using a positive discount rate, the 30-year time horizon should not undervalue the total losses by a substantial amount.

While not directly connected to the marine environment, Montenegro et al. also estimate the damages that will have to be repaired from the dredging activities—the landfill quarrying having been done. These are again likely to be significant underestimates because she cannot measure the impacts during the dredging on things such as water quality, increases in runoff due to soil disturbance, flooding, loss of soil stability, and increased erosion. Her proxy for damages is the costs of remediating the damage. Using the experience of another reclamation project in Cebu, she arrives at a present value of the loss to be US$ 19.5 million. Summing all of the damage estimates yields a present value of just under US$60 million for the 30-year period.

Damage estimates from this sort of study are valuable in assessing the trade-offs associated with land reclamation; the benefits to society from the reclamation could be estimated, and the present value of net benefits could be calculated. The case study did not have the capacity to perform a social benefit-cost study. However, she compared the social costs to the private benefits and found that under all of the different scenarios examined, the net benefits of the reclamation project were negative. Further data collection and analysis would be needed to perform the social benefit-cost computation. The study reinforces the urgent need to systematically estimate the environmental costs associated with large-scale coastal reclamation projects. Her estimates find that even when underestimating the total losses by measuring only the use values, these costs are not trivial. The hope is that knowledge of the damage costs will help improve decision-making regarding the scale and location of reclamation projects.

Lessons Learned from the Case Studies

1. Existing regulatory policies that focus on controlling activities that pollute and degrade the environment through permitting, performance, or emission standards are typically found to be inadequate in controlling pollution and reducing land degradation. The reasons are complex and include a lack of monitoring and enforcement, conflicts of interest in which the regulators are also beneficiaries of the economic activity they are charged with overseeing (e.g., local regulators also own fish farms), inadequate penalties for noncompliance, overlapping and potentially conflicting jurisdiction among government agencies, and more.
2. Greater awareness of the impact of economic activities on the environment needs to be communicated, not just to those who are creating the pollution and environmental degradation but also to those who are harmed. In some cases, such as the aquaculture sector, the polluter and pollutee are the same organization. This suggests privately inefficient operations, even without taking into account the unpriced externalities associated with the fish farms. Surely, conditions can improve with greater knowledge about the impacts of the activities generated.

3. Private actions will never be enough to address losses of ecosystem goods and services associated with pollution and land degradation. While the cases show that there are many different forms of regulation already in place, they are often ineffective. There is a role for incentive-based policies, but to date they are infrequently used because of factors such as opposition from stakeholders, information demands to implement the policies, and lack of familiarity by regulators with the concept and benefit of pollution taxes and tradable permits.
4. Understanding of the relationships between economic activities and their impact on the environment is often low and requires coordination of research between economists and scientists to provide the baseline information needed, as well as an understanding of correlations and any causal relationships that can be ascertained.
5. It is vital to estimate environmental consequences of large-scale development before the development is allowed to proceed. Doing so may help decision-makers decide whether to approve such developments, change their scale and location, or build in mitigation of strategies from the beginning, not after the damages have occurred.
6. Efforts to restore damaged ecosystems can be very cost-effective, but the barriers to these projects—in the forms of agency institutional capacity to identify and process eligible sites, regulatory complexity and constraints, and funding—need to be addressed to accelerate the return of vital coastal marine ecosystems such as mangrove forests.

References

Brown K, Daw T, Rosendo S, Bunce M, Cherrett N (2008) Ecosystem services for poverty alleviation: marine and coastal situational analysis. UK Department for International Development/Natural Environment Research Council/Social Research Council, London/Swindon/New York

Ferrer AJG, Hopanda JC, Orquejo MQ, Moscoso ADE, Sadaba RB Reversion of disused fishpond lease agreement areas to mangrove forests in region VI, Philippines. EEPSEA research report no 2011-RR9. Economy and Environment Program for Southeast Asia, Singapore

Montenegro LO, Diola AG, Remedio EM (2005) The environmental costs of coastal reclamation in Metro Cebu, Philippines. EEPSEA research report no 2005-RR5. Economy and Environment Program for Southeast Asia, Singapore

Pornpinatepong K, Kiripat S, Treewanchai S, Chongwilaikasaem S, Pornsawang C, Chantarasap P, Chandee C, Jantrakul P (2010) Pollution control and sustainable fisheries management in southern Songkhla Lake. EEPSEA research report. EEPSEA, Thailand

Sumalde Z, Francisco K, Penales M (2003) Pollution-induced fish kills in Bolinao: effects of excessive aquaculture structures and overstocking. In: Francisco H, de los Angeles M (eds) Economy and environment: selected reading in the Philippines. EDM Press, Manila

UNEP (2006) The state of the marine environment-trends and processes. UNEP/GPA-UNEP, The Hague

Chapter 8
Pollution Control and Sustainable Fishery Management in Southern Songkhla Lake, Thailand

Kunlayanee Pornpinatepong, Sakchai Kiripat, Sinad Treewanchai, Sukampon Chongwilaikasaem, Chotima Pornsawang, Pathomwat Chantarasap, and Chantip Chandee

Abstract This study explored resource-usage policy recommendations for sustainable fishery in Songkhla Lake, Thailand. The investigation began with problem recognition: decreasing fishery production. The profile of the existing usage and the economic value of the lake were identified and demonstrated in the study through a simplified conceptual model. The model illustrated the economic aspect of the ecosystem using theoretical and hypothetical concepts together with the available information. The policy options were specified next, after understanding the actual situation, using stakeholders' opinions and expert judgment. The policy options concerning the management of the better level of water quality control were the first priority. By selecting appropriated criteria (public acceptability, legal feasibility, implementation complexity, effectiveness, capital costs, operating costs, transaction costs, impacts, equity, and decentralization), the analysis and evaluation were conducted under three alternative policy options, namely, the command and control system, emission charge system, and tradable discharge permit system. The results of this study are valuable for policy recommendations that aim to maintain fishery production on a sustainable level. In addition, the outcome is useful for understanding the general situation of resource exploitation in Songkhla Lake.

Keywords Songkhla Lake • Pollution control • Fishery management • Tradable permit • Pollution charge

K. Pornpinatepong (✉) • S. Kiripat • S. Treewanchai • S. Chongwilaikasaem • C. Pornsawang • P. Chantarasap • C. Chandee
Faculty of Economics, Prince of Songkla University, Songkhla, Thailand
e-mail: kunlayanee.p@psu.ac.th; sakchai.k@psu.ac.th; sinad.t@psu.ac.th; sukampon.c@psu.ac.th; chotima.po@psu.ac.th; pathomwat.c@psu.ac.th; chantip.ch@gmail.com

N. Olewiler et al. (eds.), *Marine and Coastal Ecosystem Valuation, Institutions, and Policy in Southeast Asia*, DOI 10.1007/978-981-10-0141-3_8

8.1 Introduction

Songkhla Lake is a lagoon in south Thailand that is made up of three interconnected parts from the north to the south—Thale Noi, Thale Luang, and Thale Sap Songkhla or the Southern Songkhla Lake. The overall basin covers an area of 8729 km^2 in three provinces: Phatthalung, Nakhon Si Thammarat, and Songkhla. Southern Songkhla Lake is facing a significant decline in its once thriving fishery industry. Two main factors have led to this decline: water quality degradation and an increase in fishing activity. This study examines the pollution problem and how to address it, with the aim to extend the learning to all of Songkhla Lake. A second paper (see Part III of this volume) examines policy options that could help achieve a sustainable fishery.

First, we provide a simple conceptual model to describe the existing usage and economic value of Southern Songkhla Lake. Next, the policy options are examined, taking into consideration stakeholders' opinions and assessments by economics and fishery experts.

The results and recommendations of this study will be valuable to decision-makers attempting to formulate practical policies aimed at sustaining fishery production in Southern Lake. In addition, the study will be useful to those interested in understanding the value of the fishery industry in Songkhla Lake.

8.1.1 Research Problem

Southern Songkhla Lake, shortened to Southern Lake, is located in Songkhla Province, on the southeast coast of Thailand, and covers an area of about 180 km^2 with brackish and salt water. Southern Lake empties into the Gulf of Thailand through a small channel that also serves as the harbor entrance to the town of Songkhla. Tidal currents carry seawater into the lake all year long, which mixes with freshwater from several streams and small rivers that flow into the three lakes. Southern Lake has a season-dependent salinity—its water is brackish to salty in the dry season (February–April) when freshwater input is minimal but is mostly fresh during the wet season (November–March) when freshwater input is high.

The lake plays a unique role because it acts as the lungs for the inland harbor at the town of Songkhla. Due to the tidal currents, the depth of the harbor and its channel are maintained at 6–8 m. The seasonal fluctuations in the water level of the lakes wash away new sediments and remove the isolated patches of water hyacinth, which die because of the saline water. The open lagoon is essential for a stable and healthy ecosystem, supporting all forms of aquatic life in the lakes, because open water, wind, and waves put oxygen in the water effectively. Any change in the quality of the water, whether from natural or man-induced actions, is likely to affect the whole system.

Songkhla Lake is facing serious water pollution as a result of human activity and industrial development. In 2001, domestic wastewater (from Hat Yai City, the most populous area with the largest business center of the Songkhla Lake basin) containing 3375 kg biological oxygen demand (BOD) was discharged into the lake daily. Pesticides and fertilizers from agricultural activities, nutrients from shrimp farming and livestock waste, and oil and grease from shipping and other activities in the harbor also accumulate in the lake (ONEP 2004). In 2005, the Department of Pollution Control reported 1567 factories around Songkhla Lake, half of which are large factories and nearly two thirds of which are around Southern Lake. Their accumulated discharge is having serious detrimental impacts on the water quality, and the direct impacts on the fishery industry increase yearly. Although a sewage treatment plant was opened in Hat Yai in 1999, it has only had a small impact in mitigating pollution.

The shrimp fishery is the most profitable economic activity in Southern Lake. On average, the income from the shrimp catch is almost ten times that of other catches. Presently, however, the shrimp catch in Southern Lake has decreased severely in terms of both amount and number of species. Sirimontaporn et al. (1999) reported that there were 30 species of shrimp found in Songkhla Lake in the early 1980s, most of which were found in Southern Lake, but a later study in 2002 found only 11 species (Mabuntham 2002). A related study found that the shrimp catch from standing traps, the most common local fishing trap, decreased from 3.6 kg/day in 1996 to 0.9 kg/day in 2003 (ONEP 2004).

Although there has been a great deal of concern over these problems, they persist. This study was thus undertaken to determine appropriate policy options to manage water quality more effectively as a prerequisite to developing a more sustainable fishery industry in Southern Lake.

8.1.2 Objective of the Study

The objective of this research is to examine the impact of water pollution on fishery production and to develop appropriate water quality control policies that will ensure a sustainable fishery in Southern Lake.

8.1.3 Study Site and Scope of the Study

Of the three lagoons in Songkhla Lake, Southern Lake was selected as the study site because it is richest in biodiversity, and it is home to the most intensive economic activities and the highest population density. The shrimp fishery was selected as the indicator of both the environmental quality and economic value of the lake because shrimps are the highest-priced product in the seafood market and they are most abundant in Southern Lake.

8.1.4 Methodology

The study is a thorough review of the existing situation by means of data collection, construction of a profile of the existing situation, and estimation of the relationship between shrimp productivity and sources of pollution. The data and preliminary analysis are reported in Sect. 8.2.

Three main categories of information are included: biological conditions, physical conditions, and economic activities related to fishery production. The biodiversity of the lake was determined with the help of an experienced biologist. The historical trends and current physical conditions were also investigated, focusing on water pollution and the major sources of pollution. Using these data along with consultations with ecologists and discussions with focus groups of experienced fishermen, a profile of the existing biological, physical, and economic system of Southern Lake was constructed in the form of a simple conceptual model.

We analyze the links between pollution from economic activities surrounding the lake and government regulatory policies on water quality and fishery productivity and assess the ability of potential alternatives to existing policies to improve conditions in the lake and, hence, outcomes for fishers. The draft policy alternatives were presented to the various stakeholders, specifically the polluters; community leaders and representatives from related government agencies; and external agencies, such as NGOs and academic institutions. Once the stakeholders had a chance to study the draft policy alternatives, focus group discussions (FGDs) were conducted to investigate how they perceived the situation and to obtain their comments and suggestions. With this feedback, the details of each alternative were adjusted, and criteria for assessment were formed. Finally, the research team and other experts interpreted the results and ranked the policy alternatives.

8.2 The Southern Lake: Functions, Status, and Recent Trends

This section provides a baseline from which to investigate policy options for reducing pollution as a first step in achieving a sustainable fishery in the Songkhla Lake system. A profile of the lake system is presented in terms of biophysical properties and existing usage, followed by discussion of the causes of the current environmental degradation affecting fishery production in Southern Lake.

8.2.1 The Ecosystem of Southern Lake

The characteristics of the Southern Lake ecosystem are quite unique, especially the salinity for its seasonal variations and the distance it reaches from the mouth of the lake. In the rainy season, freshwater and nutrients from the watershed area are discharged into the lake, flushing the seawater outward to the Gulf of Thailand. However, in the dry season seawater intrudes well into the lake bringing the larvae of sea animals—various shrimp species, in particular—into the lake, which then serves as a nursery for these animals. At the same time, this seawater intrusion also kills certain aquatic weeds, which might otherwise become overabundant, thus helping maintain the equilibrium of the lake's ecosystem. The combination of freshwater, brackish water, and seawater all in the same body gives Southern Lake a highly diverse and rich ecosystem that can sustain fishery resources and marine animals all year long.

According to the Encyclopedia of Southern Thailand Culture (Sirimontaporn et al. 1999), 450 fish species and 30 shrimp species could be found in Songkhla Lake at the time of publication. The lake also serves as an important nursery ground for many economically important species of fish, crabs, and shrimps (Sookchareon 1965; Choonhapran 1994; Sirimontaporn et al. 1999; Mabuntham 2002).

8.2.2 *The Socioeconomic Structure of the Southern Lake Area*

Southern Lake has always been rich in fishery resources, and fishing is the major occupation. People also collect forest products and cultivate food crops, such as rice and fruits. With the exceptions of Hat Yai and Songkhla—the bigger cities in the area with large business centers—these natural resources are the main sources of livelihood and food around Southern Lake. The total population around the lake is approximately 800,000, distributed in 9 of the 16 districts in Songkhla Province. In recent years, government policy has encouraged industrial operations and the population to increase, thereby contributing to environmental degradation and changes in social and cultural norms, such as increased migration to towns and dependence on wage income. Southern Lake's economic structure remains related to natural resources, primarily through its fishery and agricultural production sectors. However, in recent years, the fishery sector's economic contribution to the province has decreased, while those of most other sectors have increased (Office of the National Economic and Social Development Board 2006).

8.2.3 *The Economic Value of Southern Lake*

The economic value of a lake can be understood using the concepts of use and nonuse values. Use values would be the values of goods and services directly provided by the lake system, on-site to the local people. Examples of such use values are the lake's fisheries, vegetation, and mariculture (fish farming) because they provide food and income from sales of products derived from the resources, as well as nonpecuniary or recreational benefits that come from enjoying the natural environment. Songkhla Lake also provides indirect use values, for example, habitats for a diversity of aquatic life which contribute to the fish populations that can be caught by people living many miles away from the Lake.

Nonuse or passive-use values refer to the willingness of the public to pay to maintain some goods and services provided by the lake even if this use is not actually possible or available (Freeman 1993; Bateman et al. 2002). For instance, someone who never visits, or ever intends to visit, the lake may hold a passive-use value because she values nature in general. Nonuse values can be of various types, including existence values, altruistic values, and bequest values.

In addition to use and nonuse values, there may also be "option values" for resources that are not currently used, as people may be willing to pay now to have the option of using them in the future (Bateman et al. 2002). Option values for Southern Lake could be the indigenous species of fish that might draw ecotourists or other recreational activities.

The total economic value of a lake can be measured as the sum of all the types of economic values it carries. Calculation of all the different types of values provided by Southern Lake would be quite challenging. For simplification, this study mainly considers the cause-and-effect relationship between water quality and the economic value of the lake in terms of the income households around the lake derive from fishers. More specifically, the productivity of the shrimp fishery was selected as the single collective measure of the economic value of the lake.

8.2.4 Deteriorating Water Quality

Water quality is one of the critical factors affecting the productivity of Southern Lake. This section examines the sources of pollution, the monitoring of the water quality, and the current water quality situation.

The major sources of pollution in Southern Lake are the industrial sector, agricultural sector, and lakeside or nearby communities. The factories that release wastewater into the lake are mainly from the rubber and food industries. The Industrial Office of Songkhla reported that in 2003, at least 75 factories had released wastewater into a canal flowing into the Southern Lake (ONEP 2006). The major sources of agricultural pollution are shrimp farms, pig farms, crop farms, and rubber plantations, which contaminate the lake with wastewater, pesticides, fertilizers, and other toxic substances. The other important sources of pollution are human communities around the lake. The Office of Environmental Policy and Planning (ONEP 2006) reported that in 2005, the most important source of BOD was local communities.

Data on water quality parameters were collected according to seasons and catchment areas. The Office of Environmental Policy and Planning (ONEP 2006) sets the type of parameters for water quality measurement which can be combined into an index. Since 1998, five monitoring stations have been set up in areas of special attention around Southern Lake to collect data three times a year according to the season: November (heavy rainy season), March or April (summer), and July or August (light rainy season). These data are used to assess changes related to the quantity of rainfall, which has a major influence on lake conditions, especially salinity because it affects the proliferation and diversity of all aquatic species.

The eutrophication[1] in Songkhla Lake arises from high nitrogen and phosphorus levels, which result primarily from the release of wastewater from human activities in the catchment areas. In an extensive study from 1999 to 2003, NICA (2005) reported that the nitrogen levels at the U-Tapao river mouth was three times higher than in other areas and the amounts of phosphorus at both the U-Tapao and Pawong river mouths were higher than in other areas as well. In November 2001, the amount of chlorophyll A[2] was 164 mg/L, which indicated serious eutrophication. Again, in June 2004, the amount of chlorophyll A in Southern Lake was 339 mg/L (Laoongsiriwong et al. 2004), which indicated that eutrophication was severely damaging aquatic survival, and sudden fish die-off was regularly occurring in Southern Lake.

8.2.5 A Framework for Estimating the Relationship Between Fishery Productivity and Water Quality

Figure 8.1 illustrates the framework used to estimate the relationship between the water quality of Southern Lake that leads to environmental degradation and the economic activities in the watershed, in general, and fishery productivity in Southern Lake, in particular. The model identifies three factors affecting fishing catch: the first factor is water quality, measured by a water quality composite index (WQCI), which is a measure of pollutants present in the water; the second is fishing effort, measured by the total number of items of fishing equipment used and the number of fishing households; and the third is the salinity, salts measured in parts per thousand (ppt) units of water. The model also includes positive and negative factors affecting water quality, such as rainfall, which dilutes water pollution, and human activities, such as settlement, industry, and farming, which discharge wastewater into the water basin.

[1] Eutrophication is the situation that results when a body of water has a large supply of nutrients which encourages plankton overgrowth and DO decreases, which affects the survival of aquatic life.

[2] Chlorophyll A, a measure of primary production of plankton in a water column, is an indicator of eutrophication; more than 10 mg/L of chlorophyll A in a water body usually indicates eutrophication (Nedwell et al. 2002).

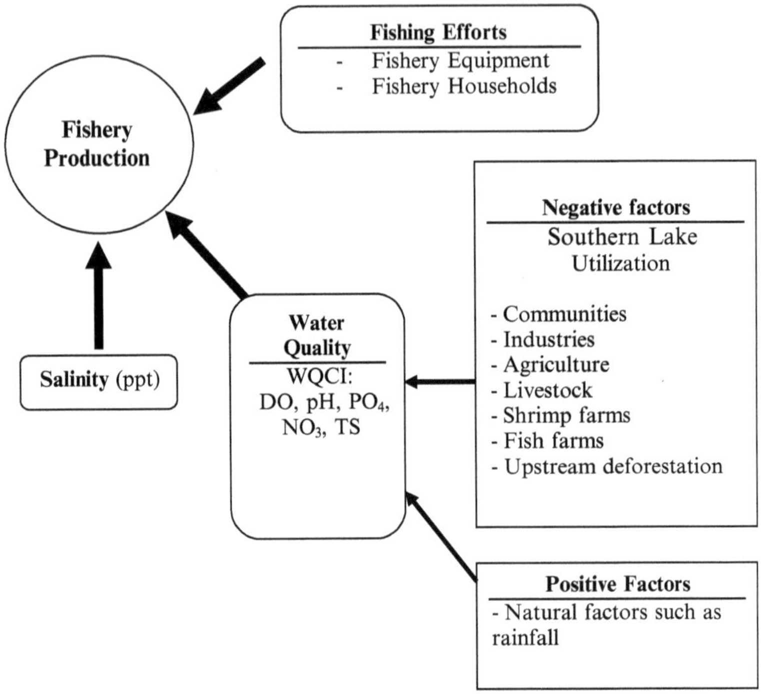

Fig. 8.1 A framework for analyzing fishery productivity in Southern Lake

8.2.6 The Empirical Model

The following production function provides a framework to estimate the impact of water quality, salinity, and fishing effort on shrimp harvests:

$$Q = f(WQCI,\ S,\ E) \tag{8.1}$$

where

Q = fishery productivity (pink shrimp catch)
WQCI = water quality composite index
S = salinity
E = fishing effort

Each of the variables is defined in more detail below.

Fishery Productivity (Q) Pink shrimp was selected to indicate the fishery productivity of Southern Lake for the following reasons:

1. Pink shrimps are the species of shrimp caught in the largest numbers in the lake (Choonhapran et al. 1996).

2. They are native to Southern Lake and are not augmented by any artificial means (NICA 2005). Therefore, as the proxy for representing the overall fishing catch, pink shrimp reflects the natural conditions of Southern Lake.
3. Data are available from NICA surveys on seasonal shrimp catches in Southern Lake.

Water Quality Composite Index (WQCI) The WQCI consists of five factors, DO, pH, PO_4, NO_4, and TSS, selected from the general list of water quality parameters[3] specified and used by the Department of Pollution Control. The WQCI was calculated on the basis of the composite index method; each of the parameters was weighted according to its importance to water quality and health of aquatic animals. The index is described by Eq. 8.2, below.

$$\begin{aligned} WQCI = {} & 0.4DO + 0.25[1/(\text{absolute value of}(7 - pH))] \\ & + 0.125(1/PO_4) + 0.125(1/NO_3) + 0.1(1/TSS) \end{aligned} \tag{8.2}$$

where:

DO = Dissolved oxygen (mg/L); a DO value of less than 4 mg/L causes the death of most aquatic animals.

pH = A measure of acidity or alkalinity of a solution (MPN/100 mL); a measure of pH in water of 7 indicates a neutral solution, which is healthy for aquatic life-forms. A pH value below 7 means the water tends to be acidic, and a value over 7 denotes water that is alkaline, neither of which is healthy for aquatic life.

PO_4 = Phosphate in the water (mg/L); phosphate content that is too high leads to eutrophication. The main sources of phosphate are fertilizers and detergents.

NO_3 = Nitrates in the water (mg/L); in a lake, high NO_3 can cause eutrophication similar to phosphates. Their sources are farming, domestic waste, and industrial waste.

TSS = Total suspended solids (mg/L); high concentrations of impenetrable material (suspended solids) obstruct both the breathing and growth of aquatic animals. An increase in suspended solids comes, for instance, from deforestation in the upstream watershed.

The relationship between WQCI and catch quantity (Q) is positive, that is to say, a higher WQCI would give rise to a higher catch quantity.

Salinity (S) The relationship between salinity and shrimp productivity is quadratic because, all other factors being equal, shrimp productivity will increase with higher salinity but decrease when the salinity becomes too high. Southern Lake is an open lagoon with brackish to salty water in which the salinity is not too high. For the

[3] The general water quality parameters specified by the Department of Pollution Control are DO, pH, PO4, NO4, TSS, TU, BOD, and fecal coliforms. The five we include in the index are the most highly correlated with the health of the shrimp fishery.

relevant range of salinity, the relationship between salinity and shrimp productivity can be represented using a positive linear relationship.

Fishing Effort (E) Unfortunately, there was no formal data available for fishing traps after 2003. However, most of the fishermen and government officers who were interviewed indicated that the number of fishing traps had reached a saturation point by 2003 in Songkhla Lake and had not changed significantly between 2003 and the writing of this report in 2007. Thus, in the regression analysis carried out in this study, it was assumed that fishing effort is constant and, hence, is not an explanatory variable in the estimation of shrimp harvests over time. A fishery in which there are changes in effort either over time or across catchment areas would include fishery effort and other variables that may impact fish stocks and, hence, harvests in estimation of a fishery's productivity as illustrated in Eq. 8.1.

Data Sources and Description The study data are primarily from secondary sources, specifically, from two government agencies: the National Institute of Coastal Aquaculture (NICA 2005) and the Office of Environmental Policy and Planning, Division 16 (ONEP 2007). The seasonal shrimp catch data was collected from one of eight docking sites around Southern Lake for the years 2004–2006. The WQCI was calculated using seasonal water quality data for the years 2004–2006 taken at two stations in and around Southern Lake—U-Tapao river mouth (UT) and Pag-ro (PR).[4] This information formed the basis for tracing which sources of pollution, coming from what sectors, primarily affected shrimp production and, by extension, the overall productivity of the lake. Finally, the salinity data measured by Pag-ro station was selected for the analysis because this station is located a good distance from the mouth of the lake, so if significant salinity intruded as far as Pag-ro station, it could be assumed that much of the lake would also have the same salinity. Pag-ro was also selected because its WQCI is not correlated with those of other stations.

Empirical Results The analysis checks for multicollinearity between variables and leads to the following three variables selected for the analysis: WQCI at U-Tapao river mouth station (WQCIUT), WQCI at Pag-ro station (WQCIPR), and the salinity at Pag-ro (SPR). The shrimp productivity equation was estimated using ordinary least squares (OLS) in semilog form. A number of different functional forms were examined for the salinity variable because there is no underlying theoretical argument supporting a specific function, and while there are some differences in significance and coefficient levels, they are minor (See Table 8.5 in the appendix for these results). The linear form is shown in Eq. 8.3 below, where ** denotes significance at the 0.01 level and * at the 0.05 level, with t-statistics shown in parentheses under each estimated coefficient.

[4] There are five stations monitoring water quality. Three of the five are highly correlated with each other. We thus used one of those (U-Tapao river mouth) and one that was not correlated with the others (Pag-ro) for the WQCI variables. Both of these stations also represent the areas of severe water quality degradation.

$$LN\_SHRIMP = 5.368^{**} + 0.034\ WQCI_{UT}{}^{*} + 0.10\ WQCI_{PR}{}^{**} + 0.185 S_{PR}{}^{**} \quad (8.3)$$
$$(3.875) \quad (24.238) \quad (8.446) \quad (13.571)$$

Based on Eq. 8.3, an improvement in water quality by one unit at the U-Tapao river mouth ($WQCI_{UT}$) would result in an increase of 3.4 % in shrimp productivity, while the same level of improvement at Pag-ro ($WQCI_{PR}$) would result in an increase of 10 % in shrimp productivity. In addition, one ppt increase in salinity at Pag-ro would lead to an increase of 18.5 % in shrimp productivity. Thus, these results suggest that efforts to improve water quality would have a significant impact on shrimp productivity.[5] A more comprehensive estimation of the impacts of water quality on fish harvests is an important topic for additional research.

8.3 Management Options for Water Quality Control

The government can respond to the negative externalities of pollution in at least two ways: It can regulate polluter behavior directly (command and control policies) or provide incentives (market-based policies) that effectively price pollution, leading to reduced levels of emissions. The following sections first describe the existing policies and then discuss additional policies the government could consider.

8.3.1 Water Pollution Management in Thailand

The Ministry of Natural Resources and Environment (MONRE) is the main governmental agency responsible for environmental quality and management at the national level. Besides MONRE, the core governmental organizations involved in environmental management include the Office of the National Economic and Social Development Board (NESDB),[6] the Department of Public Works and Town and Country Planning (DTCP), and the Department of Local Administration (DLA). The regulatory agencies in charge of monitoring water quality attempt to limit pollutant loads to current levels and reduce them as soon as possible in the future.

All the stakeholders must comply with the various acts and regulations listed in Table 8.6, in the Appendix. Stringent regulation applies to pollutants known to

---

[5] Because the number of observations covers only 3 years with three seasons (rain, dry, and before rain), multivariate cointegration testing was performed using these data to determine if a long-run equilibrium exists for the variables. The result supports the existence of a long-run equilibrium with a statistically significant p-value (at the 0.0073 level).

[6] NESDB is under the Office of the Prime Minister. DTCP and DLA are under the Ministry of Interior.

reduce water quality by increasing BOD, COD, and total Kjeldahl nitrogen (TKN). These are typically in the form of minimum concentration-based discharge requirements (see Table 8.7 in the Appendix).

8.3.2 Command and Control Policies

Since 1961, the Thai government has developed 5-year National Economic and Social Development Plans that establish the overall priorities and policy framework for the country's development. In the seventh plan (1991–1995), the government started to promote sustainable economic development that protects the environment, which continues to the present day. For instance, the ninth plan (2001–2005) focused on the management of natural resources and the environment by enhancing the participation of all sectors in society, empowering communities, incorporating strong and useful regulations, and encouraging practical research.

Limitations of Command and Control Policies The present investigation finds that command and control policies for Southern Lake have the following limitations:

- Command and control policies do not encourage industries to improve their production processes and wastewater treatment systems to reduce pollution emissions.
- Concentration-based discharge standards alone are not sufficient to control pollution, and there should also be volume-based controls over waste discharge.
- Monitoring procedures are somewhat ineffective due to insufficient numbers of enforcement officers, which limits enforcement to random inspections every 2 months or so for large and medium factories and farms, as well as the general neglect of smaller operations, such as backyard pig farms.
- The punishment for those who exceed pollution standards is quite light and not strictly enforced. For instance, polluters who dispose of effluents at a higher rate than the set standards are asked to pay an emission charge that is always lower than the cost of pollution treatment.
- Command and control policies do not encourage participation by all stakeholders, such as manufacturers and communities, as they are not made sufficiently aware of their responsibilities in environmental protection.

8.3.3 A Market-Based Instrument: Effluent Charges

Although the command and control system is currently employed in water pollution management in Thailand, there are plans for changes to help control pollution more effectively. Toward this end, the Department of Industrial Works conducted a study in 2007 on "The Application of Economic Instruments in Industrial Pollution Management," which concludes that emission charges are the most appropriate economic instruments for minimizing pollution in Thailand and that they should be used in hazardous waste management by related government agencies. Emission charges (EC) are estimated using data from the pollution a factory discharges and pollution load coefficients,[7] such as the pollutant load per unit of product or unit of raw material, for industries that generate significant BOD loads, for example.

The funds collected from emission charges are to be utilized as a circulating fund to assist or support implementation of industrial pollution management measures and environmental quality monitoring of industrial activities. All factories that discharge wastewater to their surroundings are to be obliged to pay ECs, and the factories are divided into two types: "flat rate" type and "variable rate" type.

The flat rate-type factory is classified as one having a load of influent[8] of less than 100 kg BOD per day. Most factories are medium to small scale that discharge relatively small amounts of wastewater, and ongoing effluent monitoring is not required.

The variable rate-type factory is classified as having a load of influent greater than 100 kg BOD per day. Most of such factories are relatively large scale, generating considerable volumes of wastewater, and effluent monitoring is required.

8.3.4 Certification and Education Programs

Government agencies use other methods besides regulation and legislation to attempt to control the pollution in Southern Lake including:

1. *Environmental promotion programs for entrepreneurs*: These programs are introduced to small entrepreneurs in parallel with environmental legislation—mainly various "green-technology" operations—which have the goal of reducing pollution by focusing on production lines. A factory joining this program will be certified by the government as a "production-friendly business" to improve its reputation in the local community.
2. *Public awareness and education programs*: The concept behind these programs is to encourage the environmental agencies and communities located around the lake to collaborate in being vigilant about any threats to the quality of the water.

---

[7] The coefficients were divided by the type and production scale of the factory. See Department of Industrial Works (2007).

[8] BOD load of influent is the amount of BOD in wastewater produced from the production process. Effluents are the wastewater from factories and other activities discharged into the environment with or without treatment.

8.3.5 *Tradable Discharge Permits for Water Quality Control in Southern Lake*

It is difficult to define an optimal emission fee level because of the need to estimate marginal damages and marginal abatement costs for each type of polluter. If the fee were too low, it would not provide sufficient incentive for pollution control; even if the fee was at an optimal level, there would always be distortions existing in behavior of both the enterprises and the regulatory bureaus (Yun 1998).

A tradable discharge permit (TDP) system, if properly designed, can provide more flexibility than EC for firms to meet their discharge targets. It offers industrial firms the option to meet effluent standards by purchasing effluent credits/permits from other firms, which can reduce their effluents at a lower cost. This concept is very applicable to the Southern Lake watershed because the local industries have different abatement costs and there are likely to be gains from trading effluent credits/permits. However, a TDP system might be limited by the high costs of monitoring and transactions between many small firms. Also, its structure (e.g., the issue of grandfathering versus auctioning permits) is challenging for governments to design.

A promising way to initiate a TDP for the Southern Lake watershed would be to first introduce a TDP system for large domestic and foreign firms, as well as public utilities, such as public wastewater treatment facilities. In addition, the implementing agency can work with the local industry associations, community organizations, and related government agencies to set up a monitoring and enforcement system. An illustration of how a TDP system would work in the U-Tapao catchment area, a sub-watershed located in the Southern Lake basin is found in Pornpinatepong et al. (2010).[9]

8.3.6 *Technological Options to Improve the Water Quality in the U-Tapao Catchment Area*

The cost-effectiveness of technological options for sewage treatment systems is examined using the available literature and working with engineers and specialists in the field. In this study, cost-effectiveness (measured per kilogram of BOD treated) is analyzed for three levels of wastewater treatment—primary, secondary, and tertiary—for each technical option and for each major point pollution source, namely, industry, farm, and domestic treatment facilities.

Cost-Effectiveness Analysis (CEA) of Technological Options Because of the limitations of cost-effectiveness analyses (CEAs), which ignore scale effects, an

[9] U-Tapao was selected for an initial demonstration of this system for its high population density and large variety of firms, both domestic and foreign. See Pornpinatepong et al. (2010) for details about the TDP program such as how to allocate rights and establish trading rules.

attempt is made to categorize all the different cases according to treatment plant size. The list of case studies, together with their treatment processes, is given in Table 8.8 of the Appendix. Capital, operating, and land use costs are included in the analysis. Capital costs include initial construction and equipment costs and are considered as having been incurred in a single year. For the cost of land use, the revenue of growing rubber trees was selected for the opportunity cost of land per year, which was THB 15,510/rai (or USD 2675.6/ha).[10] This figure comes from interviews with experienced rubber tree farmers in the U-Tapao area. The effectiveness of each sector was estimated in terms of the capacity to reduce BOD loading and in the differences between influent and effluent BOD loads. The cost-effectiveness ratio was then calculated by dividing the total cost per year by the amount of reduction in BOD loading per year. The smaller the number, the more cost-effective the technology. The results are shown in Table 8.1.

Two types of methodologies were identified as affecting treatment costs for basic technology options. First, land use applications had higher fixed costs and lower operating costs, while technology-based applications, such as the upflow anaerobic sludge blanket (UASB) or activated sludge (AS), had lower land use costs but higher operating costs. With high technology options, however, there was no particular cost pattern. The results show that CEAs of treatment technologies are sensitive to the size of the plant and the effluent load. For businesses with lower influent loads or smaller amounts available to invest in individual treatment plants, treatment technologies are less effective in reducing emissions, while technological treatments are more cost-effective solutions for small-scale polluters that can share a common treatment plant.

For domestic treatment facilities at the municipal level, the CEAs show that the cost of treating BODs is very high, even at full capacity, and large-scale domestic treatment facilities tend to be ineffective due to underutilization and the requirement for administrative management. Smaller-scale domestic treatment facilities are recommended because they are easier to operate effectively and offer more flexibility for future development.

8.4 Economic Evaluation of Policy Options for a Water Quality Management Program: A Case Study of U-Tapao Catchment Area

Using the data from the previous sections, the focus turns to the formulation of policy alternatives to improve water quality in the case of the U-Tapao catchment area of Southern lake. The U-Tapao catchment area was selected as the representative site for Southern Lake because it contains a high density of polluters, a good

[10] I rai equals 0.16 ha and the exchange rate in 2007 was 1USD equals THB 36.35.

Table 8.1 Cost-effectiveness analysis of technological options for sewage treatment plants in each sector

Sector	Capital costs (1000 THB)	Land costs (1000 THB)	Operating costs[a] (1000 THB)	Total costs (1000 THB)	Effluent (BOD (kg/year)	Effectiveness[a] (THB/kgBOD)	Biogas
Factories: rubber (CRSFs)/food/others							
Basic technology							
Yoong Thong[b]	526.9	182.6	55.5	765.0	498.2[c]	1,535.4[c]	Yes
Model 1994[b]	531.9	109.4	1,800.9	2,442.3	50,000[d]	48.9[e]–4,901.9[c]	–
					498.2		
Model 1995[b]	342.6	109.4	1,106.3	1,558.3	50,000[d]	31.2[e]–3,127.8[c]	Yes
					498.2[c]		
High technology							
UASB[f]	1,339.9	35.8	1,437.5	2,813.2	492,750.0[c]	5.7[c]	Yes
AS[f]	693.2	51.1	5,274.0	6,018.3	492,750.0[c]	12.2[c]	–
Pig farms							
A (large)[g]	502.8	268.7	111.0	882.5	120,656.2	7.3	Yes
B (medium)[g]	26.4	2.9	55.5	84.8	13,578.0	6.2	Yes
C (small)[g]	5.3	2.4	55.5	63.2	2,894.1	21.8	Yes
Communities							
Small scale: villages							
Sadao[h]	125.0	0.7	89.0	214.7	2,428.2[c]	88.4[c]	–
Klong Ree[h]	60.0	23.3	60.7	144.0	15,300.0[c]	9.4[c]	–

Large scale: municipalities							
Songkhla[h]	15,000.0	2,728.7	6,000.0	23,728.7	1,022,000[e]	23.2[e]–331.7[c]	–
		(176 rai)			71,540[c]		
Hat Yai (dry season)[h]	93,368.2	31,628.2	8,966.7	133,963	3,777,750[e]	35.5[e]–108.8[c]	–
		(2040 rai)			1,231,875[c]		
Hat Yai (rainy season)[h]	93,368.2	31,628.2	8,966.7	133,963	3,189,932[e]	42.0[e]–52.7[c]	–
		(2040 rai)			2,542,833[c]		

USD 1 = THB 36.23 in 2007
[a]Calculation by researcher
[b]ORRAF 2007
[c]Current situation (from survey in 2008)
[d]Assumptions by an expert
[e]Full capacity
[f]Klinpikul, 2008, Industrial manager. Chotiwat Factory, Songkhla, Personal communication. March 2008
[g]Dumrungwattanapokin 2004
[h]Hatyai Local Government 2008

monitoring plan overseen by responsible agencies, and communities that are aware of the problem.

8.4.1 Construction of Comprehensive Policy Options

The environmental problems and related policy problems facing Songkhla Lake and Southern Lake have been clearly defined in previous sections. The policy goal here is to effectively achieve the desired total water pollution reduction targets through sustainable water resource utilization. Table 8.2 lists the specific policy problems and potential options.

From this list, specific policy options have been formulated:

Policy Option 1: Strengthen the enforcement of the existing command and control system (CAC). This takes the status quo situation and increases monitoring and enforcement.

Policy Option 2: Implement a modified version of the emission charge system (ECS) from the Department of Industrial Works (2007) described above.

Policy Option 3: Implement a tradable discharge permit (TDP) system.

These three policy alternatives have been assessed based on four basic assumptions: (1) BOD loading was used as the basis for water quality control standards for all three policies; (2) under the ECS option, firms that discharge more than 100 kg BOD/day would be required to participate in the EC program using a variable rate, while firms that discharge up to 100 kg BOD/day would be charged a flat rate; (3) under the TDP, firms that discharge more than 100 kg BOD/day would be required to participate in the TDP program, while firms that discharge up to 100 kg BOD/day would be allowed to discharge their wastewater through municipality sewage treatment plants; and (4) all three policy options address household discharge.

Table 8.2 Policy problems and potential options to improve water quality in Southern Lake

Policy problem	Approaches to improve water quality
Inappropriate water quality control standards	Define an appropriate standard for BOD concentration and volume
Ineffective incentives to reduce pollution	Design incentive mechanisms (e.g., emission charges, TDP)
Unclear measures for controlling small-sized firms	Construct and manage wastewater discharge facilities for small-sized firms
Unavailable domestic discharge control measures	Construct a combined sewage treatment plant for each community (for households, small firms, and farms)
Ineffective monitoring	Restructure the monitoring system. Public participation by providing an independent environmental protection organization

8.4.2 Criteria Selection

Table 8.3 describes the criteria used to assess each policy option.

8.4.3 Evaluation of Policy Options

Table 8.4 presents a summary of assessment of the options using the criteria listed in Table 4. The advantages, disadvantages, and implications of each policy are discussed below.

The CAC Policy CAC has several advantages, namely, high public acceptability and legal feasibility, relatively low implementation complexity, no initial investment requirements, and lower operating costs than those of the other two options. However, analysis of the CAC shows several disadvantages when compared to the TDP and EC system: low effectiveness, low equity, and low decentralization. Under CAC, it would be quite difficult to solve water pollution control problems because of serious disadvantages to some stakeholders.

The EC Policy Although similar to the TDP in theoretical concept, the practical effectiveness of the EC policy was found to be lower than the TDP because of various limitations. Firstly, it is difficult to define an optimal fee level that provides sufficient incentives for pollution control. Secondly, implementation is complex because additional resources are required to collect the charges. There is also a higher administrative burden in having to deal with polluters who do not want to, or are unable to, pay the required fees, thus leading to potential conflict, deceit, and corruption. Therefore, if the EC were to be applied to the U-Tapao catchment area, there is no guarantee that it would achieve its goals. However, an EC has high legal feasibility because the legislative framework to enable its implementation is already in place. Moreover, it has high public acceptability.

The TDP Policy The main advantage of the TDP is it is highly effective because it provides great flexibility for firms to meet their discharge targets using market mechanisms. Because firms in the U-Tapao catchment area have a large range of abatement costs, the potential gains from trading effluent permits are potentially significant. This is important because it can stimulate firms to seek and develop the most effective sewage treatment technologies for their businesses. Also, with market operations, a permit trading system will take some of the administrative burden out of the hands of the government officers. Moreover, having the highest outcome in terms of effectiveness, the TDP is expected to provide the most positive social impacts. However, there are also some disadvantages with the TDP. Firstly, there are problems with legalities, as implementing the TDP would require additional legislative mechanisms. Secondly, in terms of implementation complexity, the TDP requires a well-organized system from the earliest stages, primarily identification of BOD-carrying capacities in the areas where the TDP will be

Table 8.3 Criteria to assess the three policy options

Criterion	Description	Measure(s)	Evaluation methodology (ies)
1. Public acceptability	The acceptance level among key stakeholders, local administrative authorities, and government agencies	The degree of acceptance (strongly disagree/disagree/agree/strongly agree)	Focus group discussions
2. Legal feasibility	The existing legislative framework to permit the policy options	Easy/somewhat easy/somewhat difficult/difficult	Historical data/expert judgment
3. Implementation complexity	Extent of regulatory and legislative requirements	Degree of complexity (very high/high/low/very low)	Historical
4. Effectiveness	Extent to which the proposed policy options address the six factors in proper water quality control management	1. Provision of incentive mechanisms for pollution reduction	Expert judgment/focus group discussions
		2. Interaction among and coordination of stakeholders for water quality improvement	
		3. The outcome effectiveness of the water quality control policy	
		4. Appropriate water quality control standards	
		5. The control of all sizes of firms	
		6. Availability of domestic discharge control (high/moderate/low/none)	
5. Capital costs	Monetary costs of the initial investment in required equipment	1. Number of activities needed	Expert judgment
6. Operating costs	Monetary costs of operating the option on a yearly basis	2. Comparing the cost of each activity	
7. Transaction costs	Monetary costs incurred in firm to firm interaction		
8. Impacts			
8.1 Positive impacts	Impacts on aquatic environment, natural fishery, and aquaculture fishery	The change in the economic value of the lake (fish productivity/recreational value)	Expert judgment
8.2 Negative impacts	The loss of competitiveness due to higher production costs	The impact on output price	
9. Equity	Polluter pays principle	Number of dischargers included in the system (high/moderate/low/none)	Historical data
10. Decentralization	Firms make decisions on pollution reduction	Number of firms that potentially reduce pollutants and set up plans for pollution reduction	Historical data/expert judgment

Table 8.4 Summary of the evaluation results of the three policy options

Criteria	CAC	ECS	TDP
1. Public acceptability	Agree	Agree	Disagree
2. Legal feasibility	Easy	Somewhat easy	Somewhat difficult
3. Implementation complexity	Low	High	Very high
4. Effectiveness	Moderate	Moderate	High
5. Capital costs	Low	High	Higher
6. Operating costs	Low	Higher	High
7. Transaction costs	Low	High	Higher
8. Impacts			
8.1 Positive impacts	Low	High	Higher
8.2 Negative impacts	Low	Low	Low
9. Equity	Low	High	High
10. Decentralization	Low	Moderate	High

applied; a well-defined initial permit distributional system; and a trading system. Thirdly, with this complexity, the TDP requires a greater initial investment and also has higher transaction costs. Fourthly, the TDP was found to have the lowest rate of public acceptability due to lack of understanding of this type of market system. Therefore, in order to apply the TDP, considerable education of the communities involved would be required.

8.5 Conclusions and Policy Recommendations

The major objective of this study was to develop appropriate policy alternatives for a sustainable fishery in Southern Lake, and the impact of water pollution was the first consideration. After review and synthesis using secondary data, statistical analysis indicates that the water quality changes in the lake have had significant impact on fishery production. The findings from our investigation of technological options and management policies are summarized below.

8.5.1 Technological Options to Improve Water Quality

The results of the cost-effectiveness analysis (CEA) show that the effectiveness of a wastewater treatment technology is sensitive to the size of the plant and the volume of the influent load—the wastewater flowing into the plant for treatment. Firms must take into account the appropriate size, technology, and consistency of the amount of BOD loading when deciding what technology to use. The case studies indicate that an individual treatment plant is less effective for small businesses with a lower influent load; it was more cost-effective for small-scale polluters to share a common treatment plant.

For domestic treatment facilities at the municipal level, the CEA showed that large-scale domestic treatment facilities tend to be ineffective due to underutilization, and such operations require additional administrative management. Smaller-scale domestic treatment facilities are recommended because they are easier to operate effectively and also offer more flexibility for later development.

8.5.2 Policy Recommendations

Policy recommendations out of this research are designed to address the problems identified for Songkhla Lake as follows:

- With inappropriate water quality control standards, the major problem was high concentrations and volumes of major pollutants such as BOD.
- With ineffective incentives for pollution reduction, introduction of incentive mechanisms, such as an EC system or TDPs, is necessary.
- To counter the lack of domestic discharge control measures, small-scale, shared sewage treatment facilities for each community should be set up to deal with the combined waste from households, small firms, and farms.
- As the monitoring system was shown to be ineffective, it needs to be restructured to incorporate incentive mechanisms and public participation.

Based on the literature and survey results, it is evident that firms and farms have the potential to improve their wastewater treatment technologies but that there is no incentive to do so. Market-based instruments in pollution control systems have the potential to reduce water pollution adjunct to the existing command and control policies, but there is little actual experience with them in Thailand, and their specific design and implementation would be challenging. Further studies are required to implement appropriate new systems, particularly in special areas such as the Songkhla Lake basin.

This analysis suggests that applying a tradable discharge permit (TDP) system to large-sized firms and farms would be an effective means of reducing pollution in the Songkhla Lake watershed. This is because there is a wide variety of industry in the region, and they face a range of abatement costs. The potential gains from trading effluent credits or permits are therefore likely to be significant. However, there would also be some disadvantages with a TDP. Implementing a TDP would require various legislative issues to be resolved among other complexities. As such, a TDP scheme would require a greater initial investment and have higher transaction costs than other schemes analyzed in this study. A TDP was also shown to have a low public acceptability rating, as people do not currently understand how such systems function. Therefore, in order to implement a TDP policy in a community situation, much outreach and education work would be required.

For small- and medium-scale firms and farms, an effluent charge approach could be explored where sufficient similarities across the firms and farms allow a uniform tax rate to be applied. Whichever combination of policy options is ultimately selected, further studies will be needed to determine their feasibility and effectiveness.

Appendix

Table 8.5 Functional forms tested for the estimation of shrimp productivity

	Models ($WQCI_{UT}$, $WQCI_{PR}$, S_{PR})		
Variables	1. Linear (S_{PR})	2. Quadratic (S_{PR})	3. Time lag (S_{PR})
LN_shrimp			
R_sq	0.987	0.987	0.9863
F-value	99.979^{**}	58.691^{**}	35.9347^{*}
Autocorrelation test			
D	2.701	2.466	2.6834
du ($\alpha = 0.05$)	2.287	–	–
dl ($\alpha = 0.05$)	0.368	–	–
	Indecisive zone		
Coefficients			
Constant	5.368^{**}	5.364	5.3599^{**}
$WQCI_{UT}$	0.034^{*}	0.034^{*}	0.0324
$WQCI_{PR}$	0.1^{**}	0.106	0.1023^{*}
S_{PR}	0.185^{**}	0.147	0.1837^{*}
S_{PR}^2	–	0.002	–
SPR_Lag	–	–	0.006

**Significant at the 0.01 level; *significant at the 0.05 level

Table 8.6 Water pollution legislation

Legislation	Regulated activities	Responsible ministry(ies)	Remarks
Enhancement and Conservation of National Quality Act (ECNQA 1992)	Regulates specified point sources for wastewater discharges into public water resources, or the environment, based on effluent standards	Ministry of Natural Resources and Environment (MONRE)	Key environmental law in Thailand
Factories Act of 1992	Limits levels of effluents discharged and restricts concentration levels of chemical and/or metal pollutants	Ministry of Industry (MOI)	Issues ministerial regulations and notifications to govern any or all types of factories
Public Health Act of 1992	Regulates nuisance activities related to water pollution such as odors, chemical fumes, wastewater discharge systems of buildings and factories, or animal feedlots that cause harmful health effects	Ministry of Public Health (MOPH)	Decentralized implementation by local government agencies
Building Control Act of 1979	Regulates discharge of water pollution from buildings	Ministry of Interior (MOINT)	Decentralized implementation by local government agencies
Fisheries Act of 1947	Prohibits dumping or discharging of hazardous chemicals into water resources reserved for fishing	The Ministry of Agriculture and Cooperatives (MOAC)	

Source: partly from WEPA 2007

Table 8.7 Effluent standards in Thailand

Parameter (unit)	Industries	Pig farms		Shrimp farms	Housing estates	
		Large farms (note 3)	Medium and small farms (note 3)		100–500 units	>500 units
BOD (mg/L)	Not more than 20 mg/L depending on the receiving water body or type of industry. The PCC may allow BOD levels to exceed 20 mg/L but not 60 mg/L	60	100	20	30	20
COD (mg/L)	Not more than 120 mg/L depending on the receiving water body or type of industry. The PCC may allow COD levels to exceed 120 mg/L but not 400 mg/L	300	400		–	–
TKN (mg/L)	Not more than 100 mg/L depending on the receiving water body or type of industry. The PCC may allow TNK levels to exceed 100 mg/L but not 200 mg/L	120	200		35	35

Source: partly from WEPA 2007

Notes:

(1) *BOD* biological oxygen demand, *COD* chemical oxygen demand, and *TDK* total Kjeldahl nitrogen

(2) *PCC* Pollution Control Committee

(3) Based on Standard Methods for the Examination of Water and Wastewater (APHA, AWwA, and WEF)

(4) Large farms = >5000 units; medium and small farms = ≤5000 units

Table 8.8 Technological options for sewage treatment plants in each sector

Technologies	Factories: rubber (CRSFs)[a]/food/others					Pig farms			Communities				
	Basic technology			High technology		Model/size			Villages		Municipalities		
	Yoong Thong[b]	Model 1994[c]	Model 1995[c]	UASB[d]	AS[d]	A[e] large	B[e] medium	C[e] small	Sadao[f]	Klong Ree[f]	Sonkhla[f]	Hat Yai (dry)[f]	Hat Yai (rain)[f]
1. Primary treatment													
Coarse/fine screen grit chamber				x	x				x	x	x	x	x
Rubber trap	x	x	x										
pH neutralization		x	x	x	x								
Grease trap tank/ skimming sedimentation													
Primary sedimentation tank				x	x				x	x	x		
Equalization pond				x					x	x			
2. Secondary treatment													
Anaerobic pond/ facultative pond	x						x	x		x		x	x
Anaerobic covered lagoon	x		x			x							
Fixed dome							x	x					
UASB				x									
3 Tertiary treatment													
Oxidation pond						x	x	x		x		x	x
Aerated lagoon (AL)/ activated sludge (AS)		x	x	x	x				x		x		

4. Final treatment													
Constructed wetland	x			x	x					x		x	x
Oxidation pond	x	x	x								x		

"x" is mean technology use

Source: [b]Pollution Control Department 2008

[d]Klinpikul, 2008, Industrial manager. Chotiwat Factory, Songkhla, Personal communication. March 2008

[f]Hatyai Local Government 2008 (personal communication)

Notes: [a]CRSFs stand for cooperative rubber sheet factories

[c]Model 1994 and Model 1995 are the sewage treatment plant model for rubber factories specified by Office of the Rubber Replanting Aid Fund (ORRAF 2007)

[e]The size of pig farms specified by the number of pigs in the farm which are the number of pigs >5000 pigs is large-sized farms, the number of pigs = 500–5000 pigs is medium-sized farms, and the number of pigs <500 pigs is small-sized farms (Pollution Control Department 2007)

References

Bateman IJ, Carson RT, Day B, Hanemann M, Hanley N, Hett T, Jones-Lee M, Loomes G, Mourato S, Ozdemiroglu E, Pearce OBE DW, Sugden R, Swanson J (2002) Economic valuation with stated preference techniques. Edward Elgar, Northampton

Choonhapran A (1994) Biology of Liza subviridis (Valenciennes, 1836) in Songkhla Lake and adjacent coastal areas. Thai Fish Gaz 47(3):211–226

Choonhapran A, Ratanachai C, Meechookunt A (1996) Assessment of fisheries resources in Songkhla Lake during 1994–1995. Technical paper no. 4/1996. National Institute of Coastal Aquaculture. Department of Fisheries. Ministry of Agriculture and Cooperatives, Songkhla

Department of Industrial Works (2007) A study on the application of economic instruments for industrial pollution management. Executive summery phase 3. Ministry of Industry, Bangkok

Dumrungwattanapokin T (2004) Study on efficiency and economic return for selecting wastewater treatment system concerning environmental problem solving of pig farms. National Research Council of Thailand, Bangkok

Freeman AM III (1993) The measurement of environmental and resource values: theory and methods. Resources for the Future, Washington, DC

Hatyai Local Government (2008) Wastewater treatment at Sadao and Klong Ree villagers. Water Reclamation Plant, Songkhla

Laoongsiriwong N, Tompongong P, Laoongsiriwong L, Sunbuga T (2004) Eutrophication: the effect on sea farming and the fishery in Songkhla Lake. NICA (National Institute of Coastal Aquaculture). The Fishery Resources Restoration Project in Songkhla Lake. Department of Fisheries, Ministry of Agriculture and Cooperatives, Songkhla

Mabuntham J (2002) Species diversity and quantity of aquatic animals caught with standing traps in the outer Songkhla Lake. Master of Science thesis. Prince of Songkhla University, Songkhla, 105 p

Nedwell DB, Dong LF, Sage A, Underwood GJC (2002) Variation of the nutrients loads to the Mainland U.K. estuaries: correlation with catchment areas, urbanization and coastal eutrophication. Estuar Coast Shelf Sci 54:951–970

NICA (National Institute of Coastal Aquaculture) (2005) Sea farming in Songkhla Lake. The Fishery Resources Restoration Project in Songkhla Lake. Department of Fisheries, Ministry of Agriculture and Cooperative, Songkhla

Office of the National Economic and Social Development Board (2006) Gross provincial product. Office of the Prime Minister, Thailand

ONEP (Office of Nature and Environmental Policy and Planning) (2004) Songkhla Lake basin. Ministry of Natural Resources and Environment, Bangkok

ONEP (Office of Nature and Environmental Policy and Planning). Division 16. (2006) Classification of pig farms by treatment systems. Ministry of Natural Resources and Environment, Songkhla

ONEP (Office of Nature and Environmental Policy and Planning). Division 16. Division 16 (2007) Seasonal data of water quality monitoring between 2004–2006 (unpublished paper). Ministry of Natural Resources and Environment, Songkhla

ORRAF (Office of the Rubber Replanting Aid Fund) (2007) Model 1994–1995's wastewater treatment system a cooperative rubber factory. www.thailandrubber.thaigov.net. Accessed 20 Jan 2008

Pollution Control Department (2007) An environmental quality management plan for the eastern lower south of Thailand, 2007–2011. Ministry of Natural Resources and Environment, Bangkok

Pollution Control Department (2008) Yoong Thong's wastewater treatment system. Pollution Control Department, Ministry of Natural Resources and Environment, Songkhla

Pornpinatepong K, Kiripat S, Treewanchai S, Chongwilaikasaem S, Pornsawang C, Chantarasap P, Chandee C, Jantrakul P (2010) Pollution control and sustainable fisheries management in southern Songkhla Lake. EEPSEA research report. EEPSEA, Thailand

Sirimontaporn P, Choonhapran A, Tansakul R (1999) Songkhla Lake. In: Encyclopedia of Thai culture of southern Thailand. No 7, vol 7. Thai Culture Encyclopedia Foundation, Thai Commercial Bank, Bangkok, Siam Press Management Publication. Songkhla. Thailand

Sookchareon P (1965) Larval stages of penaeid shrimp in Songkhla Lake and adjacent areas. Annual report 1965. Songkhla Marine Fishery Station. Department of Fisheries. Ministry of Agriculture and Cooperatives, Songkhla, pp 40–62

Yun P (1998) The pollution charge system in China: an economic incentive. EEPSEA research reports. Economy and Environment Program for Southeast Asia. Singapore. http://www.eepsea.org. Accessed August, 2007

WEPA (Water Environment Partnership in Asia) (2007) Water pollution legislation and responsible ministries. www.wepa-db.net. Accessed 11 Feb 2008

Chapter 9
Productivity Effects of Water Pollution Due to Excessive Aquaculture Structures and Overstocking

Zenaida M. Sumalde, Karen Lou A. Francisco, and Mildren Peñales

Abstract This study was conducted to assess the effects of excessive aquaculture structures and stocking density on water quality and aquaculture production given the policies in place in Bolinao, which is one of the major aquaculture producing municipalities in the Philippines. Results of the study showed that the average stocking density of the aquaculture operation fell within the optimal value, but there were operators whose stocking densities were more than double than that of the optimum. The aquaculture production analysis showed that stocking density affects production. Beyond a certain level, further increases in stocks result in decreasing productivity. Excessive number of fish pens and cages with overstocking may lead to fish kills due to deteriorating water quality and other associated damages that are detrimental to the aquaculture industry and to the greater community. The study suggests that several policy implications and better regulation of coastal aquaculture operation are needed.

Keywords Fish kills • Extensive aquaculture • Intensive aquaculture • Stocking densities • Water quality

Z.M. Sumalde (✉)
Department of Economics, College of Economics and Management, University of the Philippines, Los Baños, Laguna, Philippines
e-mail: zcm.sumalde@gmail.com

K.L.A. Francisco
Private Consultant, 26 Mabango St, Brgy Masaya, Bay 4033, Laguna, Philippines
e-mail: kahlou@gmail.com

M. Peñales
One Global Place, World Bank (Philippines), Bonifacio Global City, Taguig, Metro Manila, Philippines
e-mail: mhpenales@gmail.com

N. Olewiler et al. (eds.), *Marine and Coastal Ecosystem Valuation, Institutions, and Policy in Southeast Asia*, DOI 10.1007/978-981-10-0141-3_9

9.1 Introduction

9.1.1 Background

Aquaculture denotes the cultivation of edible fish and other marine species in an artificial system in both freshwater and marine environments. Historically, aquaculture production followed the traditional extensive method with low-yielding "natural ponds," where use is made of natural feeds and harvests are low but sustainable over a long period (Hagler 1997; Anderson and de Silva 1997). Increasing human populations and demand for protein have led to the transformation of aquaculture from an extensive system into intensive culture. Most aquaculture species now are grown using commercial feeds, fertilizer, and pesticides. Intensive cultivation often involves high levels of fish stocking, necessitating large amounts of feed and antibiotics to control disease. The result can be water pollution and degradation and destruction of aquatic ecosystems. In Taiwan the overuse of fertilizer and veterinary products to increase shrimp production contributes to water pollution and resulted in the decreasing resistance to disease of shrimp stock (Lin 1989). In Ranot District in Thailand, the average groundwater level in 1989 decreased from 3 to 7 m below the surface due to heavy pumping of water to supply the needs of the intensive aquaculture farms. In Taiwan and the Philippines, sinking water tables due to intensive aquaculture farming have been reported to cause sinking land levels that resulted in flooded rice fields and collapsed buildings. In addition to environmental damage, aquaculture can pose health risks, incur problems associated with land conversion, cause social conflicts, and reduce biodiversity. The trade-offs between food production and ecosystem damage are apparent, but the question is, how severe are these trade-offs and can they be reduced through effective policy?

The Philippines is a major producer of aquaculture species. In Luzon, the Lingayen Gulf is an important site for production. The industry started as early as the 1970s with fishponds in brackish water where a semi-intensive form of milkfish culture was adopted. Milkfish were given natural food, and supplemental food, in the form of breadcrumbs and commercial feeds, was given only when the natural food was exhausted. Shrimp and other fish entering the pond during inflow of seawater were also cultivated to enhance incomes (Verceles et al. 2000). By the 1980s mangroves were being destroyed to make room for more fishponds. In the early 1990s, conversion of mangroves for fishponds was prohibited, shifting the approach to milkfish operations to construction of fish pens and fish cages in the coastal water and rivers. In 1992, the Lingayen Gulf Coastal Area Management Plan (LGCAMP) was formulated, which identified zones within the Gulf area that are suitable for aquaculture. Based on the criteria, the bulk of aquaculture activity in the area is concentrated on the coastal and inland waters in four municipalities, namely: Bolinao, Anda, Dagupan, and Binmaley. The Bolinao Local Government Unit (LGU) formulated the Coastal Development Plan (CDP) in 1995, which aimed to regulate aquaculture operation. As an implementing guideline of the CDP, the

Municipal Fishery Ordinance (MFO) was passed in 1999. Bolinao is the focus of this paper; it exemplifies the problems associated with Philippine aquaculture practices and examines policy options to improve economic and environmental outcomes.

Despite the MFO, fish pens and fish cages have continued to be constructed and operated illegally. In 1997 and 1998, intense aquaculture operation was reported in Bolinao especially at the Caquiputan Channel. In 1997, 703 permits to operate fish pens and fish cages, covering 88 ha, were issued to individuals. Although it was reported that the number of permits issued decreased to 476 in 1998, inspection by the Marine Fisheries Resource Management Program (MFRMP) revealed that in 1999 there were 797 units of fish pens and fish cages installed, covering 165 ha, not including those inaccessible by boat. These figures exceed the estimated carrying capacity of 544 units (130 ha) of the Caquiputan Channel (Verceles et al. 2000). The number continued to increase so that in 2000 and 2001, there were 993 and 1170 structures, respectively. Table 9.1 summarizes the available data on the number of operations over the period, 1995–2001.

In addition to the excessive number of units, reports indicate that operators practice high stocking density and overfeeding with supplemental feeds (Boyd and Lichtkoppler 1979). The Bureau of Fisheries and Aquatic Resources (BFAR) recommended stocking density for milkfish is 1250–1500 pieces (pcs) per 100 m^2 (12–15 pcs/m^2). Reports of overstocking levels for milkfish fingerlings (3–5 cm in size) indicate that 15–50 pcs/m^2 is common. In high stocking densities, fish excreta is high in ammonia, nitrate, and nitrite, the latter binds with the hemoglobin of blood and can be fatal to cultivated and wild fish (www.twnside.org.sg/title/paccct-ch.htm). At lower levels of stocking, intensive fish culture can have positive effects; leftover commercial feeds decompose and induce the growth of algae, which can serve as feed for milkfish. Leftover feeds can also serve as extra food for wild species and coral polyps, hastening their growth. However, if the nutrient load from fish excreta and feed reaches a tipping point, algal growth becomes excessive and leads to reduction in dissolved oxygen (DO) and creates high biological oxygen demand (BOD). Very low DO level (less than 5 mg/L) is detrimental to fish growth, especially cultivated species (personal communication with Z. Catalan). Water quality monitoring done by the Marine Science Institute, University of the Philippines, in 1999, showed that DO of the waters of Caquiputan Channel in Bolinao ranged from 2.9 to 4.5 mg/L, well below the critical value (Boyd and Lichtkoppler 1979).

One outcome that can be associated with high stocking and excessive feed are the sporadic "fish kills" that have been reported in the aquaculture areas. A fish kill is a significant and sudden death of fish or other aquatic animals. The event is characterized by large numbers of aquatic animals dying over a short time usually in a clearly defined area (Jacinto 2002). While natural factors such as weather can play a role, fish kills events have occurred in specific areas where fish pens are close together with high stocking densities per pen. If there are excessive feedings, large amount of feeds are not eaten by the fish and settle at the bottom. This contributes to algal growth that competes for oxygen with the fish. Fish excreta mixed with the

Table 9.1 Number of aquaculture structures in Bolinao, 1995–2001

Year	Number of structure
1995[a]	242
1997[a]	1076 (visual survey); 703 (permits officially issued)
1998[a]	476 (permits officially issued)
1999[b]	797 (visual survey)
2000	993[c] (MFRMP, 2000); 371[d] (permits officially issued)
2001[c]	1170 (AFMA-DMEQCMA study)
January 2002[e]	1067 (LGCAMC physical count); 621 (permits officially issued)

[a]Verceles et al. 2000
[b]MFRMP (ongoing study)
[c]Agriculture and Fisheries Modernization Act-Database on Monitoring of Environmental Quality for Coastal Management and Aquaculture (AFMA-DMEQCMA)
[d]Coastal Resources Management (CRM) office, Bolinao
[e]Philippine Daily Inquirer (PDI), Feb. 9, 2002d

other nutrients increase the nutrient loading and can result in eutrophication and eventually fish kills. Unfortunately, these fish kills are considered part of the business or due to nature and are often left undocumented.[1] Another negative outcome of overfeeding is the sediments that increase turbidity in the water and reduce the health of coral reefs. Thus, excessive stocking and feeding not only injures the cultivated species but also wild species living in the surrounding areas of the fish farms.

On the positive side, municipalities earn revenues from permits issued, renewal fees, and tax payments of the aquaculture operators. For example, in 1998 the aquaculture industry in Caquiputan Channel in Bolinao was reported to have contributed 2.3 million Philippine pesos (PHP) to municipal revenues and provided employment to the community. Employment figures, however, are not recorded at the municipal office since the workers are hired on an as-needed basis and not as regular workers in the aquaculture areas.

In February 2002 a major die-off of milkfish and other species occurred in the region, and at that time it was the largest fish kill recorded in the country. What role did aquaculture practices play in the fish kill? This question is a major motivation behind this study: If the relationship between fish farming practices, water quality, and health of the ecosystem can be better understood, policies can be designed to regulate the industry, and information gleaned can be disseminated to operators as an incentive to change their practices not only to their own benefit but to the ecosystem and the local community.

[1] Section 9.3 examines fish kills in more detail.

9.1.2 Statement of the Problem and Significance of the Study

While aquaculture provides dietary protein and employment and revenues to local governments, it also is a major factor in the deterioration of water quality where there are high concentrations of operations. In the study area, the Caquiputan Channel, monitoring by the University of the Philippines-Marine Science Institute (UP-MSI) indicates that water quality has deteriorated by at least 20 % since the start of coastal aquaculture operation (Verceles et al. 2000; personal communication with the aquaculture technician and CRM manager of Bolinao). The ordinance governing aquaculture operations that limits the number of permits has not been effective and the number of operations is well in excess of the maximum permits issued. Local governments have not allocated budget to protect water quality (Verceles et al. 2000). Thus, the excessive number of pens and cages operating in the region and overfeeding and high stocking densities are significant factors in the continuous decline of water quality of the Channel.[2]

9.1.3 Objectives of the Study

This study is a comprehensive examination of the aquaculture industry at the Caquiputan Channel in Bolinao. It focuses on specific aspects of the longer paper by Sumalde et al. (2003)[3] to estimate the impact of overstocking on aquaculture productivity, examine aquaculture management practices, and identify policy options to address the problems in the industry with recommendations for future research.

[2] Other sources contributing to low water quality are nutrient runoff from agriculture and sewage waste.

[3] The key finding from that study is that water quality declines as the stocking intensity of the fish pens increases, supporting the hypothesis that environmental degradation increases with the scale of the aquaculture activity. Sumalde et al. (2003) undertook water quality sampling to measure dissolved oxygen and total suspended solids in the study region. With the help of scientists from the University of the Philippines-Marine Science Institute, samples of water were taken from the fish pens and fish cages of aquaculture operators surveyed by the authors at stations 200 m away from the structures and in a barangay without aquaculture structures. The water quality data described in this paper comes from those measurements.

9.2 Theoretical and Conceptual Framework to Estimate Aquaculture Productivity

We use a short run model for aquaculture productivity that assumes in each period that output is a function of variable inputs, each with diminishing marginal productivity. Because the fishery can be restocked each period, there is no need to include stock effects that would be integral to a capture fishery.

If there are *n* variable inputs, the generic production function is expressed as:

$$Y = Y(X_i) \tag{9.1}$$

where Y = total production and X_i = variable inputs, with i = 1, 2,. . .n.

The equation is specified as a quadratic in each variable input to capture diminishing returns and is expressed as:

$$Y_a = \alpha + \beta_1 SD + \beta_2 SD^2 + \beta_3 F + \beta_4 F^2 + \beta_5 WD \tag{9.2}$$

where

Y_a = aquaculture production (kg/100 m^2) in region a
SD = stocking density (pcs/100 m^2)
F = quantity of feeds per production cycle (bags per 100 m^2)
WD = water quality dummy

1 if DO level inside the pen is = <4 mg/L
0 if DO level inside the pen is >4 mg/L

The water quality variable is modeled as a threshold: For levels below 4 mg/L, anoxic conditions can occur, increasing mortality levels of fish (and other species). In the absence of information on the relationship between different DO levels and mortality or milkfish survival specifically in the Philippines, the 4 mg/L cutoff value for DO is based on a study on DO and marine life conducted by the US Environmental Protection Agency's Environmental Research Laboratory (USEPA-ERL).

Ideally, the model would take into account the endogeneity between water quality and aquaculture practices, and a two-stage least squares or recursive model would be estimated. As explained below, the data is limited and hence Eq. 9.2 is estimated in a single stage, simply to check whether the signs conform to theory; β_1 and β_3 should be positive, while β_2, β_4, and β_5 should be negative. Section 9.3 presents the results that confirm the hypotheses. Future research is needed to estimate a more technically appropriate model.

9.3 Research Methodology

At the time of this study, 120 multiunit aquaculture operators existed in Bolinao. A survey was used to obtain data on operations to enable estimation of the production function in Eq. 9.2 and examine the relative impact of the variable inputs on fish productivity. Letters from Bolinao Aquaculturist Association and a questionnaire were sent to the members of the Association. Total enumeration was the goal, but this was not possible due to the inability to locate all the owners and nonresponse to the survey. Only 34 questionnaires were found to be complete and considered in the productivity analysis. Information on the operators included socioeconomic characteristics of the operators; information about fish pen/cage operations, including their number and size; aquaculture practices, production, and expenses; problems related to aquaculture operations; and other related information. Key informants were also surveyed for qualitative information about the aquaculture industry in the area. Secondary information related to aquaculture operations, water quality studies, fishery guidelines, and other information were gathered from a variety of other sources.

9.4 Characteristics of the Aquaculture Sector in Bolinao

Prior to estimating the production function for aquaculture operators in Bolinao, some descriptive statistics of the sector gleaned from the survey and secondary data are presented. Table 9.2 summarizes data from the survey on the size of the operations, average stocking, and estimates of mortality—note the diversity in operations, which is apparent in the wide ranges in the data.

9.4.1 Net Income from Aquaculture Operation

The total revenue and total costs for a typical operation in Bolinao were estimated to determine net income. Total revenue from fish pen operation was obtained by multiplying total production by the market price of a kilogram of milkfish. In 2001, total revenue ranged from PHP 500,000 to PHP 3,116,667, during the first operation cycle, and averaged PHP 1,004,098. The price of milkfish earned by the individual operator ranged from PHP 45 to PHP 78/kg with average values of PHP 51–PHP 59/kg. Since the operators reported variable sizes of fish pens, net income was in terms of a fish pen area of 4000 m^2.

Total costs for a 4000 m^2 fish pen were estimated at PHP 694,120 per cropping cycle. Total fixed costs consist of the license fee, operating permit, and interest on borrowed capital, and in 2001 it equaled PHP 13,900. Total variable costs include expenses to purchase the fry or fingerlings, feed, labor, and other material inputs

Table 9.2 Stocking, production, and mortality statistics for aquaculture operators in Bolinao, 2001 (Sumalde et al. 2003)

Measure	Average for operators sampled	Range
Number of fish pen units	2	1–11
Area of fish pen (m^2)	3301	1200–6000
Stocking density	1119	625–2971
Number of structures near operator	6	1–30
Mortality in 2001 (% of stocking density)	16	0.33–60
Mortality in 1998 (% of stocking density)	14	0.17–50

and in 2001 summed to PHP 680,229. Table 9.3 provides the detailed cost data for a typical operator. Note that total fixed costs are a minimal share (2 %) of total costs, and feed incurs the major cost. Table 9.4 presents the averages and ranges of total revenue, total costs, and net income.

To calculate the average net income, the individual net incomes of all of the operators who responded were summed and divided by the number of respondents. On average, net income from a 4000 m^2 fish pen was approximately PHP 310,000. On a per site basis, the average net income ranged from PHP 227,000 to PHP 535,000, which suggests milkfish culture is lucrative.[4] However, could net incomes be even higher had the industry been managed properly with fewer operations and better management of food supply to prevent overstocking and overfeeding? That question is explored next.

9.4.2 *Perception and Problems Related to Aquaculture Operation*

The majority of operators noted that it was taking longer for their fish to grow to a marketable size (the grow-out period). The main reasons and the percentage of respondents who cited them (in parentheses) were actions by the aquaculture industry (71 %), poor quality fingerlings (3 %), and weather conditions (7 %)—they indicated a belief that cold weather coupled with deteriorating water quality slows milkfish growth. Overall, 48 % of the respondents mentioned they face low water quality and, in one part of the region, that response rate rose to 75 %. Some operators noted they reduced the number of structures, but the objective data on growth in the number of structures belies this assertion. This is evidence of a classic "commons" problem. No one is willing to individually reduce their overstocking to benefit the whole of the fishery.

[4] By way of comparison, average annual family income in Region 1, which includes Bolinao, was just under PHP 120,900 in the study year (NSO 2001).

Table 9.3 Costs incurred by a fish pen operator per cropping for a 4000 m^2 pen, Bolinao, Pangasinan, 2001

Type	Amount (PHP)
Fixed costs	
License fee[a]	5000
Permit	75
Interest on borrowed capital	8825
Total fixed cost	**13,900**
Variable costs	
Changing/washing of net	2768
Rope/nylon cord	2190
Bamboo	15,000
Raft	1500
Miscellaneous	4500
Net	20,540
Fry/fingerlings	120,895
Feeds	491,664
Labor	21,172
Total variable costs	**680,229**
Total cost	**694,129**

[a]License fee is PHP 10,000/unit/year and the operator has two cropping per year

Table 9.4 Total revenue, total costs, and net income for fish pen operators in Bolinao, 2001 (all values given in PHP, unless otherwise noted)

Total revenue, total cost, net income	Average	Range
Total harvests (kg/4000 m^2)	17,930	9000–56,667
Average price (PHP/kg)	53	45–60
Total revenue	1,004,098	500,000–3,117,000
Total costs	694,129	220,775–2,037,000
Net income	309,969	36,495–1,079,000

9.5 Effects of Aquaculture on Production

The following section provides an analysis of the effects of intensive aquaculture operations on the aquaculture industry itself and an estimation of losses due to fish kills and other damages.

9.5.1 Aquaculture Production Analysis

Using the data from the survey of 34 operations, Eq. 9.2 is estimated using ordinary least squares to see the functional relationship between fish productivity and water quality, stocking levels, and feed.[5] Results of the regression run for aquaculture production are as follows:

$$Y_a = 141.29^{**} + 2.542SD^{*} - 0.00075SD^{2**} + 2.999F^{ns} - 0.0205F^{2ns} - 40.78WD^{***}$$
$$R^2 = 0.93 \tag{9.3}$$

***significant at 0.01 level
**significant at 0.05 level
*significant at 0.10 level
Ns = not significant at 10 % level

The regression analysis indicates the signs of the variable input coefficients conform to a priori economic expectations. For the stocking density, as stocks increase, so will yield up to a certain level, beyond which yield will start to decline. The stocking density coefficient is consistent with what would be expected from a short run production analysis, the law of diminishing productivity, and fishery economic theory: As more stocks are put in a given space, overcrowding occurs and the competition for space eventually leads to slower growth and higher mortality levels due to the fish waste products reducing oxygen levels. The coefficient of the feed variable appears to have no significant effect on production. This non-significance of feed on production may imply no significant variation in feed level used by the operators, meaning yield differences do not depend on quantity of feed.[6] The coefficient of the water quality dummy is negative as expected, indicating the production frontier will shift downward. Structures with DO levels less than 4 mg/L are likely to suffer from lower productivity compared to those structures with DO levels higher than 4 mg/L. Data on water quality analysis in the sample barangays showed that there were variations in DO levels in specific sampling points, with some points having DO values greater than 4 mg/L.

While the coefficient estimates in Eq. 9.3 should not be used for predictive purposes given the endogeneity issue, they are useful in suggesting the underlying relationship. From Eq. 9.3, if stocking density per 100 m^2 of a fish pen area exceeds 1693 fry, production will start to decline. This result is quite close to the recommended stocking density of 1250–1500 fry/100 m^2 fish pen area that was identified by INTAQ (2000).

[5] Recall that a more robust model would recognize and address the endogeneity problem between water quality and aquaculture practices (feed and stocking density).

[6] Multicolinearity between feed and stocking density is also a possible problem that could be addressed with more data.

9.5.2 Losses from Fish Kills Due to Aquaculture Practices

Occurrences of fish kills had been common in Bolinao but were undocumented since most of the occurrences are site specific—that is, by specific pens or cages—not throughout the whole aquaculture area. Results of MFRMP mariculture (coastal aquaculture) monitoring showed that fish kills occurred year-round in specific fish pens in Bolinao and two other aquaculture municipalities, Dagupan and Anda, and rates of fish kills peaked during the months of April, May, and June. These fish kills were isolated occurrences, so it did not raise alarm among the local officials, despite some of them being fish pen and fish cage owners themselves (PDI 2002d).

A major fish kill occurred in Bolinao on February 1–3, 2002, and is considered the worst fish kill event in the country to date. It started in the evening of February 1, during a neap tide when tidal currents are slow, causing less oxygen-rich water flowing in from offshore areas. A few days before the event, operators had been reported to have been using aerators to introduce oxygen into the fish pens or cages. This practice is typical in fishponds where water does not flow in naturally, but not in coastal waters where water currents provide oxygen. The use of aerators is also an indication that there was high oxygen demand in the area. Two days before the fish kills, the milkfish were crowding the surface, evidently gasping for air due to low DO in the water. In Region 1, water quality testing by the Regional Director of the Bureau of Fisheries and Aquatic Resources (BFAR) recorded a 1.2–2.3 mg/L DO value a day after the massive fish kill. A level below 2 mg/L—well below the threshold of 4 mg/L used in the present analysis—can cause gill suffocation and cause fish kills. The losses in this event were estimated to be PHP 600 million (PDI 2002d). Unlike previous isolated events, this fish kill affected not only Bolinao but also other villages near it. Kills occurred in Luciente 2, Luna, and Tara, part of this study, as well as the other barangays with aquaculture operations, such as Luciente I, Salud, and Lucero.

At the time, records at the municipal CRM office listed 173 fish pens and 448 fish cages with permits issued, 621 units in all, but visual inspection done by BFAR showed otherwise: 600 fish pens and 467 fish cages or a total of 1067 units. Both counts were way above the estimated carrying capacity of 544 units (130 ha) of the water. Of these, 353 were in Guigiwanen and Luciente 2 where the fish kill started. Luciente 2 and Luna were also among the areas affected by fish kills the previous September, in 2001.

It was not only the aquaculture industry that was affected; many wild fish were also observed at the surface, gasping for air. While many died, others may have moved away from these oxygen-depleted areas. Fishers dependent upon wild stocks reported low or negligible harvests during this period. One resident said that children usually caught 5 k of fish each per day at the beach, but they came home with no fish that day (PDI 2002b). Unfortunately, there is no hard evidence available to estimate the damage suffered by the capture fishery.

The damage due to the 2002 fish kill event was so large that the inquiry into the event reached national attention. On February 5, 2002, the University of the

Philippines-Marine Science Institute (UP-MSI) issued a statement on the massive fish kills in Bolinao that indicated two causes: a bloom of dinoflagellates (plankton cells) associated with eutrophied waters and low DO levels in the municipal waters of Bolinao. Dinoflagellates are known to asphyxiate fish and a high density of them was found in the water column and in the gills of the fish. This was the first time such a bloom occurred in the Philippines.

But, what really caused the massive fish kills? The ultimate cause has to be associated with aquaculture practices, although no definitive proof can be established. The decrease in DO was not a sudden event, but is associated with the large increase in the number of fish pens and fish cages and resulting unconsumed feeds and fish wastes. According to the Regional Chief of BFAR in Region 1 "around 30 % of the feeds given are not consumed by the fish culture and are eventually thrown out into the water and seabed" (PDI 2002e). The buildup of organic matter eventually leads to eutrophication, which caused a phytoplankton bloom. The bloom led to an increased rate of die-off resulting in more organic matter on the sea bottom. Decomposition of the additional organic matter then depleted the dissolved oxygen in the water and caused further stress to the confined fish.

In light of this event, Bolinao's mayor said that the coastal development plan would be strictly implemented and that they might do away with the fish pens but continue with fish cages, as it is believed that fish pens release more pollution than fish cages where feeds and wastes that settle at the bottom cannot be removed like in fish cages (PDI 2002a). However, UP-MSI believed that for the Bolinao's water to recover from this disaster, aquaculture operations would have to be stopped all together for a period of time. On February 17, 2002, the local government suspended aquaculture operations as well as the issuance of permits (PDI 2002c).

A month after the fish kill event, the Senate conducted an inquiry and initiated Senate Resolutions 238 and 241, authored by Senators Loren Legarda-Leviste and Manuel Villar, which urged:

> The Senate Committee on Agriculture and Food and the Senate Committee on Local Government to conduct an inquiry, in aid of legislation, into the massive fish kill of milkfish, which occurred at Caquiputan Channel in Bolinao, Pangasinan, valued at PHP 600 million, with the intent of introducing remedial measures to prevent its spread to other waters and to ensure that there would be no reoccurrence of the same fish kill in Bolinao, Pangasinan and in other waters of the Philippine archipelago. (Doyo 2002)

By the date of this study (2002), no concrete legislative actions had been passed by the Senate, nor were there any resolutions from local officials or administrative orders from BFAR to address the fish kill issue.

9.5.3 Other Damages and Problems Due to Aquaculture

In addition to economic loss to aquaculture operators, capture fisheries, and government tax and permit revenues, social and environmental hazards, including health costs, can arise. In the wake of the 2002 fish kill, for example, 75 people were hospitalized due to allergic reactions after eating dried milkfish and other people who were affected complained of headache, dizziness, vomiting, difficulty in breathing, abdominal pain, nape pain, itchiness, chest pain, and skin redness (PDI 2002d). In addition, the laboratory experiments of the UP-MSI had to be suspended, so as not to be contaminated by the polluted waters. The effect on the total marine resources can also lead to "unquantifiable" costs associated with the accumulated decaying marine fauna and flora that spreads to the adjacent coastal areas.

9.6 Conclusions and Policy Implications

This study was conducted to assess the effects of excessive aquaculture structures and stocking density on water quality and aquaculture production given the policies in place at the time of study. A sample of operators and key informants, from both the aquaculture and non-aquaculture barangays, were surveyed and interviewed. Water quality was monitored in aquaculture and non-aquaculture areas to determine the deterioration in water quality due to aquaculture operations.[7] The many negative impacts from a major fish kill in 2002 are documented.

We conclude with the following observations.

- MFO guidelines were not followed: The number of aquaculture structures greatly exceeded the carrying capacity of the area. The inconsistencies between the actual number of structures and those listed or recorded at the CRM office of Bolinao indicated that most of the structures were illegally constructed, with more structures than permitted in operation.
- Aquaculture operations have positive net incomes, but regression results indicate that lower levels of stocking and feeding would increase total income from aquaculture.[8]
- The data from our survey of the aquaculture industry in Bolinao finds that the average stocking density fell within the optimal value, but there were operators whose stocking densities were more than double that of the optimum.
- The water quality analysis showed that aquaculture has an impact on the water quality in the area. The high and negative correlation between stocking density and DO values in the sampling areas implies that overstocking contributed to water quality deterioration.
- The aquaculture production analysis showed that stocking density affects production. Beyond a certain level, further increases in stocks result in decreasing productivity. Production is likely to decrease if the DO level falls below 4 mg/L.

[7] Results from the water quality monitoring are available in the longer version of this paper. See Sumalde et al. (2003).

[8] While total income is predicted to rise with optimal stocking due to the elimination of the overstocking externality, incomes of some of the operators would fall, creating an equity issue.

- Excessive numbers of fish pens and cages with overstocking may lead to fish kills and other associated damages. These are detrimental not only to the aquaculture industry itself but to the greater community as well.

The study suggests several policy implications and better regulation of coastal aquaculture operation is needed. These include strict enforcement of the MFO, penalties for violation that are high enough to make violation of the ordinances costly, and incentives for compliance to the ordinance considered (e.g., reductions in taxes, technical assistance). Authorities need to better communicate the risks associate with overstocking to aquaculture operators and to provide evidence that there are negative consequences to them as well as their community if they put too many fish in their enclosures.

Additional research is clearly needed regarding the following priorities:

1. An analysis of the technical relationship between growth of both wild and cultivated fish species and levels of DO. This would be useful to determine optimal stocking densities that reduce water pollution and the risks of fish kills.
2. A longitudinal or time series study on water quality, hydrologic characteristics, weather conditions, and fish growth rates. A technical study on the effects of aquaculture wastes on water quality, coral reefs, and spawning grounds for capture fish species to determine the survival of fish larvae and juveniles would help identify potential adverse spillovers from aquaculture.
3. A comparative study on the socioeconomic and environmental effects of different types of aquaculture operations (fishpond, freshwater aquaculture, coastal and inland aquaculture, concrete tanks, etc.) can give important information about aquaculture management.

References

Anderson T, de Silva S (1997) Strategies for low-pollution feeds and feeding. www.agri-aqua.ait.ac.th/naca/aquaasia/3vol2_1_18.htm. Retrieved 6 Mar 2002

Boyd C, Lichtkoppler F (1979) Research and development of International Center for Aquaculture. Agricultural Experiment Station, Series No. 22, p 3–10

Doyo MCP (2002) Fish pens, fish cages, and drowned fish. Human face. 14 Mar 2002. Phil Daily Inquirer

Hagler M (1997) Shrimp – the devastating delicacy. www.greenpeace.org/~oceans/shrimpaquaculture/srimreport.html. Retrieved 6 Jul 2003

INTAQ (2000) Milkfish farming in the Philippines, INTAQ technical bulletin. Integrated Aquaculture Specialist, Mandaue City

Jacinto G (2002) Fish kill in Bolinao, Pangasinan: a case study paper presented at the National Forum on Aquaculture. 14 Feb 2002. UP MSI, Quezon City

Lin CK (1989) Prawn culture in Taiwan. What went wrong? World Aquac 20(2):19–20

Municipality of Bolinao (2000) Updated socio-economic profile of Bolinao

NSO (National Statistics Office) (2001) Family income and expenditure survey. National Statistics Office, Metro Manila

PDI (Philippine Daily Inquirer) (2002a) Another fish kill hits Bolinao. 5 Feb 2002. Phil Daily Inquirer

PDI (Philippine Daily Inquirer) (2002b) Bolinao fish kill spreads, emergency state declared. 8 Feb 2002, Phil Daily Inquirer
PDI (Philippine Daily Inquirer) (2002c) Bolinao recovery plan. Fish pen operations suspended. 17 Feb 2002. Phil Daily Inquirer
PDI (Philippine Daily Inquirer) (2002d) Local execs own fish pens, cages. 9 Feb 2002. Phil Daily Inquirer
PDI (Philippine Daily Inquirer) (2002e) Bolinao fish kill: what went wrong? 6 Feb 62002. Phil Daily Inquirer
Sumalde Z, Francisco K, Penales M (2003) Pollution-induced fish kills in Bolinao: effects of excessive aquaculture structures and overstocking. In: Francisco H, de los Angeles M (eds) Economy and environment: selected reading in the Philippines. EDM Press, Manila
Verceles LF, McManus LT, Aliño PM (2000) Participatory monitoring and feedback system: an important entry toward sustainable aquaculture in Bolinao, Northern Philippines. Sci Diliman 12(2):78–87

Chapter 10
Reverting Disused Fishpond Lease Agreement Areas to Mangrove Forests in Region VI (Western Visayas), Philippines

Alice Joan G. Ferrer, Jinky C. Hopanda, Michael Q. Orquejo, Alan Dino E. Moscoso, and Resurreccion B. Sadaba

Abstract This paper evaluated the reversion of disused Fishpond Lease Agreement (FLA) areas in Region VI (Western Visayas) to mangrove forests. A two-stage and five-step evaluation process and multiple data collection methods were used. This study covered 62 disused FLA areas in the region, representing 84 % of the total 74 canceled FLAs (regardless of whether reverted or not to the Department of Environment and Natural Resources), as well as FLA areas that had been abandoned or left undeveloped. Results showed a lower number of canceled FLAs and reverted FLA areas than what should ideally be. The suitable area for reforestation was only 23 % (284 ha) of the total area of FLA areas visited. The benefit of reforesting disused FLA areas suitable for reforestation far outweighs the cost of doing so. The option to place the responsibility of paying for reforestation efforts on the lessees (through performance bond, green tax/fine, fees reflective of economic rent, and beneficial use tax) will generate higher funds and will likely positively impact other desirable goals in aquaculture; however, this strategy is relatively difficult to implement. The option to place the responsibility of funding reforestation efforts on funding organizations (Department of Environment and Natural Resources (DENR) grant, foreign fund, private sector support) is relatively easy to implement but may generate lower amounts of funds and may be

A.J.G. Ferrer (✉)
Division of Social Sciences, University of the Philippines Visayas, Miagao, Iloilo, Philippines
e-mail: aj_ferrer2005@yahoo.com

J.C. Hopanda • M.Q. Orquejo
University of the Philippines Visayas Foundation, Inc., Miagao, Iloilo, Philippines
e-mail: jinkychopanda@gmail.com; morquejo@gmail.com

A.D.E. Moscoso
School of Technology, University of the Philippines Visayas, Miagao, Iloilo, Philippines
e-mail: admoscoso@yahoo.com

R.B. Sadaba
Division of Biological Sciences, College of Arts and Sciences, University of the Philippines Visayas, Iloilo, Philippines
e-mail: sadaba09@yahoo.com

N. Olewiler et al. (eds.), *Marine and Coastal Ecosystem Valuation, Institutions, and Policy in Southeast Asia*, DOI 10.1007/978-981-10-0141-3_10

unsustainable. This study recommends (1) conducting an inventory of all FLA areas in the region, (2) formulation and approval of FLA cancelation and reversion rules and guidelines, (3) conduct of suitability assessment for reforestation of disused FLA areas in the region, and (4) actual reversion of suitable disused FLA areas to mangrove forests.

Keywords Fishpond lease • Mangrove rehabilitation • Mangrove reversion • Western Visayas • Mangrove forest

10.1 Introduction

10.1.1 Description of the Problem

Large tracks of mangrove forests have been lost and degraded in the Philippines starting in the 1960s and 1970s when the government encouraged development of aquaculture to increase food production (Genio et al. 2007). Around half of the 279,000 hectares (ha) of Philippine mangroves disappeared between 1951 and 1988 due to conversion into milkfish or shrimp ponds (Primavera 2005). In 1994, the total mangrove area covered 120,000 ha, while fishponds took up 232,000 ha (Primavera 2010). Mangroves continue to diminish due to unsustainable utilization and continued conversion to fishponds. This is despite the understanding of the multiple benefits of mangroves and the passage of laws and policies banning mangrove cutting, preventing further conversion of mangrove areas to other uses, and promoting the reversion of abandoned, underdeveloped, and undeveloped fishponds to original mangrove state.

Mangrove experts recommend that mangroves should be returned to their original habitat. The optimal area for planting mangroves is the middle to upper intertidal zone (Primavera and Esteban 2008; Samson and Rollon 2008). However, most of these areas have long been converted to brackish water fishponds. One way to bring the mangroves back in the middle intertidal zone is to recover and restore abandoned, underdeveloped, or underutilized fishponds (collectively referred to here as AUU or disused fishponds)[1] under the Fishpond Lease Agreement (FLA) (Section 23, Fisheries Administrative Order [FAO] 197) (Melana et al. 2000; Primavera 2000; Samson and Rollon 2008; Primavera and Esteban 2008). However, to this day, there has been no active effort from government agencies to revert AUU fishponds into mangrove forests. According to Melana et al. (2000) and Yao (2000), the reversion of abandoned fishponds under FLAs as practiced in the country is an extremely difficult activity that requires considerable time and resources to accomplish. For this reason, there is little practical experience with restoring disused fishponds back into mangroves.

[1] "Disused" was used by Stevenson (1997) in his paper on "Disused Shrimp Ponds: Options for Redevelopment of Mangrove." Disused shrimp ponds are unproductive and idle shrimp ponds.

In recent years, there have been calls to revert disused ponds under FLAs to mangrove forests. In July 2010, the participants of a Visayas-wide workshop on FLA cancelation and mangrove reversion echoed the message sent in 2007 to the Department of Agriculture (DA) and the Department of Environment and Natural Resources (the DENR) to implement their joint memorandum on the reversion of AUU fishpond areas covered by FLAs to mangrove forests. A similar call for reversion came out during consultations conducted in Luzon (January 2010) and Mindanao (March 2011).

This study evaluates the policy on reversion of abandoned, underutilized, and underdeveloped (or disused) fishponds under FLAs to mangrove forests in Region VI (Western Visayas), Philippines. Specifically, it assesses (1) the process of identifying disused FLA areas, (2) the process of FLA cancelation, (3) the process of reverting the jurisdiction over FLA areas from the Department of Agriculture's Bureau of Fisheries and Aquatic Resources (DA-BFAR) to the DENR, and (4) the suitability of disused FLA areas for mangrove reforestation; then it (5) determines the costs and benefits associated with reverting disused FLA areas to mangrove forests, (6) analyzes potential policies for financing the reforestation, and (7) recommends actions to improve the process of reversion of disused fishponds to mangrove forests.

10.1.2 Focusing on Region VI (Western Visayas)

Region VI is 1 of 17 administrative regions in the Philippines. Located in the middle of the country, the region consists of six provinces, with a total land area of 20,614 km^2 and a population of 6.8 million according to the 2007 Census. There are five strong reasons for choosing Region VI for the study site:

1. It has the most extensive mangrove area that has been converted to fishponds. Experts estimate that 95 % of the region's mangrove areas have been converted to fishponds, with the largest area located in the provinces of Aklan and Negros Occidental (Primavera 2010). For every hectare of mangrove, there are 5.52 ha of brackish water fishpond (or for every 4 ha of mangrove, there are 22 ha of fishpond).
2. Most FLAs issued in the country are in Region VI (BFAR-FRQD 1991)—out of 4518, 1487 FLAs (33 %) are in Region VI. The region has the largest FLA area in the country at 14,253 ha—equivalent to 24 % of the country's total of 59,556 ha—and the most number canceled or reverted to the DENR. Out of the 202 canceled FLAs (covering 6390 ha) nationwide, 63 are in Region VI (1331 ha, 31 % of the total canceled). Moreover, 21 FLA areas (478 ha) had been reverted to the DENR, representing 58 % of the number of total canceled FLAs nationwide and 36 % in terms of area.[2]

[2] FLAs that are canceled are not necessarily administratively reverted to the DENR.

3. Fishpond development in the region represents the largest acreage in the country (36,315.28 ha, 33 %), and current FLA areas have not reached even half of the total area released for development, yet the native mangroves are almost gone.
4. The region ranks second nationwide in terms of brackish water volume and area of production (BAS 2010). In terms of milkfish production, it is number one, contributing one third of the total milkfish production for years 2004–2009.
5. Many areas in the region have been identified as being prone to potential geologic hazards (landslides, soil movement, and flooding). The flood-prone areas include 64 municipalities (55 % of the nation's total) and 14 cities (88 % of the nation's total). In 2008, the region suffered its worst flooding ever recorded caused by Typhoon Frank, leaving 531 people dead, 226 people missing and presumed dead, and 2555 people injured (PDI 2010). The total damage was estimated at PHP 3.5 billion, broken down as follows: PHP 609 million for rice, PHP 157 million for corn, PHP 29 million for agricultural facilities, and PHP 2.25 billion for fisheries (Philippine Star 2008). Damage to the fishery sector amounted to PHP 1.25 billion for milkfish and PHP 1 billion for shrimp. Mangroves act as a protective barrier during typhoons and storm surges. Damages might have been far less had more mangrove areas been present.

10.1.3 Leasing of Public Lands for Fishpond Purposes

The Fishpond Lease Arrangement (FLA) governs the lease of public lands for the purpose of fishpond development; it is "an agreement entered into by and between the Secretary of Agriculture (DA) and qualified fishpond applicant for the use of public land for fishpond development purposes for a period of twenty-five (25) years" (Section 1, FAO 197). The BFAR (under the DA) is the agency responsible for the issuance of FLAs. These public lands—mangrove forestlands and other swamps—were released to the BFAR for fishpond purposes (Section 3e, JAO 01 2008) until November 1998 when the DENR ceased releasing public lands for fishpond purposes per Memorandum Order 98-17. The significant conditions under a fishpond lease include[3]:

1. No more than 50 ha for an individual lessee and 250 ha for a corporation or fisherfolk organization shall be leased.
2. A lease is for a period of 25 years and renewable for another 25 years.
3. Lease rates shall be determined by the BFAR. All fees collected shall be remitted to the National Fisheries Research and Development Institute (NFRDI) and

[3] Before 1998, FAO No. 125 Series of 1979 provided the rules and regulations governing the lease of public lands for fishpond development. From 1998, FLAs have been governed by Sections 45–50 of the 1998 Philippine Fisheries Code (RA 8550) and FAO No. 197, Series of 2000 stipulates the implementation rules and regulations.

other qualified research institutions to be used for aquaculture research development.
4. Leased area should be developed on a commercial scale and producing within 3 years from the date of the FLA approval. Areas not fully producing within 5 years from that date shall automatically be reverted to the public domain for reforestation.
5. There shall be no subleasing in whole or in part.
6. Transfer or assignment of rights under the FLA is permissible only on prior written approval by the BFAR.
7. Lessee must undertake reforestation of riverbanks, bays, streams, and seashore fronting the dike of fishpond, subject to the rules and regulations to be promulgated.
8. Lessee must provide facilities that will minimize environmental pollution to settling ponds, reservoirs, etc.

Failure of lessees to comply with any of the above conditions means cancelation of the FLA. Furthermore, to avoid cancelation, lessees must introduce improvements to the area within 180 days of the date of issuance of the lease. A particular provision that is of interest states:

> Fishpond areas covered by an existing FLA, which have been determined jointly by DA [Name in Full], the DENR and LGUs [Local Government Units] concerned, as abandoned, undeveloped or underutilized portions after five (5) years from the issuance of the FLA can be reverted to their original mangrove state and that necessary steps should be made to restore such areas to their original mangrove state. (Section 24 of FAO 197)

A total of 4518 FLAs covering an area of 59,556 ha have been issued in the country since 1973; one third (1487) are in Western Visayas (Table 10.1). Almost two thirds of the FLAs are in three regions, including VI, as well as IV (Southern Tagalog) and V (Bicol), amounting to more than half (55.5 %) of the total FLA area; Region VI has almost one quarter of the total area.

10.2 Methodology

10.2.1 Evaluation Stages, Steps, and Criteria for Analysis

The study uses a two-stage, five-step process for evaluating the FLA policy. The first stage assesses the processes of FLA cancelation and the reversion of jurisdiction over a disused FLA area from the DA-BFAR to the DENR. The second stage examines how a disused FLA area could be restored to a mangrove forest. For the second stage, the study uses the data from canceled FLAs and FLA areas in the region officially identified as abandoned and undeveloped. The disused fishponds were assessed for their suitability for reforestation. For those found suitable, the costs and benefits of reforestation were calculated. The evaluation ends with an analysis of seven potential policy options for financing reforestation.

Table 10.1 FLAs issued in the country from 1973 to present by region (Fisheries Regulatory and Quarantine Division, Bureau of Fisheries and Aquatic Resources 2010)

Philippine regions	No. of FLAs	% of FLAs	Total area (ha)	% of area
I	207	4.58	1272	2.14
II	8	0.18	78.71	0.13
III	76	1.68	519.66	0.87
IV	763	16.89	11,528.64	19.36
V	462	10.22	7,278.31	12.22
VI	1487	32.91	14,253.19	23.93
VII	452	10.00	4,530.54	7.61
VIII	211	4.76	5,207.19	8.74
IX	422	9.34	8,253.69	13.86
X	69	1.53	1,481.64	2.49
XI	137	3.03	1,435.70	2.41
XII	81	1.79	1,200.99	2.02
XIII	143	3.16	2,515.84	4.22
Total	4518	100.00	59,556.09	100.00

The origin of FLAs can be traced to FA0 14 of 1937. However, many of the areas covered by FLA that were declared "alienable" and "disposable" were sold and eventually titled. With the passage of Presidential Decree 704 in 1975, this practice stopped. Section 23 of the Decree states that "no public lands suitable for fishpond purposes shall be disposed by sale; only those fishpond sales patent already processed and approved on or before November 9, 1972, shall be given due course." There are now 17 regions because some of the 13 shown here were subdivided after 1973. The table represents all the FLAs issued since 1973

10.2.1.1 Stage 1: Return of FLA Area Jurisdiction from the BFAR to the DENR

Step 1: *Assessment of the FLA cancelation process*. The process of reverting disused FLA areas to mangrove forest starts with identifying areas for FLA cancelation. This assessment not only identifies areas for FLA cancelation but also the number canceled compared to the number that could potentially be canceled, the reasons for cancelation, and the challenges of FLA cancelation.

Step 2: *Assessment of the process of returning the jurisdiction of disused FLA areas from the BFAR to the DENR*. After cancelation of an FLA, the DA, on the recommendation of the BFAR, may either open the area for application or revert it to the DENR for restoration to mangrove forest, depending on the level of development of the area for fishpond purposes. If the FLA area is abandoned, underutilized, and undeveloped, then it qualifies for "automatic" reversion to the DENR jurisdiction. This assessment focuses on the process of turnover of disused FLA areas from the BFAR to the DENR, the number of disused FLA areas reverted to the DENR compared to the potential number of disused FLA areas, and the challenges associated with the reversion process.

10.2.1.2 Stage 2: Actual Reversion of Disused FLA Area to Mangrove Forest

Step 3: *Suitability assessment of disused FLA areas for reversion to mangrove forest*. Environmental parameters remaining in the area need to be ascertained to facilitate identification of areas suitable for mangrove reforestation and of site characteristics that bear on the cost of reforestation. The information forms the basis for computation of the costs of reforestation in the next step of evaluation.

Step 4: *Determination of the costs and benefits of mangrove reforestation*. The cost of reforestation was estimated based on "restoration or replacement cost." The cost depends on the mode of reforestation (natural regeneration or active planting), the cost of suitable species, and site characteristics. Natural regeneration includes the social cost of protecting the area from human encroachment (maintenance cost) and the breaking of installed dikes to allow water to flow freely. The recommended protection period is 5 years from planting of mangroves and makes use of local labor cost. The mangrove expert in the team recommends a ratio of one caretaker to 1 ha of mangrove. In the case of active planting, the costs from nursery to outplanting of recommended species like *Avicennia* and *Sonneratia* and other middle/landward species are estimated using data for projects undertaken in the Philippines and include nursery bagging costs, PHP 21,000 for 4500 plants; nursery shed, PHP 6000–8000; and outplanting, PHP 12,600/ha. Calculation of benefits uses the net present value of the benefits from a hectare of mangroves in Thailand, as estimated by Sathirathai and Barbier (2001), adjusted by Saplaco (2000) to apply to the Philippines.

Step 5: *Analysis of potential policies to finance mangrove reforestation*. We consider seven potential policies for payment of reforestation recommendations made by previous studies, reports, and preliminary interviews with key informants in August 2010. Adapting Holl and Howarth (2000), the potential options were classified according to the party responsible for the restoration cost: the FLA lessee or no responsible party. If the responsibility lies with the FLA lessee, then the next question is when that responsibility should be assigned, before or after the damage. The different potential financing options are as follows:

1. *Lessee is the responsible party and responsibility is assigned before damage* through the following:

    - A performance bond is imposed on all renewals (note that no new applications are allowed), and transfers of FLAs should be enforced during the whole life of the contract to ensure that lessees perform their duties and responsibilities.
    - The lease and all other fees related to FLA are to reflect economic rent; rentals, surcharges for late payments, and all other related fees should reflect the opportunity cost of the resource. This means an increase in the current rates.

2. *Lessee as responsible party and responsibility is assigned after damage* as follows:

   - A green fine is levied on damages. The green fine shall be imposed on lessees who do not comply with existing fishery laws, rules, and regulations, particularly with regard to pollution. The operation of FLA areas can cause environmental harm; in this case, accountability is secured after the damage is done.
   - Beneficial use tax is a payment that contributes to the beneficial use of public land; it should not be misunderstood as payment for the land, but rather as a cost of doing business.

3. *No responsible party*, in which case funds for restoration could be acquired through:

   - The DENR grant—Restorations paid by general taxes at the national level through the budget of the DENR. This option does not target the party responsible for the damage (i.e., lessees) but benefits the society as a whole.
   - Foreign funding support—Funds sought from foreign donors.
   - Private sector support—The private sector (volunteers, corporations with corporate social responsibility programs, youth, other organizations) may contribute directly or indirectly to mangrove restoration, for a range of reasons.

The criteria used to assess potential policies include the following:

- Feasibility refers to the level of preference ranking of stakeholders (LGU managers, aquaculture managers, environmental managers, and mangrove and aquaculture experts) as reflected in their individual assessment ranking of the options and the possibility of acceptance of the lessees. A rank of 1 is deemed "very high," 2 or 3 is "high," 4 is "moderate," 5 or 6 is "low," and 7 is "very low."
- Funds raised refer to the potential amount to be generated with implementation of the policy.
- Ease of implementation refers to the extent of regulatory, legislative, and administrative requirements for the option to be implemented.
- It is also important that the option can address other desirable societal or aquaculture goals, such as reduction in inequity, increase in production efficiency, and increase in compliance with environmental laws and regulations.
- The option must also meet the sustainability criterion that funds can be collected consistently and indefinitely.

For each criterion, there are specific indicators used in evaluating the options. Each criterion was evaluated using a simple three-point scale (high, moderate, and low). One point is assigned whenever an option is likely to meet an indicator (even modestly), and more points are given as the likelihood of achieving the objective increases. The information was based on data collected from key informants and secondary data.

10.2.2 Data Collection

The study used multiple data collection methods, including archival work, onsite visual inspection of abandoned FLA areas or those with canceled FLAs, interviews, experts' opinions, an intensive literature review, and collection and assessment of pertinent records/data from government agencies. The covered 62 FLA areas representing 84 % of the official number (74) of canceled FLAs in Region VI and FLA areas officially identified as abandoned and undeveloped. These abandoned and canceled FLAs are located in 27 sites across 19 cities and municipalities in the region's four provinces: Negros Occidental, Iloilo, Guimaras, and Capiz.

10.2.2.1 Cancelation of Fishpond Lease Agreements

Official records from the BFAR-FRQD indicate that there are 63 canceled FLAs in Region VI, covering a total area of approximately 1331 ha. Of these, only 21 FLAs (478 ha, 36 %) had been reverted to the DENR at the time of data collection. Three FLA areas (12 ha) were identified as "abandoned" and eight (209 ha) as undeveloped. These areas are spread all over the region. The average area of land covered by the 21 canceled FLAs reverted to the DENR is 22.7 ha. The average size of canceled FLAs not reverted to the DENR is 20.3 ha, and of abandoned FLA areas, approximately 4 ha. The average of undeveloped FLA areas is 26.3 ha. On average, the FLAs were canceled after 7–17 years from the date of approval, on average 14 years (the mode is more than 10 years). The most popular reason for FLA cancelation is failure to pay rental fees. This is true regardless of whether the area was reverted (95 %) to the DENR after FLA cancelation or not (89 %). Other common reasons are failure to submit required reports on the development, operation, and production of leased fishpond areas and failure to develop the areas. Of the areas covered by canceled FLAs and reverted to the DENR, two thirds were described as "devoid of improvements/abandoned" and "neither viable nor suitable for fishpond purposes."

More FLAs Should Have Been Canceled More FLAs should have been canceled or be in the process of cancelation for the following six reasons:

1. Many lessees fail to regularly pay their dues. At the rate of PHP 1000/ha, the expected total yearly collection is PHP 14.3 million. The amounts collected through the years have been falling below the expected level, even with surcharges and back rentals.
2. Many do not submit production reports. Out of 37 randomly selected active FLAs, 11 lessees have failed to make regular payments over the last 2–10 years and 26 have failed to submit regular production records for the last 2–16 years.
3. Interviews with agriculture and fishery technicians in 17 LGUs selected from the region's four provinces showed that a number of FLA areas are still in the active

list of the BFAR, but they are no longer being actively used for production purposes, have been already abandoned, or are undeveloped.
4. The number of expired FLAs is considerable: four of every ten FLAs in the active list have already expired, some more than a decade ago.
5. Subleasing is rampant and it is a violation of the terms and conditions of FLA. For instance, one interviewee was subleasing his 11 ha FLA at PHP 100,000/year for 7 years from a municipal employee who is the wife of a local politician. In many instances, the lessee and the sublessee make it appear as though they are "partners." That is, the FLA lessee partners with a person who finances the development of the fishpond.
6. Many lessees fail to comply with the obligation to introduce improvements to the FLA area within 180 days from the date of issuance of the lease. In one area of Bacolod City, the lessee had just started to develop the FLA area awarded to him 5 years earlier.

Reasons for the Low Rate of Cancelation of FLA We find that the number of canceled FLAs is lower than what it should be for a number of reasons. First, there are no approved guidelines covering FLA cancelation; the BFAR follows a practice based on FAO 197, and the steps taken and the length of time involved in FLA cancelations differ for every case. For instance, from the 29 canceled FLAs reviewed, the length of time involved in cancelation varied from 7 months to 14 years. Second is the power of connections. The BFAR recommends an FLA cancelation and the DA Secretary makes the final decision. Lessees can still appeal at the department level. Key informants indicated that the more "powerful and connected" the lessee, the higher the probability that his/her FLA will not be canceled. At the time of this study, 11 FLAs recorded as canceled by the BFAR were still pending final decision at the department level. Many FLA lessees are politicians (current and past), businessmen, or wealthy. Considered "untouchable," dealing with them is "risky," even if they blatantly violate the contract terms. Often, their FLA areas are fenced and local people also report such FLAs are also guarded by armed caretakers.

The third reason is poor monitoring and recording. According to a key informant, visual inspections of fishponds are done only when there is a need for inspection, such as during renewal of lease and transfer of rights, or when the BFAR Central Office calls for inspection of a particular FLA fishpond. Communications—usually reminders for rentals—between the Regional Office and the lessees are carried out through registered mail, which is slow and expensive. Moreover, records of each FLA area are done manually and documents are physically stored in folders. There is no computerized database to facilitate monitoring or hasten report generation.

Fourth, at the national level, the BFAR has only three staff members assigned to FLA-related work. Similarly, the Regional Office where the responsibility of identifying FLAs for cancelation rests has only two personnel dedicated to 1487

FLA areas covering over 14,000 ha across six provinces; with a limited travel budget, they are constrained to a maximum of four field visits in a month. The BFAR relies on the Provincial Fisheries Officers (PFOs) for field support, but the PFOs are also responsible for other fisheries programs—the PFOs are usually assisted by two staff members hired on contractual basis.

10.2.2.2 Reversion of Jurisdiction over Disused Fishpond Lease Agreement Areas

Once an FLA is canceled, jurisdiction over its area does not automatically return to the DENR from the BFAR; the law requires that only AUU is to be reverted automatically to the DENR. Based on copies of cancelation orders available in the individual FLA folders for 26 canceled FLAs with areas not reverted to the DENR, 11 were declared open and available to any interested applicant without prejudice to the findings of the BFAR if the area is still suitable for fishpond purposes upon determination by the BFAR. Eight were simply canceled, two were ordered canceled and to be returned to the DENR, and one was canceled with final termination to be decided after site inspection.

The official status of the 42 canceled FLAs with areas not reverted to the DENR indicates that 17 were awaiting "verification if still suitable for fishpond purposes" and 8 were identified as "vegetated with mangroves." Others were identified as "awarded to new lessee," "vacant and open to new application," or "subject of application." Notably three canceled FLA areas were identified as being with the Philippine Tourism Authority, the Development Bank of the Philippines—FLAs were once used as collateral for bank loans—and a local government unit. Two of these areas were deemed no longer suitable for mangrove reforestation.

More Areas with Canceled FLAs Should Have Been Reverted The number of reverted FLA areas in the region (currently 21) is fewer than what it should be. Three FLA areas were officially identified as "abandoned" and eight as "undeveloped" but not reverted to the DENR jurisdiction. The areas covered by at least 24 canceled FLAs should have been reverted to the DENR due to being "devoid of any development or improvement," and 8 were identified as vegetated with mangroves and 3 as having no developments introduced. The minimum number of reverted FLA areas should have been 56. These include the 21 FLAs already identified as canceled and whose areas have been reverted to the DENR and the 35 (at the minimum) that are qualified for reversion. The 35 include 24 of the 42 canceled FLAs whose areas have not been reverted to the DENR—3 were identified as abandoned and 8 as undeveloped.

Issues and Problems of FLA Area Reversion What factors give rise to less than half of the FLA areas being reverted to the DENR, and thus made available for reforestation? For one, there is no approved definition of "abandoned" and "underutilized" fishponds. One definition of "undeveloped fishpond" is found in Section 1.12 of FAO 197 and Section 46 of RA 8550. Two, there is an absence of guidelines for approving reversion of disused FLA areas. Section 14 of the Joint DA-the DENR-DILG Administrative Order No. 1, Series of 2008 provides for an interagency technical working group (ITWG) chaired by the DENR, with members from the DA-BFAR, LGUs concerned, and FARMCs. The group would be tasked with identifying FLA areas that have been abandoned for 5 years from the start date of the lease and developing and preparing guidelines for reverting AUU areas to their original mangroves. Unfortunately, no such collaboration has been created.

Third, there is in fact no formal collaboration or coordination between the BFAR (monitors FLA) and the DENR (monitors mangroves), despite the presence of orders to do so. This is true at all levels, national, regional, and local. Moreover, cancelation orders for FLAs—signed by the DA Undersecretary for Fisheries and Livestock—are not provided to the DENR; copies are provided to the lessee, Director of BFAR Regional office, Director of BFAR main office, and Chief Legal Division of DA. Key informants indicated that at the regional level, the DENR regional office communicated with the BFAR in 2002 to set up a meeting to discuss collaboration as prescribed by the FAO 197; a follow-up communication was made in 2004, and the meeting finally happened until 2010.

Fourth, there are advantages of retaining the areas of BFAR FLAs. According to the key informants, there are good reasons to recommend the DA Secretary to cancel only the FLAs. With this, the FLA areas remain open for applications to reopen it as a fishpond or continue to stay under the BFAR's jurisdiction. The popular view is that there are sectors in the BFAR wanting to retain FLA areas under BFAR jurisdiction, even if they qualify for return to the DENR. The key informants believe the BFAR holds onto FLAs because "the more FLAs, the more power" and "FLAs are a money-making machine." The money is in the legal fees and rentals and also "under-the-table" payments.

Meanwhile, the DENR is not seen as capable at managing FLA areas under its management or of restoring them to their original mangrove state. One key informant suggested that it is better to just leave the FLA in canceled status without reverting the area to the DENR, which the informant speculated the DENR would "just convert it to other uses"; the tenure instruments available at the DENR are considered "threats." The same key informant believes that FLA areas suitable for mangrove forests will regenerate naturally.

The fear that the disused area will be converted by the DENR to other uses is not without basis. Another key informant (from the DENR) affirmed this by pointing to "political influences" as reasons. A case in point is the application for an FLA in an area that was reverted already to the DENR: one applicant secured certification that the DENR poses no objection to fishpond development. Although, in this instance, the actions of both agencies are questionable (the BFAR, for entertaining new applicants, despite the previous ruling of reverting it to the DENR and the DENR for issuing a certificate), it highlights that the development future of an FLA area, even when already reverted to the DENR, can be changed.

Fifth, the DENR is seen as being inactive and ineffective. A key informant from the DENR disclosed that the department has done nothing with respect to reversion of AUU FLA areas.[4] Another key informant described the entire reversion process as having "zero governance." Still another key informant from the DENR said there are no records of reverted FLA areas available at the DENR Central Office; this was echoed by a key informant at the regional level. However, this is only partly true as a memo exists that states a "list of areas covered by Fishpond Lease Agreements which have been canceled for cause and reverted to the jurisdiction of that Department" was sent and acknowledged by DENR.

10.2.2.3 Suitability of Disused Fishpond Lease Agreement Areas for Mangrove Reforestation

Not all disused fishpond areas are suitable for mangrove reforestation (see Table 10.2). Only 23 % (284 ha) of the total area (1240 ha) visited in Western Visayas were found suitable because they are in the middle of an intertidal area near a freshwater source and adjacent to areas with mature mangroves. About 82 % of the FLA areas visited for this research were inundated, 10 % were vegetated with mangroves, 5 % were still being used as fishponds despite the cancelation order, a small portion (1.3 %) is part of a river, and nearly 1 % have squatters or have become a filled area (thus, severely altered).

10.2.2.4 Cost and Benefit of Reverting Fishpond Lease Agreement Areas to Mangrove Forests

The costs and benefits of the ecosystem services provided by mangroves were computed for a span of 15 years. It made use of values transferred from Sathirathai and Barbier (2001), adjusted to current period and local prices. The said study estimated the value of protecting the remaining mangrove area in Tha Po Village in Surat Thani Province in southern Thailand. The local direct use value of mangrove is the income generated from the forests in terms of various wood and nonwood products. Market prices were used if the extracted products were sold and surrogate prices when the products were only for subsistence. For the ecological values of the mangrove area, the production function approach was used in estimating the value of offshore fishery linkages, while replacement cost method was used to value coastline protection and stabilization. The values were discounted at three discount rates: 5 %, 10 %, and 15 %. The analysis was done for two scenarios. One scenario allows harvesting of mangrove products, such as wood for fuel, medicines, and others, thus represents the direct use values from a mangrove forest. The other scenario bans harvesting and reflects the mangrove's functions in protecting

[4] It should be noted the DENR-CMMO (main office) partnered with the NFR and other stakeholder groups in organizing consultations on AUU reversion to mangroves in Luzon and Mindanao.

Table 10.2 Summary assessment of the suitability for mangrove reforestation of disused FLA areas

Province and number of FLA areas visited	Total area of visited FLA areas (ha)	Area for reforestation considering all areas that can be reforested based on environmental parameters (ha)	Area for reforestation by natural regeneration or active planting (ha)	Area for reforestation by natural regeneration only (ha)	Area for reforestation by active planting only (ha)
Negros Occ. (32)	189	84	41	43[a]	0
Iloilo (16)	640[b]	134	0	0	134
Guimaras (3)	35	*6*[c]	0	0	*6*[c]
Capiz (14)	376[d]	60	60	0	0
All (62)	1240	284	101	43	140

Areas in hectares are estimates and numbers are rounded to the nearest unit. The total shown (62) represents all the FLAs that were located

[a]Needs hydrological restoration

[b]Excluding 1 FLA area (40 ha) not visited because the exact location was not found

[c]Taken care of by the Zoological Society of London

[d]Excluding 2 FLA areas (47 ha) not located

ecosystem services in terms of indirect use values (protection against storm surge). The computations are summarized in Tables 10.3 and 10.4.

The benefits outweigh the costs of reforestation over the 15-year duration at all three discount rates. The net benefits are higher with reforestation by natural vegetation, even while sustainable harvesting is allowed. The results support the argument that restoration of disused FLA areas into mangrove forest is in the social interest whether or not harvesting is allowed. The net present values (NPV) are very close for both scenarios, suggesting that a ban on harvesting would be feasible, particularly because indirect benefits have likely been underestimated given they do not explicitly include damages from climate change, damages that mangroves help mitigate.

10.3 Potential Financing Options for Mangrove Reforestation

10.3.1 Feasibility

The feasibility indicator is based on consultation with experts. Mangrove and aquaculture experts prefer the options where the responsibility of paying for reforestation rests with lessees (e.g., performance bonds, higher leases, green

Table 10.3 Net present values (NPV) at different discount rates for two scenarios (in USD) with reforestation through natural regeneration

Discount rate (%)	NPV benefit without harvesting[a]	NPV benefit with harvesting[b]	NPV cost	B-C ratio with harvesting	B-C ratio without harvesting
5	33,807–34,244	34,610–35,046	5637	6.00–6.08	6.14–6.22
10	20,415–20,678	20,899–21,163	4511	4.53–4.58	4.63–4.69
15	18,544–18,784	18,984–19,224	4603	4.03–4.08	4.12–4.18

Note: 1 USD = P43.31

[a]Using values from Sathirathai and Barbier (2001), indirect use value only

[b]Using direct and indirect use values from Sathirathai and Barbier (2001); values were adjusted according to the formula of Saplaco (2000)

Table 10.4 Net present values (NPV) at different discount rates for two scenarios (in USD) with reforestation through active planting

Discount rate (%)	NPV benefit without harvesting[a]	NPV benefit with harvesting[b]	NPV cost	B-C ratio with harvesting	B-C ratio without harvesting
5	33,807–34,244	34,607–35,046	6957	4.86–4.92	4.97–5.04
10	20,415–20,678	20,898–18,983	5831	3.50–3.55	3.58–3.63
15	18,544–18,784	21,163–19,224	5924	3.13–3.17	3.20–3.25

[a]Using values from Sathirathai and Barbier (2001), indirect use value only

[b]Using values from Sathirathai and Barbier (2001), direct and indirect use values. Values were adjusted following the formula of Saplaco (2000)

taxes; see Table 10.5). Likewise, the environment managers prefer green taxes/fines and higher leases and have a low preference for DENR grants. In contrast, the aquaculture managers much prefer DENR grants in combination with performance bonds, higher leases, and private sector support. The LGU managers much prefer beneficial use taxes (as they provide other benefits for the LGU), green taxes/fines, and foreign funding grants. It is noted that the experts and the LGU managers have also very low preference for DENR grants, which may reflect their dismay with the DENR's ongoing reforestation projects. Based on previous efforts to raise the lease rate, the probability of lessees accepting any option that will require them to pay is very low. The options where the responsibility of paying for reforestation lies with lessees received moderate to low rates of preference in terms of feasibility. Aggregating all of these viewpoints, the only option with consistently high feasibility is private sector support.

10.3.2 Ease of Implementation

The options that require the lessees to pay for reforestation appear to be relatively more challenging to implement than those that do not, using the criteria provided in

Table 10.5 Feasibility of options

Option	Preferential ranking of				Likelihood of accepted by lessees	Indicator summary
	Mangrove and aquaculture experts	Environment managers	Aquaculture managers	LGU managers		
Performance bond	High	Low	Very high	Low	No	Low
Lease reflective of economic rent	Very high	High	High	Low	No	Moderate
Green tax/fine	High	Very high	Low	High	No	Moderate
Beneficial use tax	Moderate	Low	Low	Very high	No	Low
The DENR grant	Very low	Low	High	Very low	Yes	Low
Foreign funding grant	Low	Moderate	Very low	High	Yes	Moderate
Private sector support	Moderate	High	Moderate	Moderate	Yes	High

Table 10.6 Ease of implementation

	Needs legislation	Requires a proposal for submission	Need study to determine optimal rate/amount	Requires additional inputs	Entails low enforcement and monitoring cost	Indicator summary
Performance bond	Yes	No	Yes	No	Yes	Moderate
Lease reflective of economic rent	No	No	Yes	Yes	No	Low
Green tax/fine	Yes	No	Yes	Yes	No	Very low
Beneficial use tax	No	No	No	No	No	High
The DENR grant	No	Yes	No	No	Yes	High
Foreign funding grant	No	Yes	No	No	Yes	High
Private sector support	No	Maybe	No	No	Yes	Very high

Table 10.6. Imposing a green fine was found to be highly complex to implement, while imposing a performance bond and raising the lease were found to be moderately complex.

Two options—performance bonds and green fines—require national legislation to be implemented. On the other hand, the Fisheries Code of 1998 (Sections 9, 14, and 16) stipulates imposing rental and application fees and a cash bond deposit. The optimal levels for a green fine or performance bond need to be determined through scientific studies. In order to determine who are to be charged with green fines, information on each FLA area is needed. Such information can only be collected by actual area visits.

The revenue powers of the local governments can be invoked (1991 LGC, Sections 129–130, Section 186, Section 205d) to impose a beneficial use tax. A local ordinance is needed before implementation. The assessment for real property (1991 LGC, Sections 218a, 220) and tax rate can follow. Similarly, there is a need to visit the FLA areas to calculate fair market value. Section 13.d of FAO 197 stipulates, "Annual rentals shall be set at levels that reflect resources rent accruing from the utilization of resources." To implement this, political will and additional resources (e.g., budget and personnel) are needed for enforcement and monitoring. A credible empirical study to support an increase in rents and to what amount is necessary.

The other options only require special skills in drafting, packaging, and submitting grant proposals. This is especially true in seeking funding grant from the DENR, foreign funding agencies, and the private sector.

Except for performance bonds, which are one-time payments at the time of FLA renewal, the other options that require lessees to pay for reforestation would entail high enforcement and monitoring costs to ensure regular payments. The rest of the options do not entail monitoring and enforcement cost.

10.3.3 Likely Impact on Other Desirable Goals

The options requiring lessees to pay for mangrove reforestation will attract lessees who are serious in fishpond production. If so, positive impacts will likely be incurred, such as reduced inequity (fewer lessees who treat having an FLA as an end in itself, hold the land without development); increased production efficiency, which ensures enough revenues to pay the fees and turn a profit; and increased compliance with environmental laws (to avoid fines). All options push for compliance with environmental laws. The options to seek grants may not impact increasing fishpond or aquaculture production, in general; their viability is based on the least resistance from the industry. Seeking private sector support may have a positive impact on inequity, since, in some sense, it would provide a venue for the rich to help the poor (Table 10.7).

Table 10.7 Likely impact on other desirable goals

Option	Reduction in inequity	Increase in production efficiency	Increase in compliance with environmental laws and regulation	Indicator summary
Performance bond	Yes	Yes	Yes	High
Lease reflective of economic rent	Yes	Yes	Yes	High
Green fine	Yes	Yes	Yes	High
Beneficial use tax	Yes	Yes	Yes	High
The DENR grant	No	No	Yes	Low
Foreign funding grant	No	No	Yes	Low
Private sector support	Yes	No	Yes	Moderate

10.3.4 Funds Generated

Ideally, the cost of bonds or taxes should cover the "worst case scenario" or the level that would deter the lessees from defaulting. For these options to work, they have to generate a sufficiently high amount of revenues to ensure that a disused FLA area can be rehabilitated as a mangrove forest.

The country has seen many reforestation efforts in the past. There had been six major externally funded mangrove rehabilitation projects in the country from 1957 to 2006 (Primavera and Esteban 2008). Funds came from the Overseas Economic Cooperation Fund (OECF), the World Bank, Asian Development Bank, and Japan Bank of International Cooperation. The World Bank funded the latest project on Community-based Resource Management Program (CBRMP) for USD 38 million. On the other hand, one funding source not yet explored in the country for projects, such as mangrove reforestation, which will benefit the environment, is the Global Environment Facility.

The DENR grant is taxpayers' money. This means that restoration efforts funded through the DENR are taxpayer-supported. Presently, the DENR funds reforestations projects but not beyond PHP 200,000 (USD 4619) per reforestation site. Similarly, although there are reforestation projects initiated by or with the help of the private sector, the donation is typically only for a specific reforestation project and for the labor component (Table 10.8).

Table 10.8 Amount of funds generated

Option	Estimated amount to be generated and some comments	Indicator summary
Performance bond	PHP 88.6 million[a] (USD 2.0 million)	High
Lease reflective of economic rent	PHP 96.3 million annually[b] (USD 2.2 million)	High
Green tax/fine	Expected to be high given the rampant violation, particularly for not observing a greenbelt, generating pollution, and other violations of fishery laws	High
Beneficial use tax	PHP 11.4 million annually[c] (USD 263 thousand)	High
DENR grant	PHP 200 thousand (USD 4.5 thousand)	Low
Foreign funding	For example, WB: PHP 38 million under the CBRMP	High
Private sector support	Usually one-time sponsorship of tree planting by corporations. Also include volunteer service on actual outplanting of mangroves	Low

Numbers are rounded to the nearest unit
[a]7032 ha of 671 expired FLAs × PHP 12,600/ha = direct outplanting cost
[b]14,253 ha × FSP recommendation of USD 156/ha per year (1991). Currently, this is equivalent to PHP 6700/ha per year, which is also the lowest rental rate for private fishponds existing in the Region VI
[c](PHP 40,000/ha assessed value*0.02)* 14,253 ha of total area

10.3.5 Sustainability

Funds are needed to ensure continuous reforestation, maintenance, and protection of mangrove areas. Among the seven options, there are those that are one-time payments (performance bonds), payments when needed (green fines), granted when qualified (the DENR grant, foreign funding, and private sector support), and regular payments (lease and beneficial use tax) (Table 10.9).

10.3.6 Evaluation Summary

The optimal option is one with the following characteristics: highly feasible, easily implemented, high positive impacts on other desirable objectives, and able to generate a high amount of funds on a sustainable basis. Table 10.10 shows that no single option meets all of these criteria—there are trade-offs, which are common when dealing with complex problems. The trade-off is largely between the amount of revenue raised versus implementation complexity and feasibility. For example, in a situation in which lessees pay for reforestation, the beneficial tax and lease rates based on economic rents are promising. Seeking grants from funding agencies is less likely to raise revenue than bonds and taxes, but would be relatively easier to implement.

Table 10.9 Sustainability

Option	Possibility of regular payments
Performance bond	Low
Lease reflective of economic rent	High
Green tax/fine	Low
Beneficial use tax	High
The DENR grant	Low
Foreign funding grant	Low
Private sector support	Moderate

Table 10.10 Summary matrix evaluating the seven options

	Feasibility	Ease of implementation	Likely impact on other desirable objectives	Amount generated	Sustainability
Performance bond imposition	Low	Moderate	High	High	Low
Lease reflective of economic rent	Moderate	Low	High	High	High
Green tax imposition	Moderate	Very low	High	High	Low
Beneficial use tax	Low	High	High	High	High
The DENR grant	Low	High	Low	Low	Low
Foreign funding grant	Moderate	High	Low	High	Low
Private sector support	High	Very high	Moderate	Low	Moderate

10.4 Conclusions and Recommendations

10.4.1 Conclusions

The current state of the mangrove resources in the Philippines calls for urgent intervention to prevent further deforestation and degradation. One intervention is to bring the mangroves back to their natural habitat, the middle intertidal coastal areas, which can be done by reverting disused FLA areas back to mangrove forests. Despite the policies issued on this, efforts toward this end have been very slow.

The entire FLA policy is problematic. The processes of FLA cancelation and reversion of jurisdiction over disused FLA areas from the BFAR to the DENR are fraught with problems (institutional, economic, and political), leading to fewer canceled FLAs and reverted FLA areas than feasible. If the process of reverting disused FLA areas to mangrove forest is to work, some critical changes have to be made. Such changes would have to start in the way the FLAs are enforced and canceled and the way the jurisdiction over the areas is transferred from the BFAR to the DENR. A number of basic things need to be done, such as formulation and approval of guidelines on the FLA cancelation and reversion of area jurisdiction.

Disused FLA areas do not automatically qualify for reforestation. A suitability assessment should be made before any decision to revert to the original mangrove state; this will allow creation of a priority list for rehabilitation. If the disused FLA area is not suitable for reforestation, its best use must be determined and pursued.

Reforestation does not always mean active reforestation. Given suitable environmental conditions, an area may vegetate naturally. Of the total area (284 ha) suitable for reforestation, 43 ha were found suitable for natural regeneration, leaving 140 ha for active planting and another 101 ha that can either naturally vegetate or be actively planted.

For disused FLA area suitable for reforestation, the costs will reflect the characteristics of the site. For example, active planting incurs higher costs than natural regeneration. Aside from protecting the area from encroachment (required in natural regeneration) and maintaining it, costs are incurred for nursery establishment and outplanting when active planting is necessary.

The evaluation of the financing options illustrates the trade-offs among the options. The options where the responsibility for paying for reforestation rests on lessees (performance bonds, green taxes, fees reflective of economic rent, and beneficial taxes) would generate high amounts of revenue. They would work jointly with institutional reforms to create the right incentives to have a more efficient and timely process of restoring fishponds that no longer provide food to mangroves. These fees would lead unprofitable FLAs to cease operation quicker than the current system. Their drawbacks take the form of opposition from the aquaculture industry and complexity in implementation. On the other hand, the options where the responsibility to pay for reforestation cannot be identified (the DENR grant, foreign fund, private sector support) are somewhat easier to implement, but may generate lower amounts and not be sustainable.

10.4.2 Recommendations

A smaller incentive package (higher rentals, shorter lease periods, and a smaller maximum area than presently prescribed) is needed for existing FLAs to help reduce the inefficient use of land characterized by abandoned licenses. It would also attract and retain lessees who are serious about fishpond production and contribute to the goals of reducing income inequality and improving food production and security.

Since issuance of new FLAs is no longer allowed, a better screening process for renewal and transfer of leases is important. Approved guidelines on renewal and transfer of leases must be observed. Also, the high number of expired FLAs deserves attention. The decision on what to do with expired FLAs should be evaluated in terms of efficiency, equity, food security, and environmental impact.

A survey of all FLA areas in the region is needed to establish the potential for returning the mangroves to their natural habitat (i.e., the middle intertidal region). This survey could be scaled up to the national level at a future date. The main objective is to identify the FLAs for cancelation and to have their areas reverted to the DENR. Once the FLA areas are returned to the DENR, suitability assessment for reforestation should follow immediately, and mangrove reforestation should be carried out in suitable disused FLA areas.

In support of the reversion of disused FLA areas to mangrove forests, institutional changes are required:

1. Approved guidelines on FLA cancelation and on the reversion of area jurisdiction from the BFAR to the DENR. Once an FLA is canceled, the reversion of jurisdiction to the DENR should be automatic (i.e., it should not be open for application as a fishpond).
2. A database of all FLA areas, including all types of fishponds and mangrove areas in the region. There should be electronic record keeping of FLAs and regular monitoring of the FLA areas.
3. A review of related laws, policies, rules, and regulations on fishponds related to FLAs is needed, with the view to harmonize.
4. Coordination among the DENR, BFAR, and concerned LGUs, as specified in JAO 2008: The National Task Force on AUU reversion to mangrove forests with membership from these agencies should already be created and should start to work.

LGUs may be given bigger roles in FLA implementation, especially since they have better access to local information (such as whether an area is actively used for fishpond purposes) than the DENR or BFAR. Moreover, they could assume bigger roles in mangrove reforestation, particularly in the reversion of disused FLA areas in their jurisdiction.

The national government should embark on a nationwide program for mangrove reforestation. The program could reside under the National Greening Program launched by the national government with various agencies of the government, with the DENR and DA as implementers.

Lastly, a Mangrove Reforestation Fund could be set up for this purpose; funds could be from different sources such as the lessees (by increasing rental and other fees, imposition of green fine, beneficial use tax) and funding agencies. Credible empirical studies are needed to determine correct rentals, fines, or taxes.

References

BAS (Bureau of Agricultural Statistics) (2010) Fisheries statistics of the Philippines 2007–2009. BAS, Quezon City

BFAR-FRQD (Bureau of Fisheries and Aquatic Resources-Fisheries and Quarantine Division) (n.d.) Fisheries Administrative Order No. 125-1, s. 1991. http://www.bfar.da.gov.ph/legislation/fao/fao14.htm. Retrieved 4 Aug 2010

Genio EL Jr, Rejesus RM, Pomeroy RS, White A, Smith B (2007) Factors affecting fisherfolk's support for coastal resource management: the case of local government initiated mangrove protection activities. Ocean Coast Manag 50(2007):808–882

Holl KD, Howarth RB (2000) Paying for restoration. Restor Ecol 8(3):260–267

Melana DM, Atchue J III, Yao CR, Edwards R, Melana RR, Gonzales HT (2000) Mangrove management handbook. Department of Environment and Natural Resources, Manila, Philippines through the Coastal Resource Management Project, Cebu City, Philippines, 96 p

PDI (Philippine Daily Inquirer) (2010) Gov't releases P600M for typhoon Frank rehabilitation. Inquirer Visayas. http://newsinfo.inquirer.net/breakingnews/regions/view/20100530-272924/Govt-releases-P600M-for-typhoon-Frank-rehabilitation. Retrieved 5 Oct 2010

Philippine Star (2008) Crop damage reaches P3.3 B. The Philippine Star. http://www.philstar.com/Article.aspx?articleId=69235. Retrieved 5 Oct 2010

Primavera J (2000) Development and conservation of Philippine mangroves: institutional issues. Special issue – the values of wetlands: landscape and institutional perspectives. Ecol Econ 35 (2000):91–106

Primavera J (2005) Mangroves, fishponds, and the quest for sustainability. Science 310 (5745):57–59

Primavera J (2010) Paper presented at the seminar-workshop on FLA cancellation and mangrove reversion, 28–30 Jul 2010, Grand Hotel, Iloilo City

Primavera JH, Esteban JMA (2008) A review of mangrove rehabilitation in the Philippines: successes, failures and future prospects. Wetl Ecol Manag 16:345–358

Samson MS, Rollon R (2008) Growth performance of planted mangroves in the Philippines: revisiting forest management strategies. Ambio 3(4):234–240, Royal Swedish Academy of Sciences

Saplaco R (2000) VALUASIA: benefits transfer for Southeast Asia. Unpublished undergraduate thesis. College of Economics and Management, University of the Philippines Los Banos, Laguna

Sathirathai S, Barbier EB (2001) Valuing mangrove conservation in Southern Thailand. Contemp Econ Policy 19(2):102–122

Stevenson NJ (1997) Disused shrimp ponds: options for redevelopment of mangrove. Coast Manag 25(4):423–425

Yao CE (2000) Saving the mangroves of Bohol. Over seas: the online magazine for sustainable seas. 3(11). Downloaded from http://www.oneocean.org/overseas/200011/saving_the_mangroves_of_bohol.html on 25 Jan 2010

Chapter 11
Environmental Trade-Offs from Coastal Reclamation: The Case of Cebu, Philippines

Lourdes O. Montenegro

Abstract Coastal land reclamation—the process of creating new land by infilling coastal areas—is becoming a popular way of expanding the space available for economic activity in East and Southeast Asia. Although this method can provide more land for economic activity, these reclamation projects are not without their environmental trade-offs. These trade-offs are not often explicitly considered or valued when evaluating project feasibility. As a result, there is an absence of a clear conception of the types and magnitudes of the trade-offs that could be used to guide policy decision-making and implementation. This study fills this gap by investigating the case of a proposed large-scale reclamation project in Cordova, a town in Cebu, Philippines. Environmental impacts and costs were identified and monetized for use in a social cost-benefit analysis, which accordingly assessed whether the project would generate net social value to the appropriate reference community. The total potential environmental costs were estimated to be almost USD 60 million (PHP 3.3 billion). Among the environmental costs valued in the study, damage to corals and externalities from landfill quarrying were the most significant. The prevention of damages to the coral reef alone would reduce the estimated environmental costs by more than half. The relatively large value of forgone recreational benefits from the coral reef could be avoided if the reclamation activity would not be in the coastal areas where fair to good coral cover exists. This potentially high environmental cost also suggests an alternative development strategy that the municipality could pursue: one that has more to do with sustainable ecotourism than with attracting new industry.

Keywords Environmental valuation • Coastal reclamation • Cost-benefit analysis • Project evaluation

---

L.O. Montenegro (✉)
Lee Kuan Yew School of Public Policy, National University of Singapore, Singapore, Singapore
e-mail: mailmontenegro@gmail.com

N. Olewiler et al. (eds.), *Marine and Coastal Ecosystem Valuation, Institutions, and Policy in Southeast Asia*, DOI 10.1007/978-981-10-0141-3_11

11.1 Background

Coastal land reclamation—the process of creating new land by infilling coastal areas—is becoming a popular way of expanding the space available for economic activity in East and Southeast Asia. Singapore, Hong Kong, and many coastal cities in Japan, China, Taiwan, and South Korea have changed the outlines of their coastlines to create new industrial, institutional, and even agricultural lands. For example, it has been estimated that South Korea has reclaimed more than 62,000 hectares (ha) of its coastal area since the Second World War (Hwang 1999; Moores and Bräunlich 1999). In Hong Kong, about a tenth of the country's developed area is reclaimed land, including one project that provided space for a Disney theme park, and more is planned (Jiao 2000).

While providing more land, these reclamation projects are not without their environmental trade-offs. Unfortunately, these trade-offs are not often explicitly considered or valued when evaluating project feasibility. As a result, there is an absence of a clear conception of the types and magnitudes of the trade-offs that could be used to hopefully guide policy decision-making and implementation. This paper fills this gap by investigating the case of a proposed large-scale reclamation project in Cordova, a town in Metro Cebu, Cebu Province, Philippines.[1] Environmental impacts and costs were identified and monetized for use in a social cost-benefit analysis that assesses whether the project generates net social value to the appropriate reference community.[2]

11.1.1 *The Cordova Reclamation Project: History, Description, and Stakeholders*

The Municipality of Cordova is located in the south eastern part of Mactan Island, Cebu Province, Philippines. Its 846 ha land area is divided into 13 barangays (or villages) where all except one have access to the coast. An important portion of the local economy revolves around artisanal fishing and harvesting or gleaning of marine products. According to the Department of Agriculture's Provincial Office,

[1] Metro Cebu is the Philippines' second largest urban center and is composed of the cities of Cebu, Mandaue, Lapulapu, and Talisay, as well as the municipalities of Minglanilla, Naga, Compostela, Liloan, Cordova, and Consolacion. These cities and municipalities lie within an area that has a radius of about 25 km.

[2] Important issues in a cost-benefit analysis are to what scale and from whose perspective is the study done. Is the perspective at the local community level, regional, national, or international? The outcomes can be quite different, and this is the case for a major reclamation project. Net benefits at the community level may be negative, while those at the national level, positive. Thus, a cost-benefit study is not necessarily a definitive methodology with which to make decisions if there are significant divergences between net benefits, which depend on the jurisdictional level of the analysis.

the town's 175 km^2 of municipal fisheries produced more than 800,000 metric tons of fish in 1990. Most of the fishing activities in the area are small-scale family affairs, usually concentrated on the 17.4 km^2 intertidal reef flat and ledge bordering the town's south, south eastern, and western coasts (University of San Carlos 1998). There is a 14 ha marine sanctuary in a narrow channel between Cordova's eastern coast from the western side of Gilutongan Island. This sanctuary, which falls within Cordova's jurisdiction, is very popular among local and international tourists who visit for recreational diving and snorkeling. In 2002, the municipality collected some PHP 2 million in total user fees from more than 47,000 visitors; the user fees contributed to more than half of Cordova's municipal government budget for that year.

Although there are rival proposals to reclaim Cordova's foreshore area, the analysis presented in this chapter is limited to the 2700 ha reclamation proposal from the Malayan Integrated Industries Corporation (MIIC), a developer from Manila. Among the various proposals, it is the only one containing a sufficiently detailed description, which enabled the identification and monetization of impacts. At the time of this research, the proposal was on hold, and there were prospects for undertaking a much smaller project covering an area less than 5 % of that proposed by MIIC.[3] The large-scale project is the subject of this paper but estimates are provided per hectare, when possible, so that alternative scales could be assessed.

Under the MIIC proposal, all construction costs are to be assumed by MIIC and thus represent private costs. The developer has offered to provide the local and national government 30 % of the newly created land. The project costs are estimated at PHP 37.5 billion in 1997 prices (MIIC 1997). The proposed reclamation would require approximately 170 million m^3 of landfill to create 2700 ha—an area three times larger than Cordova's land area and even larger than its intertidal flats. The reclaimed land is to be divided up and used for a variety of purposes. The MIIC proposal envisions close to a quarter of the area to be set aside for industrial use, while more than 30 % is earmarked for residential and tourism purposes.

It is estimated that the reclamation project would enable Metro Cebu to absorb an additional 100–120 exporting firms. These would be expected to provide employment for about 50,000 workers. Some of these workers would come from other islands in the region, as well as from neighboring regions in the Southern Philippines (Mactan Island Integrated Master Plan 1995 cited in Schema 1996). Estimation of the social benefits of the project would thus have to examine the net new employment (i.e., providing jobs for people who would otherwise be unemployed versus simply a redistribution of employment from one region of the country to another or the use of imported workers). If implemented, reclamation work would be accomplished within 14 years from the start of the project. The project

[3] On February 17, 2014, the Cebu Daily News reported that the MIIC reclamation proposal has been revised into a 3,500 ha joint venture with Hyundai Heavy Industries. In contrast to the 1997 proposal, the current project now includes an international seaport, a 12-lane circumferential highway, and a bridge connecting Cordova with Cebu City, in addition to infilling 800 ha more of coastal area.

plan has three phases, at a rate of almost 200 ha per year (MIIC 1997). The MIIC reported that it expects to earn a financial internal rate of return (FIRR) of 55–69 % on the sale of its 70 % share of the newly created land. The local and national governments project the revenue from the project to be between PHP 10 billion and PHP 16 billion from the sale of their 30 % share of the newly created land, which represents a social benefit.

The private developer proposes to obtain landfill materials for the Cordova Project by dredging either the offshore seabed around Talisay City in the southern part of Metro Cebu or other offshore areas around the neighboring island province of Bohol (MIIC 1997). However, many experts are skeptical about the technical feasibility of sourcing landfill offshore given that previous projects ended up relying on land-based quarries. Given the scale of the proposed Cordova Reclamation Project (CRP), there is a multitude of stakeholders and a variety of interested groups and entities. These represent many different levels of jurisdiction, from local to national. At the local level, there is the immediate Cordova community. This is made up of people who are, in varying degrees, dependent on the potentially affected coastal resource for their livelihoods (a profile of this group is discussed in the next section). They are, however, in a position to gain potential employment as construction workers during the reclamation development and as employees of the firms that locate on the newly created land. Another local stakeholder is the local government unit (LGU). It is in a position to gain increased revenues from the sale or lease of its share of any newly created land. It could also gain increased tax revenues from any new firms that locate in the area.

More broadly, the project could provide additional employment prospects to the regional population. Many of these people may, however, also be negatively affected by pollution and other externalities from the widespread quarrying for landfill materials, which is expected to take place in many upland areas. Direct construction costs are expected to be borne by private investors who will come from within the region or from Manila. From a national perspective, the scale of benefits from the project may be smaller than expected as it may only serve to divert economic activity from other parts of the country. However, for the purposes of this study, the analysis of social costs and benefits is limited only to those affecting populations at the local and regional levels.

A survey of a stratified random sample of 377 Cordova households was conducted in April to May 2003, as part of this analysis. The survey collected information on demographics, income, education, employment, coastal resource-based livelihood activities (e.g., fishing and reef flat gleaning), and perceptions regarding the proposed reclamation project. The results of the survey indicate an overwhelming majority of households are aware of the project. When asked what they thought were the positive effects of coastal reclamation, about 38 % of respondents said that they see no positive effects at all. About 46 % are of the opinion that the project would only have positive effects in terms of job creation and economic growth. About 16 % mentioned positive effects but qualified their responses by highlighting the negative effects that they feel the reclamation would have. The dominant concern expressed by the survey respondents is that, while jobs might be created during the construction process, these jobs would only be temporary. There is also the worry that any new firms that might locate to the new land would only hire people with certain educational attainment (e.g., secondary and tertiary education) and not benefit the entire community.

11.2 Environmental Impacts of Coastal Reclamation

Coastal reclamation, especially as it is carried out in the Philippines and in other parts of East Asia, affects both coastal and upland environments. The most apparent environmental impacts arise from adverse effects on the coastal environment. Reclamation covers up and destroys intertidal reef areas and the marine environment associated with them. Related dredging operations can also damage adjacent corals.

A number of recent studies have shown that coastal land reclamation is one of the major factors responsible for the loss of intertidal flats and other coastal resources in East and Southeast Asia (Kawabe 1998; Hwang 1999). Intertidal reef areas are often the first targeted for reclamation because they are, from an engineering perspective, convenient and cost-effective to landfill. However, intertidal reef flats are unique and highly productive ecosystems and these are also categorized as marine wetlands. Like all wetlands, they provide a variety of valuable ecosystem services such as (1) shoreline protection and storm damage buffer zones, (2) fisheries and fish nurseries, (3) nutrient cycling and storage and related pollution control, (4) commercial goods, and (5) extended food web control (Pearce and Turner 1990). Many of the ecological goods and services that are provided by such wetlands are, however, seldom marketable and may take the form of a public good. In this case people's preferences for them, as reflected by the choices they make (i.e., their "willingness to pay" in terms of other goods), do not show up in the market. Such goods and services seldom have a market price and their values are therefore underestimated.

Coral reef ecosystems are also readily affected by reclamation activity since corals are very sensitive to any increase in the concentration of suspended particles in the water in which they live. Dredging activities related to reclamation work result in heavy siltation and also cause anoxic conditions in surrounding waters. This contributes to the smothering of corals. Considering coral reefs take a very long time to recover and regenerate, this impact is usually regarded as permanent (USC 1991).

Coral reefs rival tropical rainforests in terms of the amount of biological diversity they support. They are highly valuable ecosystems that provide a variety of goods and services. They are, for example, crucial in the maintenance of productivity in the fisheries sector. They are also admired for their beauty and often attract a variety of tourists and sports people. The Philippines sits within the so-called coral triangle at the center of biodiversity in the marine tropics. Yet, its coral areas are greatly threatened—less than a third can be classified as being in good or excellent condition (World Bank 2000).

Environmental impacts also result from landfill quarrying activities associated with reclamation. Landfill quarrying may have adverse on-site and off-site impacts on climate, air, noise, topography, groundwater, land use/vegetation, wildlife, surface water regime, and soil stability and erosion rates (GSECI 2002). The plurality of these impacts and the lack of the biophysical data needed to model their effects make them harder to quantify and monetize.

11.3 Estimating the Forgone Benefits from the Affected Coastal Area

Part of the environmental costs of reclamation activities arise from the forgone benefits of lost and damaged coastal ecosystems, such as intertidal flats and coral reefs. Since the reclamation scenario examined in this study involves the total and irreversible destruction of these ecosystems, the total economic value (TEV) approach was adopted. The TEV concept encompasses not just the use values of a resource or ecosystem but their nonuse values as well. Three categories of forgone benefits were valued in this study: (1) on-site fishery, (2) reef flat gleaning, and (3) the recreational value of the affected coral area.

The monetary values reported in this chapter are to be understood as lower bound estimates of the costs from the loss of the coastal ecosystem. First, the analysis was limited to a 30-year time period. This means only the net present value (NPV) of the flow of forgone benefits for the next 30 years is reported, thereby potentially underestimating the value of ecosystems that can theoretically provide goods and services for a very long period of time when used sustainably. Second, nonuse values are excluded and only use values are measured.

11.3.1 On-Site Fisheries

Since the products of on-site fisheries are sold in markets, a price-based approach to valuation was used. For those few products that were not sold in organized markets, surrogate market prices of the closest substitutes were used. The value of the loss of on-site fishery output took into consideration the potential response of fishermen to the loss of their traditional fishing grounds. Hence, this valuation was understood as being made up of two components: (1) lost fishing output and (2) the increased cost of fishing at more remote sites outside the reclamation area (off-site). Lost fishing output was measured as the loss in net fishing income. The increased cost of fishing off-site included the amortized capital costs of acquiring a larger motorized fishing vessel and additional fuel costs.

Almost a quarter of all households surveyed had at least one member engaged in fishing. The south eastern coast facing the Gilutongan reserve was shown to be the most heavily dependent on fishing. Around 83 % of fishing respondents spend three-fourths of their fishing time within or adjacent to the intertidal flat. Almost all fishing activities in the affected area are small-scale family affairs that can be categorized as artisanal. The majority (97 %) of fishing respondents either do not use fishing vessels (25 %) or use fishing vessels that are below 3 metric tons in capacity (72 %). Slightly more than 50 % of the fishing vessels were reported to be non-motorized, indicating that the majority of fishermen do not have the immediate capacity to venture to more distant and more abundant fishing grounds, such as those found near the waters of Bohol Island. At 2003 prices, it would cost approximately PHP 36,000 (USD 643) for a fisherman to purchase a motorized fishing vessel adequate for the journey to fishing grounds in nearby provinces like Bohol. Annual fuel costs for a fishing journey to neighboring provinces would amount to about PHP 19,000 (USD 340) more than the costs for fishing in local waters.

Respondents reported that they catch 51 different types of fish, crabs, squids, octopi, and eels from the intertidal flat. The four kinds of marine products most commonly caught are fish from the families Siganidae (spinefoot/rabbitfishes), Callionymidae (dragonets/mandarinfish), and Labridae (wrasse) as well as common varieties of crabs. Prices for spinefoot/rabbitfishes in three Metro Cebu wet markets ranged from PHP 70 to PHP 100/kg, at the time of the survey (2002). On the other hand, dragonets were selling for an average of PHP 35/kg, and crabs and wrasse, for an average of PHP 40/kg.

The average annual revenue from fish catches from the Cordova flat for each fisherman ranged from PHP 33,896 to PHP 46,608, in 2002 nominal pesos. Aggregating over the 1300 fishers working in the area, this sums to an annual fish revenue of about PHP 44 million to PHP 61 million (USD 786,000 to USD 1.1 million). As mentioned previously, this is a lower bound estimate.

Taking into account all fishing-related costs at constant 1997 prices—including an imputed shadow wage rate of about PHP 9 per hour (which is 35 % of the legislated minimum for areas outside Metro Manila)—the annual net fishing income from the Cordova flat is estimated to range from PHP 17.2 million to PHP 29.9 million. Cordova fishermen worked an average of 8 h/day, which translates to a shadow wage rate of about PHP 73, an amount approximately the same level as that paid to workers in the area whose education is high school and below.

To determine the benefits from on-site fishing that will be forgone if the CRP goes ahead, it was necessary to first set up a "without the project" scenario. For this study, it is assumed that if the reclamation project does not go ahead, then the Municipality of Cordova would continue to implement its current Coastal Resource Management Plan (CRM), which serves dual purposes. Firstly, it lays the basis for estimating the size of future harvests. Secondly, it highlights the alternative policy directions available to the local government.

A review of data on fish trap yields in Sumilon Island—another site in Cebu province where a form of coastal resource management has been implemented since 1977—suggested that a 5 % annual increase in fishery production (and gleaning harvest) is a realistic assumption to make for the area. In their article on the experience of coastal resource management efforts in Sumilon Island, Russ and Alcala (1994) reported positive changes in the yield of reef fish from traps throughout the period 1976–1986, ranging from 6.25 % to 38.30 %. The authors attributed the increases in yield to biomass spillover from the marine reserve established in the area. Projecting over a 30-year period, a 5 % annual increase in fish production is plausible given that improvements begin from a very low production baseline in what was formerly a highly productive coastal area.[4]

Taking this into account, a 5 % annual increase in on-site fishery production for the 30-year period is assumed in the "without the project" scenario, and half of the net income from on-site fishing is assumed to be forgone or lost in the "with project" scenario. As previously discussed, the other forgone benefits from on-site fishing caused by the CRP are assumed to take the form of incremental fishing costs (e.g., fishermen who shift to more distant sites). These are estimated as the amortized capital costs of acquiring a motorized fishing vessel and the additional fuel expenses for longer fishing journeys. Using these assumptions, the net present value (NPV) of the on-site fishery, at an 8 % discount rate, is estimated at approximately PHP 290 million or USD 5 million over a 30-year period.

11.3.2 Reef Flat Gleaning

Reef flat gleaning refers to an activity that is relatively distinct from fishing and which generally involves collecting marine products (normally by hand), such as shellfish, mollusks, sea cucumbers, sea urchins, and others. It is a highly labor-intensive activity that is usually done on the intertidal flat, adjacent to the coast.

Most reef flat gleaning activities undertaken by Cordova residents are for food or to supplement family incomes. None of the respondents reported reef flat gleaning as their primary means of income, although 34 % reported gleaning as their secondary type of income. Of this 34 %, a majority (90 %) concentrate their gleaning efforts on the Cordova intertidal flat. It is estimated that 1500–1800 people engage in some form of gleaning, and the majority are women. The south eastern coast has the highest number of households in which at least one member is engaged in some form of gleaning activity. Since the products of reef gleaning are also sold in markets, a price-based approach to valuation is also applied.

Each gleaner harvests approximately 460 kg of marine products each year. At an average price of PHP 20/kg (USD 0.36/kg), this amount is worth about PHP 9,180 (USD 164) annually. This roughly translates to an aggregate value—net of imputed labor costs—between PHP 8.9 and PHP 10.9 million (USD 160,000–195,000) annually. The value of the loss of reef flat gleaning is taken to be the value of the total loss of the harvest (less harvest cost). Using the same "without the project"

[4] A sensitivity analysis could be done at lower and higher annual rates of fish production.

scenario as used for estimating forgone on-site fishing benefits, the net present value of reef gleaning products, over a 30-year period at an 8 % discount rate, is estimated to be approximately PHP 207 million (USD 3 million).

11.3.3 Recreational Value of Affected Coral Area

The coral reefs around Mactan Island are considered relatively unique on a global scale because they are adjacent to Cebu, a major urban area that is home to almost two million people. The reefs are said to have significant economic potential (Ross et al. 2003). As such, the forgone recreational value of the coral area is an important factor in the costs of reclamation. The length of the eastern coast of Mactan Island is the site of various medium- to high-priced resorts that cater to both domestic and foreign tourists. Across the Gilutongan Channel are several smaller islands that are only 15–30 min away from Mactan Island by motorboat. These islands are a popular destination for domestic and foreign "island hopping" tourists.

The results of an underwater assessment, conducted as part of this research, revealed that the coral reefs of Cordova are typical fringing reefs, which are generally in fair condition with high biological diversity. Live coral cover ranges from approximately 12 % to 70 %. Most of the coral life forms are of the encrusting, branching, and massive types. Good coral populations were found to be located only in the southern reef edge. The coral reefs are poorly developed in the south western part of the study area, which faces Cebu Harbor, across a narrow channel. This poor coral cover has been attributed to the area's high levels of suspended particles in the water column (USC 1998) and to pollution from industries, household effluent, and oil discharges from Cebu City. The total area of the coral reef adjacent to the Cordova mainland is estimated to cover 9.35 km^2 or 935 ha. Of this area, an estimated 640 ha of coral in the south and south eastern portion of the reef flat is classified as having fair to good cover and could potentially be developed for recreational diving and snorkeling. It is this coral area that would be affected by the proposed reclamation project.

A comparison of the coral reef fringing the Cordova mainland with the coral area within the 14 ha Gilutongan Marine Sanctuary (mentioned in Sect. 11.1) shows that both can be classified as fair and are relatively similar in terms of live coral cover. The Gilutongan Sanctuary is a very popular dive and snorkeling destination with approximately 3367 visitors/ha/year, or more than nine visitors/ha/day. Since 2000, the Cordova local government has been collecting accreditation fees from dive operators and user fees from visitors. Around 30 % of the revenues from fees go to the local barangay, while 70 % go to the municipality, which assumes most of the operation and management costs. Annual revenues are now close to PHP 2 million and contribute to a very substantial portion of the local government budget.

The value of forgone recreational benefits from the 640 ha coral reef area that would be affected by the reclamation work was estimated by applying the unadjusted unit-value benefit transfer. Two key assumptions were made concerning the "without the project" scenario. First, in lieu of reclamation, Cordova is assumed to manage the relatively healthy coral areas along its southern reef edge in ways similar to its current management of the Gilutongan sanctuary. Second, projections of annual visitor density for the next 30 years are assumed to mirror the historical growth of visitors to the Gilutongan sanctuary (per hectare). These projections were then multiplied by the mean willingness to pay estimate for coral areas in Mactan Island reported by Arin and Kramer (2002). Using a "payment card" elicitation format, Arin and Kramer (2002) sampled 129 dive tourists at three Philippine dive destinations, including Mactan Island, and found that the mean amount tourists are willing to pay to enter a coral area is USD 5.5. If, as suggested by the Gilutongan experience, full capacity for a coral reef area is in the range of 3500–4000 visitors/ha/year, then a 640 ha coral reef at full capacity can accommodate 2.2–2.5 million visitors annually. Based on the projected number of visitors within a 30-year period, the net present value of forgone recreational benefits—with the reclamation project—provided by the coral reef is estimated at PHP 1.7 billion (USD 31.6 million).

11.4 Estimating the Damage from Landfill Quarrying

Landfill quarrying is another important channel through which reclamation affects the environment. Valuation efforts were focused on the impact of landfill quarrying on (1) surface water regimes (e.g., changes in surface water runoff and flood/peak volumes) and (2) soil stability and soil erosion (e.g., the occurrence of landslides and mass wasting and increases in downstream sediment transport). The cost of mitigating the adverse impacts of landfill quarrying on these two environmental aspects was taken to be the value of these impacts. Although mitigation cost is far from ideal as a method of valuing environmental damage, it can still provide a reasonable and accessible approximation.

The experience of the 300 ha Cebu South Reclamation Project (CSRP), the most recently completed reclamation project in Metro Cebu, was examined in order to estimate the environmental costs of landfill quarrying. It was assumed that the Cordova project would use similar types of landfill material and sources as the CSRP. The primary landfill material used for the CSRP was diorite sourced from small-scale quarries (defined by Philippine law as quarries of less than 5 ha) in the upland areas of Cebu Province and the nearby islands of Leyte and Bohol.

Due to data and resource limitations, the impact study of quarrying was limited to examination of ten quarry sites. All ten sites examined are located in Naga, a town in the southernmost portion of Metro Cebu and the place from which most of the landfill material for the CSRP was sourced. The topographic, geological, and hydrological makeup of Naga is typical of upland areas in Cebu. An expert assessment of the physical impacts of the quarrying, as well as the type and extent of the mitigation measures, was carried out for each of the ten sites. The mitigation measures consisted of (1) planting vetiver grass at each of the quarry sites and

(2) the construction of gabion dams (check/protection dams) to minimize downstream impacts.

The cost of implementing mitigation measures for each of the ten quarry sites was estimated, and this calculation was in turn used in the mitigation cost estimates for the landfill quarrying linked to the Cordova project itself.[5] The initial cost of implementing mitigation measures for all ten sites studied was estimated at PHP 9.4 million (USD 230,700) or an average of PHP 1.3 million (USD 23,000) per quarry site.

To maintain the effectiveness of installed gabion dams, regular maintenance is required. Maintenance estimates were adapted from actual per-unit construction and maintenance costs reported for four existing gabion dams located in the uplands of Metro Cebu. The cost of maintenance for the ten sites under review is estimated at some PHP 188,800 or USD 3370.

The Cordova reclamation—based on the MIIC proposal—will need approximately 170 million m^3 of landfill material, some 13 times more than the volume used for the CSRP. It will therefore require an average of 12 million m^3 of landfill annually for the entire 14-year construction period. A project of such scale would require about 1100 small-scale mountain quarries to fulfill its landfill requirements if, as suggested by the Naga case, each site produces an average of 150,000 m^3. That translates to around 81 new quarry sites per year. It is reasonable to expect that these quarry sites will be located within Cebu province and on neighboring islands in the region in order to minimize transport costs.

In calculating the potential mitigation costs for the damages that would result from the increase in quarrying activities necessary to supply the Cordova project, the following assumptions were made: (1) the landfill would be sourced from small-scale quarries similar to CSRP; (2) gabion dams would be installed immediately upon the start of quarry operations; (3) vetiver grass would be planted immediately after the site is abandoned or exhausted; and (4) the gabion dams would be maintained regularly for the entire 30-year period. At a discount rate of 8 %, the net present value of these mitigation costs is estimated at more than PHP 1 billion or around USD19 million.

[5] If the mitigation costs are borne by the developer, they would be counted as private costs in a benefit-cost study, not social costs. However, in this context, they represent damages that would not be incurred without the project. If the developer did not pay the costs, they would be included as social costs of the project.

11.5 Environmental Costs of the Proposed Cordova Reclamation

The environmental cost of the Cordova reclamation, in present value terms, is estimated at about USD 59.8 million or PHP 3.3 billion. This is equivalent to almost 15 % of the USD 404 million (approximately PHP 22 billion) direct reclamation development costs reported in the MIIC reclamation proposal. Almost 86 % of the total environmental cost is attributed to damages from landfill quarrying and to the coral reef. The estimated value of damage to corals alone makes up some 53 % of the total. The quantified and monetized impacts are summarized in Table 11.1. These numbers are lower bound estimates given that the analysis excludes other environmental impacts such as pollution arising from construction and economic activities on reclaimed land or valuing damages using willingness to accept compensation by the affected parties rather than willingness to pay.

11.6 Economic Benefits from the Proposed Cordova Reclamation

There are two types of benefits expected to be derived from the reclamation project: private financial benefits and social benefits. It is important to distinguish one from the other. Private financial benefits accrue to the private investor and developer who earn revenue from the sale of newly created land. These financial gains from the sale of land are not counted as benefits to society as a whole. For the Cordova project, it is estimated that the net present value of private financial benefits would be positive and range from PHP 41.5 billion (USD 741 million) to PHP 81.9 billion (USD 1.5 billion) at 1997 prices (MIIC 1997).

Social benefits, on the other hand, refer to the benefits that accrue to society as a whole. These benefits are complex and depend on the share of land that goes to the region, how it is used, levels of incremental tax revenue, net new jobs, whose benefits are being measured, and so on. The MIIC proposal did not define or report these social benefits, so they could not be included in the present research.

11.7 Concluding Remarks

This research illustrates the urgency of systematically incorporating explicit measures of environmental costs into the appraisal process of large-scale coastal reclamation projects. It shows that these costs are not trivial. Their inclusion may affect conclusions relating to the net social benefits of such projects. Furthermore, these costs may represent large losses from the perspective of less-advantaged members of society. Knowing what and how large these environmental costs are

Table 11.1 Estimated environmental costs of the proposed Cordova Reclamation Project

Impact	Present value (30 years, 8 % discount rate)	Valuation approach
Loss of on-site fishery	USD 5 million	Forgone fishing rents
Loss of reef flat gleaning	USD 3.7 million	Forgone net income
Damage to corals	USD 31.6 million	Forgone recreational benefits
Damage from landfill quarrying	USD 19.5 million	Mitigation costs
Total	USD 59.8 million	

can help guide the design and implementation of such reclamation projects in a way that lessens their impacts.

The total potential environmental costs are estimated at almost USD 60 million (PHP 3.3 billion). It is thought that this is a lower bound figure since it does not include the various negative externalities that arise temporarily from reclamation construction. It also does not include the impact of increased levels of pollution caused by economic activities on the created land. Of the four categories of environmental costs valued in this study, damage to corals and externalities from landfill quarrying are the most significant. The avoidance of damage to the coral reef would alone reduce estimated environmental costs by more than half.

Small-scale quarrying for landfill does not necessarily translate to small-scale impacts and costs. It is also clear that the number of such small-scale quarries—estimated at more than a thousand in this study—needed to fulfill the CRP's landfill requirements may be much too large and widely dispersed for an environmental agency to consistently monitor, inspect, and regulate them. This would be particularly true if this responsibility fell solely on an understaffed local or provincial authority.

The study estimates the forgone benefits from the loss of on-site fishing, reef gleaning, and recreational tourism from the coral reefs. Losses are most significant for the barangays located on the town's southern and south eastern coast. The losses associated with coral damage were larger than those for fishing and reef gleaning. However, this study concludes that the damages to fishers and gleaners would nevertheless constitute a significant loss of livelihood for thousands of Cordova residents. Losses from tourism, on-site fishing, and reef gleaning could be minimized by avoiding the reclamation of the southern and south eastern coastal area where there is good coral cover and the bulk of fishing and gleaning households are located. Although reclamation work on a smaller scale may yield smaller economic benefit, it would require smaller landfills and consequently, incur smaller external costs. The potentially high environmental cost associated with the loss of coral also suggests an alternative development strategy that the municipality could pursue: one that has more to do with sustainable ecotourism than with attracting new industry.

References

Arin T, Kramer RA (2002) Divers' willingness to pay to visit marine sanctuaries: an exploratory study. Ocean Coast Manag 45:171–183

GSECI (Ground Structures Engineering Consultants, Inc) (2002) Final environmental impact assessment report: Peters' Mine Concession, Georgetown

Hwang M (1999) Coastal land-use change by reclamation of tidal flats along the western coast of the capital region in Korea. Seoul National University, Seoul

Jiao JJ (2000) Preliminary conceptual study on the impact of land reclamation on groundwater flow and contaminant migration in Penny's Bay. Department of Earth Sciences, University of Hongkong, Hongkong

Kawabe M (1998) To enhance the environmental values of Tokyo Bay – a proposition for integrated coastal zone management. Ocean Coast Manag 41:19–39

MIIC (Malayan Integrated Industries Corporation) (1997) Final feasibility study report for the Cordova reclamation and development project (executive summary). Cordova Mayor's Office, Cordova

Moores N, Bräunlich A (1999) The saemankum project. Common Wadden Sea Secretariat (CWSS), Wilhelmshaven

Pearce D, Turner K (1990) Economics of natural resources and the environment. John Hopkins University Press, Baltimore

Ross M, Ross N, Green S, Amores A, Carina J, Menguito T (2003) Experience from improving management of an "Urban" marine protected area: Gilutongan Marine Sanctuary, Municipality of Cordova, Cebu, Philippines. World Resources Institute, Washington, DC

Russ GR, Alcala AC (1994) Sumilon Island reserve: 20 years of hope and frustrations. ICLARM Q 17(3):8–12

Schema Konsult Inc (1996) Mactan Island integrated master plan study. Volume 1, Draft final report. 235 p

USC (University of San Carlos) (1991) Environmental impact assessment of the proposed reclamations in Lapulapu, Cordova and Consolacion. University of San Carlos – Area Research Training Center, Cebu City, Philippines

USC (University of San Carlos) (1998) Baseline marine resource assessment and monitoring for the proposed Cordova Reclamation Project. Marine Biology Section, University of San Carlos, Cebu City

World Bank (2000) Philippines environment monitor. World Bank Country Office, Manila

Part III
Institutions and Policies in Fisheries Management

Alice Joan G. Ferrer

The Importance of Fisheries in Southeast Asia

Fishery resources are important to people as a source of food and livelihoods. Fish is an important source of animal protein, minerals, and essential fatty acids for about 950 million people worldwide (USAID 2005). In 2009, fish accounted for 16.6 % of the world's intake of animal protein and 6.5 % of all protein consumed (FAO 2012). As of 2010, about 55 million people depended on the primary production sector of the fisheries (FAO 2012).

World fishery production is growing at a steady pace, owing primarily to the increase in aquaculture, which compensates for the stagnation and decline in capture fisheries that started in the 1990s (FAO 2012; Dey and Kanagaratnam 2007). From 1961 to 2009, the annual rate of average production grew by 3.2 %, while the annual population growth for the period was 1.7 %. In 2010, fishery production was registered at 148 million tonnes and valued at USD 217.5 billion, with 128 million tonnes for food (FAO 2012). Moreover, employment in fisheries grew 2.1 % annually from 2005 to 2010, outpacing the annual population growth of 1.2 % during the same period (FAO 2012).

Asia, the most populous region of the world, is the biggest producer and consumer of fish. It accounts for about two-thirds of the global consumption (FAO 2012). In 2009, per capita consumption in Asia was estimated to be 20.7 kg. Asia is home to about 87 % of the world's population that is dependent on fisheries for livelihood and as source of income. The region is also most significant in terms of small-scale fisheries employment, output, and economic value (FAO 2014).

A.J.G. Ferrer (✉)
Division of Social Sciences, University of the Philippines Visayas, Miagao, Iloilo, Philippines
e-mail: aj_ferrer2005@yahoo.com

To many Southeast Asians, fisheries are integral to their way of life. Annually, about 21 million tonnes of fish production—one-quarter of the total global production—takes place in the 10 member states of the Association of Southeast Asian Nations (ASEAN) that include Indonesia, Malaysia, the Philippines, Singapore, Thailand, Brunei, Vietnam, Laos, Myanmar, and Cambodia (Garces et al. 2008). In 1995, the "center of marine biodiversity" was identified as being in Southeast Asia (SEA), particularly in the Philippines and in Indonesia (Carpenter and Springer 2005); these two countries are the leading capture fishery producers in the region. Komatsu (2013) reported that as of 2009, Indonesia ranked first in total fish production volume in Southeast Asia, accounting for 33.8 % of the region's total, while the Philippines ranked a far second at 16.3 %. The next in the top 5 producing countries are Vietnam (15 %), Myanmar (13.8 %), and Thailand (10.9 %).

The trend for SEA nations has been a decline in production from coastal areas, while operations in the high seas as well as in the coastal areas of adjacent nations have been expanding (Komatsu 2013). It was estimated that in South Asia and SEA, coastal fish stocks have declined 5–30 % over the last five decades, negatively impacting fisher incomes, fisheries employment, revenues, trade, and social stability (ADB 2014). With declining production and increasing population, food security in Southeast Asia is threatened.

Key Issues Affecting the Fisheries in Southeast Asia

A number of challenges threaten the future potentials of fishery resources that are vital to food security, SEA economies, and the well-being of coastal communities in SEA. The key management issues and challenges facing the coastal fisheries in SEA are multiple and interrelated. Various sources (such as Silvestre and Pauly 1997; Silvestre et al. 2003; Salayo et al. 2006; Garces et al. 2008; ADB 2014) identified some of these management issues and challenges. The management issues include (1) inadequate information and research, (2) institutional weakness and constraints, (3) inadequate infrastructure, (4) inadequate fisheries and development policies, (5) weak interagency coordination, and (6) weak law enforcement and poor compliance to regulations. The challenges, meanwhile, include (1) overfishing; (2) overcapacity; (3) habitat and environmental degradation; (4) intersectoral and intrasectoral conflicts; (5) destructive fishing; (6) poverty among small-scale fishers; (7) illegal, unreported, and unregulated fishing; (8) increasing population; (9) inequitable distribution of benefits; (10) climate change; (11) rising fuel cost; (12) globalization of trade and market access; (13) pollution (particularly domestic waste); and (14) coastal and aquaculture development. Dominating these issues are depletion and degradation of coastal fishery resources, mainly caused by overfishing, property rights disputes, and weak governance.

Need for Effective Fisheries Management

As early as the 1990s, when world fishery production began to stagnate, many called for more effective environmental management (FAO 2001–2014). Existing fisheries management tools and strategies implemented in SEA can be classified as input controls (limits on the hours or days when fishing is permitted, limits on the number of fishers, licenses or permits, surveillance efforts on fishing activities, technology limits, bans on some fishing gear, boat size limits, engine horsepower limits, bans on the use of multiple gear per boat, protection of critical fish habitats, bans on fishing in spawning aggregation areas, zoning, and allocation of fishing areas), output controls (catch quotas or total allowable catch, fish size limits, limiting bycatch,[1] and discards), conservation measures (seasonal closures, fishing bans related to reproduction of fishes or migration runs, fish habitat restoration, stock enhancement and restocking, and ban on species), subsidies (by government, such as free gear or boats, discounted gas prices, tax cuts, and gear buyback programs), and traditional fisheries management, which designates "sacred areas" (ADB 2014: 152).

The Role and Importance of Institutions and Policies in Fisheries Management

Fishery resources are common-pool resources characterized by a lack of or limited excludability and rivalry or subtractability (Ostrom et al. 1994). That is, they are used commonly by all people (limited excludability), but the fish caught by one fisher is no longer available to another fisher (rivalry). Without effective institutions to manage fishery resources, utilization becomes uncontrolled and exploitative, leading to overharvesting and declining quality.

"Institutions" is defined as:

> The set of working rules that are used to determine who is eligible to make decisions in some arena, what actions are allowed or constrained, what aggregation rules will be used, what procedures must be followed, what information must or must not be provided, and what payoffs will be assigned to individuals dependent on their actions. (Ostrom 1990: 51)

That is, "rules, norms or strategies create incentives are constituted and reconstituted by human interaction in frequently occurring or repetitive situations" (Crawford and Ostrom 1995: 582). Rules can be selected and enforced by the government to guide behavior of the people in a given situation or developed and enforced by the people themselves. Considering the rules, individuals adopt strategies that have consequences not only for themselves but also for others. As

[1] Bycatch refers "to that part of the catch which is not the primary target of the fishing effort. It consists of both fish which is retained and marketed (incidental catch) and that which is discarded or released" (Clucas 1997).

individuals become more knowledgeable of the actions of everyone in a situation, they may change norms and strategies, leading to changes for themselves and the relevant environment (Crawford and Ostrom 1995; Ostrom 2000, 2003).

Institutions are sometimes known as the "rules of the game" (Torell and Salamanca 2002) or "rules in use" (Rudd 2002) that influence incentives and behavior. They can be formal (laws, regulations, policy, or procedures) or informal (social norms or habits) (Rudd 2002). Formal institutions require a third party for enforcement, while informal institutions rely on mutual agreements or relations of power and authority between actors in the social group (Torell and Salamanca 2002).

"Organizations" are sometimes referred to as "institutions" (Imperial 1999). An organization, however, is different for it is defined as a "set of institutional arrangements and participants who have a common set of goals and purposes, and who must interact across multiple action situations at different levels of activity" (Polski and Ostrom 1999: 4). Organizations can be viewed as the "players" of the "rules of the game" they have developed (FAO 2014).

Institutions are important in management because they define the rights and rules of resource use; they determine access by groups of people, members of an organization, or individuals in a community. Rights are actions that are authorized, while rules refer to prescriptions that forbid, permit, or require acts performed in relation to a right (Ostrom 1990).

Analysis of the institutions underlying fisheries and coastal resources management in SEA is important. Improved management and efficiencies in the fisheries sector require formal institutions that promote sustainable practices and behaviors to protect the resources and the people dependent upon them. This is a challenge for institutional entities with authority and responsibility for the management and utilization of resources.

Tools and Institutions in Fisheries Management

Five studies undertaken for Economy and Environment Program for Southeast Asia (EEPSEA) in the Philippines (Javier 2003; De Guzman 2004; Ferrer 2009), Malaysia (Sidique et al. 2014), and Thailand (Pornpinatepong et al. 2013) all focused on tools and policies to achieve sustainable management of fisheries and coastal resources. The tools include designating marine protected areas (Javier 2003; De Guzman 2004); employing a set of management tools including marine protected area (Ferrer 2009); establishing artificial reefs (Sidique et al. 2014); and setting fishing effort controls, including individual transferable quotas (ITQs) (Pornpinatepong et al. 2013). The institutional arrangements implementing these tools were also examined. There are four broad types of institutional arrangements used to implement the tools and manage the fisheries (Imperial and Yandle 2005): (1) the "leviathan" or centralized bureaucratic arrangement or command-and-control model based on government regulation, (2) market-based arrangements, (3) community-based arrangements that rely on self-regulation of fisheries by communities and user groups, and (4) comanagement, which relies on shared management between government agencies and user groups. This section briefly introduces the EEPSEA studies.[2]

Marine Protected Areas

Designating a marine protected area (MPA) is a common management tool for fisheries management in Southeast Asia. As defined, an MPA is "any area of intertidal or subtidal terrain, together with its overlying water, and associated flora, fauna, historical and cultural features, which has been reserved by law to protect part or all of the enclosed environment" (International Union for the Conservation of Nature [IUCN] 1994; Kelleher 1999). There are evidences of positive biological impacts of MPAs (Adan 2004; Alcala et al. 2004; Dalby and Sorensen 2002; Gell and Roberts 2003; Hilborn et al. 2004; Razon et al. 2012; Russ et al. 2003, 2004; White et al. 2006).

The EEPSEA study by De Guzman (2004) provides further evidence on the positive impacts the community-managed marine reserve in Baliangao, Misamis Occidental, has had in enhancing the ecological conditions and the economic profitability of the fishery in Danao Bay, Philippines. Biological data collected for the study show improved habitat quality and greater biodiversity and biomass in the marine reserve and adjacent areas. Also, the study documents increases in fish catch per unit of effort and income of fishers who fish in areas adjacent to the marine reserve. This information is a significant contribution to the literature on marine protected areas, which, according to Christie (2004), is dominated by biological studies and lacks data on the social impacts.

In the Philippines, the 1998 Fisheries Code stipulates that at least 15 % of the municipal waters[3] should be reserved for fishery refuge or sanctuaries. The most recent data indicates there are more than 1,800 MPAs in the country (Cabral et al. 2014); most, however, are small. According to Weeks et al. (2010), 90 % of the 985 MPAs in the Philippines covered 14,943 km^2 as of 2008, but at less than <1 km^2 each, and the combined no-take area (1,459 km^2) in Tubbataha Reef and Apo Reef natural parks constitutes 85 % of the total no-take area. Moreover, managing these marine protected areas sustainably is not just a challenge; many management systems are altogether dysfunctional.

The issue of functionality and sustainability of MPAs demands finding the best institutional arrangement. This question is the main focus of the EEPSEA study by Javier (2003). Javier's study focuses on the role of governing institutions in biodiversity conservation in selected marine protected areas in the Philippines. It is one of the earliest studies to fill the information gap on the relationship between

[2] Visit www.eepsea.net for the full research reports of the studies.

[3] Water area within 15 km from the shoreline

type of institutional arrangement and effective resource management. The study evaluates three different institutional arrangements: Department of Environment and Natural Resources/nongovernmental organization-led MPAs, local government/NGO-led MPAs, and community-managed MPAs. It looks at five essential property rights: use rights, exchange rights, distribution rights, management rights, and rights to authority with the goal of identifying the "governing institution" most effective at managing MPAs. Key performance indicators (biophysical, socio-economic, and institutional) were used to measure the impacts of management schemes on the performance of the marine protected areas. These indicators were used to rank the criteria for evaluating MPAs. The result was a composite index for determining the best MPA among the study sites. This composite index was determined through the arbitrary ranking of selected key performance indicators as ranked according to highest value (5) and lowest value (1). The study concluded that an MPA can perform well, regardless of type of governing institution, if there is effective implementation of the conservation plan, strict enforcement of laws, and efficient sourcing of implementation funds. The study contrasts the commonly held idea that greater efficiency can be achieved and undesirable distributional implications prevented with comanagement or community-based resource management as the sole management tool (USAID 2005).

Single Versus Combinations of Management Options

This brings to light the fact that MPAs, or any single management tool, cannot be viewed as a singular panacea to overfishing and other threats to the fishery resource, neither can any single management tool restore fishery health nor ensure the sustainability of a coastal fishery (De Guzman 2004; Javier 2003); an MPA has to be implemented with other complementary policy instruments and fisheries management strategies if it is to bring about positive net benefits. Similarly, Ferrer (2009) evaluated 10 potentially viable management options that address multiple threats (overfishing, degraded resources, low income, weak law enforcement, and inappropriate management) to the fisheries of the Visayan Sea. The study aimed to develop more effective fisheries management and help the fishery and fishers who depend on the resource. The study highlights the trade-offs between options and recommends implementation of MPAs in combination with other tools and strategies by a single management body.

Reducing Fishing Effort

Reducing fishing effort would allow the fishery resources to rest, populations of exploited species to recover, and the habitats modified by fishing to regenerate. In Thailand, Songkhla Lake is characterized by overfishing and stock depletion leading to declining catch and fishing incomes. Fishing effort control is a potential solution; in what way fishing effort could be reduced is the question. To provide evidence to the effectiveness of this regulation, Pornpinatepong et al. (2013) investigated fisher behavior using game theory in response to five combinations of management policy scenarios: open access, external regulations with individual transferable quotas (ITQs) or individual quotas (IQ), and comanagement with ITQs or IQs. Comanagement with ITQ was found to have the highest potential for maintaining fishery production sustainable level in Songkhla Lake.

Artificial Reefs

In Malaysia, the government introduced artificial reefs (ARs) to conserve and enhance natural fish stocks in marine waters in response to marine resources declining considerably over the past few decades, primarily due to overfishing. ARs were expected to rehabilitate degraded fish stocks and secure the livelihoods of the fishers along the coastal villages. Sidique et al. (2014) examined the actual impact of ARs on fisher households' income in Terengganu (districts of Besut, Setiu, and Kuala Terengganu). Three regressions were carried out for each different set of independent variables entered in the model: Model 1, basic catch value model, used only demographic and fishing effort variables; Model 2 incorporated the AR effect on the value of the catch and operating costs; and Model 3 considered district-level effects by including variables on fishing congestion and the number of large ARs. The study found the incomes of those fishing near the ARs are higher than those fishing far from ARs, suggesting the positive impact of ARs on the fishery resource.

Lessons for Fisheries Management

The fishery resources have inherent characteristics (limited excludability and rivalry) that warrant management. Fisheries management, however, is not easy and needs information that helps it adjust to changing conditions. The five EEPSEA studies provide valuable lessons that can inform policy to improve fisheries management:

1. Marine protected area benefits extend to neighboring areas, not only within the MPAs. Spillover of biomass to nearby areas of a marine protected area was documented by the study of De Guzman (2004).
2. Regardless of the governing institution, MPAs work under the right conditions. The study of Javier (2003) concluded that the type of institution governing the MPA does not matter—as long as it is efficient and effective in implementing its conservation plan, enforcing its laws, and sourcing its funds—MPAs can perform well. Furthermore, the study provides evidence that community-based

resource management schemes are no more effective than other institutional arrangements. The key elements for success of the MPA are good leadership, adequate human power to support technical aspects, monitoring, enforcement, availability of funding, and provision of livelihood support to the community.

Although De Guzman (2004) concluded that community-based management regimes can improve management of fishery resources, that one cannot generalize the effectiveness of a particular governing institution: one institution may perform well at one site, while another may do better at another site.

3. MPAs work well when they are networked and complemented by other management tools. While three of the EEPSEA studies (Javier 2003; De Guzman 2004; Ferrer 2009) show that establishing an MPA is far better than not having established one, its impact may not be large enough to create a truly sustainable and economically viable fishery area. They recommended networks of MPAs that are supported by other complementary fisheries management tools.
4. There are trade-offs in the use of a combination of management options. Ferrer (2009) provided evidence of the trade-off among the management options. The challenge is in finding the right mix of the management tools, which depends on the priorities set by the fisheries management.
5. Artificial reefs work, but work better when complemented by other management tools and work best when incorporated with MPAs. The study of Sidique et al. (2014) showed that the government policy to deploy ARs in Terengganu, Malaysia, was successful; ARs are impacting fisher incomes positively. However, the positive impacts of ARs will likely be undermined by the absence of clear demarcation of AR sites and the fishery remaining open access. Other management measures to complement the AR must be implemented to ensure sustainable use of fisheries in the AR sites.
6. Collaboration of stakeholders is important in fisheries management. Pornpinatepong et al.' (2013) study of Thailand provided evidence that comanagement of fishery resources leads to better results than external regulation, in terms of both sustainability of the resource and reduction of fishing resource extraction. Unfortunately, the comanagement process is more complex to manage. It requires greater cooperation among the fishers and strong solid support from the government. When left entirely to the community to manage, it may not succeed due to noncooperative behavior by other resource users who prefer open access to the resource.

References

Adan WR (2004) Multiple small-scale marine sanctuaries in municipal waters: the Magsaysay example. In: Turbulent seas: the status of Philippine marine fisheries. Coastal Resource Management Project, Department of Agriculture-Bureau of Fisheries and Aquatic Resources, Cebu City, pp 232–236

Alcala AC, Russ GR, Maypa AP (2004) Evidence for fishery enhancement effects of marine reserves in central Philippines. In: Turbulent seas: the status of Philippine marine fisheries.

Coastal Resource Management Project, Department of Agriculture-Bureau of Fisheries and Aquatic Resources, Cebu City, pp 215–218
Asian Development Bank (ADB) (2014) Economics of fisheries and aquaculture in the coral triangle. Asian Development Bank, Mandaluyong City
Cabral RB, Aliño PM, Balingit ACM, Alis CM, Arceo HO, Nañola CL Jr, Geronimo RC, Partners MSN (2014) The Philippine marine protected area (MPA) database. Philipp Sci Lett 7 (2):300–308
Carpenter KE, Springer VG (2005) The center of the center of marine shore fish biodiversity: the Philippine Islands. Environ Biol Fish 72:467–480
Christie P (2004) Marine protected areas as biological successes and social failures in Southeast Asia. Am Fish Soc Symp 42:155–164
Clucas I (1997) A study of the options for utilization of bycatch discards from marine capture fisheries. FAO fisheries circular no 928. FAO, Rome
Crawford SES, Ostrom E (1995) A grammar of institutions. Am Polit Sci Rev 89:582–600
Dalby J, Sorensen TK (2002) Coral reef resource management in the Philippines: with focus on marine protected areas as a management tool. University of Copenhagen, Botanical Institute, Department of Physical Ecology. MS thesis, 155 p
De Guzman AB (2004) A fishery in transition: impact of a community marine reserve on a coastal fishery in northern Mindanao, Philippines. EEPSEA research report no 2004-RR6. Economy and Environment Program in Southeast Asia, Singapore
Dey MM, Kanagaratnam U (2007) Community based management of small scale fisheries in Asia: bridging the gap between fish supply and demand. WorldFish Center, Penang
FAO (Food and Agriculture Organization of the United Nations) (2012) The state of world fisheries and aquaculture. FAO, Rome
FAO (2014) Small-scale fisheries, 2008–2014, website. Asia. FI institutional websites. In: FAO Fisheries and Aquaculture Department [online]. Rome. Updated. http://www.fao.org/fishery/topic/16634/en. Accessed 19 Jan 2014
Ferrer AJG (2009) Evaluation of fisheries management options for the Visayan Sea: the case of northern Iloilo. EEPSEA research report no 2009-RR5. Economy and Environment Program for Southeast Asia, Singapore
Garces LR, Pido MD, Pomeroy RS (2008) Fisheries in Southeast Asia: challenges and opportunities. In: Pandya A, Laipson E (eds) Transnational trends: Middle Eastern and Asian views. The Henry L. Stimson Center, Washington, DC
Gell FR, Roberts CM (2003) Benefits beyond boundaries: the fishery effects of marine reserves. Trends Ecol Evol 18(9):448–455
Hilborn R, Stokes K, Maguire JJ, Smith T, Botsford LW, Orensanz MMJ, Parma A, Rice J, Bell J, Cochrane KL, Garcia S, Hall SJ, Kirkwood GP, Sainsbury K, Stefansson G, Walters K (2004) When can marine reserves improve fisheries management? Ocean Coast Manag 47:197–205
Imperial MT, Yandle T (2005) Taking institutions seriously: using the IAD framework to analyze fisheries policy. Soc Nat Resour 18:493–509, http://www.envs.emory.edu/faculty/YANDLE/S%26NR-July2005.pdf. Accessed 15 Jan 2014
IUCN (International Union for the Conservation of Nature) (1994) Guidelines for protected area management categories, CNPPA with the assistance of WCMC. IUCN, Gland/Cambridge
Javier MEP (2003) Do institutions affect the performance of marine protected areas? Evidences from the Philippines. EEPSEA research report no 2003-RR5. EEPSEA, Singapore
Kelleher G (1999) Guidelines for marine protected areas. IUCN, Gland/Cambridge, UK, xxiv +107 pp
Komatsu M (2013) Special study on sustainable fisheries management and international trade in the Southeast Asia and Pacific Region. ADBI working paper 438. Asian Development Bank Institute, Tokyo
Ostrom E (1990) Governing the commons, the evolution of institutions for collective action. Cambridge University Press, Cambridge/New York
Ostrom E (2000) Collective action and the evolution of social norms. J Econ Persp 14(3):137–158

Ostrom E (2003) How types of goods and property rights jointly affect collective action. J Theor Polit 15(3):239–270
Ostrom E, Gardner R, Walker J (1994) Rules, games, and common pool resources. The University of Michigan Press, Ann Arbor
Polski MM, Ostrom E (1999). An institutional framework for policy analysis and design. In: Workshop in political theory and policy analysis working paper W98-27. Indiana University, Bloomington, Feb 1999
Pornpinatepong K, Chantarasap P, Seneerattanaprayu J, Hemtanon W, Saelim P (2013) Response of fishermen to fishing control policies in Southern Songkhla Lake, Thailand: a field experiment. WorldFish (ICLARM) – Economy and Environment Program for Southeast Asia (EEPSEA), Laguna
Razon BC, Liao LM, Nakagoshi N, ナカゴシ, ノブカズ, 中越信和 (2012) Success and failure of marine protected area management affecting the fish catch by adjacent fishermen in Sarangani Bay, Mindanao, Philippines. South Pac Stud 33(1):1–23
Rudd MA (2002) An institutional framework for designing and monitoring ecosystem-based fisheries management policy experiments. http://dlc.dlib.indiana.edu/dlc/bitstream/handle/10535/136/Rudd_%28OMRN%29.PDF;
jsessionid=AC1572CF3FE616BF44E36039BC3E7A77?sequence=1. Accessed 30 Dec 2013
Russ GR, Alcala AC, Maypa AP (2003) Spillover from marine reserves: the case of Naso vlamingii at Apo Island, the Philippines. Mar Ecol Prog Ser 264:15–20
Russ GR, Alcala AC, Maypa AP, Calumpong HP, White AT (2004) Marine reserve benefits local fisheries. Ecol Appl 14(2):597–606
Salayo ND, Ahmed M, Garces L, Viswanathan K (2006) An overview of fisheries conflicts in South and Southeast Asia: recommendations, challenges and directions. NAGA, WorldFish Center Quarterly, vol 29, no 1 and 2, p 11, Jan–Jun 2006. http://aquaticcommons.org/9481/1/overview.pdf. Accessed 14 Jan 2014
Sidique SF, Noh KM, Islam GMN, Noh AFM (2014) Economic impacts of artificial reefs: the case of fisher households in peninsular Malaysia. WorldFish (ICLARM) – Economy and Environment Program for Southeast Asia (EEPSEA), Laguna
Silvestre G, Pauly D (1997) Management of tropical fisheries in Asia: an overview of key challenges and opportunities. In: Silvestre G, Pauly D (eds) Status and management of tropical coastal fisheries in Asia. ICLARM conference proceedings 53, Manila, Philippines, pp 8–25, 208 p
Silvestre GT, Garces LR, Stobutzki I, Ahmed M, Santos RAV, Luna CZ, Zhou W (2003) South and South-East Asian coastal fisheries: their status and directions for improved management: conference synopsis and recommendations. In: Silvestre G, Garces L, Stobutzki I, Ahmed M, Valmonte-Santos RA, Luna C, Lachica-Aliño L, Munro P, Christensen V, Pauly D (eds) Assessment, management and future directions for coastal fisheries in Asian countries. WorldFish Center conference proceedings 67, Penang, pp 1–40, 1120 p
Torell M, Salamanca AM (2002) Navigating the institutional landscape: introduction and overview. In: Torell M, Salamanca AM (eds) Institutional issues and perspectives in the management of fisheries and coastal resources in Southeast Asia. ICLARM, Penang, ICLARM – The WorldFish Center and Swedish International Development Cooperation Agency (Sida)
USAID (United states Agency for International Development) (2005) Review of the status, trends and issues in global fisheries and aquaculture, with recommendations for USAID investments, report of USAID SPARE fisheries and aquaculture panel. USAID, Washington, DC
Weeks R, Russ GR, Alcala AC, White AT (2010) Effectiveness of marine protected areas in the Philippines for biodiversity conservation. Conserv Biol 24(2):531–540
White AT, Gomez E, Alcala AC, Russ G (2007) Evolution and lessons from fisheries and coastal management in the Philippines. In: McClanahan TR, Castilla JC (eds) Fisheries management: progress towards sustainability. Blackwell Publishing, Oxford, pp 88–111

Chapter 12
A Fishery in Transition: Impact of a Community Marine Reserve on a Coastal Fishery in Northern Mindanao, Philippines

Asuncion B. de Guzman

Abstract Establishing marine reserves or marine protected areas (MPA) is a popular management tool to ensure the sustainability of nearshore fishery against the threats of overfishing and other environmental impacts. Fish biomass in MPAs builds up rapidly in the absence of fishing and eventually contributes to fishable biomass in the areas surrounding the reserve through adult migration or "spillover." Improved ecological conditions allow MPAs to deliver more socioeconomic benefits to resource users and thus provide higher economic rent. Accordingly, this study investigated the impacts of a community-managed marine reserve in Baliangao, Misamis Occidental, on the fishery resources and fishers of Danao Bay over a 1-year period in order to demonstrate the ecological and economic benefits of a well-managed MPA.

The assessment of the biodiversity and physical conditions of the various ecosystems in Danao Bay showed improved coral and fish diversity. Monitoring of fish catch and fishing effort was conducted in six coastal villages across Danao Bay in order to obtain estimates of catch per unit of effort (CPUE), monthly and annual yields, and costs and revenues from capture fisheries. The results showed that fish production had moderate seasonality, with higher CPUE during the calmer summer months (February–May). To demonstrate whether or not the artisanal fishery of Danao Bay was generating rent, a cost-revenue analysis and estimation of the net and net present values of fishing activities were carried out. The estimated annual revenues suggest that some gear types are more profitable than others, known as that appears to be a function of CPUE and fishing costs. The calculated total annual net operating values (ANOV) and net annual values (NAV) for each gear showed differential profitability of fishing gear, with some earning zero or negative rents (e.g., gill net and handline) while a few others earning positive rents (e.g., fish corral). Net annual values are considered to be producer surplus, a profit over costs that accrues to the gear owner, and also represent the annual value of contribution of the marine ecosystem to the artisanal fishery of the bay. Summing up the NAV for all gear results in a range of values for total NAV, with a midpoint

A.B. de Guzman (✉)
Gaia Resource and Environmental Consultancy Services, Poblacion, Naawan, Misamis Oriental, Philippines
e-mail: sonydeguzman@gmail.com

N. Olewiler et al. (eds.), *Marine and Coastal Ecosystem Valuation, Institutions, and Policy in Southeast Asia*, DOI 10.1007/978-981-10-0141-3_12

of PHP 104,317.12 (at a 10 % discount rate). This sum indicates the amount of fishery rent generated by the bay's fishery resources. Many gears appeared unprofitable; however, the positive midpoints of the total NAVs indicate that some fishery rent is being earned, although it is probably below the rent that could be earned if the fishery were managed more efficiently. These results indicate that the coastal fishery surrounding the marine reserve in Danao Bay could be on a transition toward becoming a viable, sustainable industry.

Keywords Marine protected areas • Artisanal fishery • Biomass spillover • CPUE • Economic rent

12.1 Introduction

Tropical fisheries are characterized by such complexity, diversity (multi-gear, multispecies, many fishers), and vulnerability that its management is a significant challenge. The open access to most fishery grounds has resulted in overfishing and contributed to negative ecological and economic impacts on communities. Key nearshore fishery habitats, such as coral reefs, sea grasses, and mangroves, are particularly vulnerable to degradation because of their public accessibility.

In response to overfishing and other threats faced by the capture fisheries, marine reserves or marine protected areas (MPA) are established to ensure the sustainability of fish stocks that support reef fisheries. The concept of marine reserves is founded on the premise that fish population levels recover once fishing stops (Holland and Brazee 1996). Many fishery scientists believe that considering the alarming levels of overexploitation of many reefs, marine reserves may be one of the few management options available to maintain a critical spawning stock biomass needed to sustain reef fisheries (Roberts and Polunin 1993; Russ and Alcala 1996). The popular theory is that fish biomass builds up rapidly, in the absence of fishing in the MPA and, given limited space, eventually migrates to nearby fishing grounds. This biomass migration or "spillover" contributes to fishable biomass in the areas surrounding the reserve (Alcala 1999).

Data available on the ability of marine reserves to enhance coastal fisheries in surrounding areas through biomass spillover is meager and invariably relies on circumstantial evidence. The pioneering work of Russ and Alcala (1996) in Apo Island, central Philippines, provided some early evidence of spillover of adult fish from a reserve to fished areas. A study by Rodwell and Roberts (2000) also showed, through simulation, that full protection in Mombasa Marine National Park and Reserve would lead to an increase in total fish biomass and that the movement of adult and larval fish from the reserve would increase the total fishery catch of the surrounding areas.

Meanwhile, the most likely selling point of establishing an MPA as a fishery management tool is its potential to deliver socioeconomic benefits to resource users by sustaining fish stocks through improved ecological conditions. A major policy question that guides research in evaluating effectiveness of marine protected areas is: Do MPAs help enhance the fish catch of the surrounding fishing grounds? Enhancement of fisheries is understood as the generation of positive economic rent or profits by the fishers. Moreover, such fishery rent should be sustained over the long term, rather than generated in pulses against periods when profits are zero. Marine reserves help enhance the catch of fish population and, consequently, economic value of unprotected areas outside the reserves. This is achieved through (a) improved yields through larval recruitment and adult spillover, (b) decreased fishing costs resulting from reduced and focused fishing effort and the use of less capital-intensive fishing gear, and (c) improved management involving the community in protection and advocacy, thus increasing compliance with fishery regulations by fishers outside the marine reserve.

Sustaining profits, however, depends on the ability of fishery managers and local institutions to effectively regulate fishing activities within the fishing grounds. One management option that is gaining popularity among municipal fishers is to stop open access fishing by excluding nonresident fishers from the municipal fishing ground, issuing fishing permits, and establishing "no-take zones" and periodic closures to protect spawning and recruitment.[1] The establishment of MPAs can transition an open access scenario to effective fisheries in surrounding areas.

An economic analysis of fisheries in the surrounding areas of the MPAs would generate critical data to demonstrate to policy makers and local communities the necessity of protecting their coastal ecosystems. Although valuation methodologies are available, hardly any analysis has been carried out on marine reserves in the Philippines. It is equally important to analyze the existing institutional arrangements that lend support to the management of MPA projects and how these and social relations among various sectors of the fishing community can influence the success or failure of a project.

Studying the impacts of the Baliangao Marine Reserve in Misamis Occidental, on the fishery resources and fishers of Danao Bay is a means of demonstrating the ecological and economic benefits of a well-managed MPA. The research aims to (a) determine if the marine reserve contributed to the improved ecological condition of the bay, (b) evaluate the economic profitability of the municipal fishery, and (c) determine how existing institutional arrangements influenced the management of the marine reserve and the surrounding fishery.

[1] Recruitment is the entry of young fish into the population or fishing ground as a result of reproduction.

12.2 Methodology

12.2.1 *Baliangao Marine Reserve in Danao Bay*

Danao Bay is a shallow embayment of about 2000 ha fronting the Bohol Sea, along the northern shore of Misamis Occidental, and it straddles the municipalities of Baliangao and Plaridel. Around 2001, Danao Bay had an estimated 430 artisanal[2] fishers, where 260 work part time and 170 work full time. About 167 nonresident fishers, dubbed "strikers" (Heinen and Laranjo 1996), further intensify fishing pressure on the fish resources and are often suspected of engaging in illegal fishing operations.

The marine reserve in Baliangao was established in 1991 and was originally known as the Misom Sea Sanctuary under the partnership of the local government of Baliangao and Pipuli Foundation, Incorporated[3] and was renamed the Baliangao Wetland Park after the rich mangrove forest was incorporated into the protected area (Heinen and Laranjo 1996). It was declared a national protected area under the National Integrated Protected Areas System (NIPAS) Act of 1992 (RA 7586) on November 22, 2000. It is presently called the Baliangao Protected Landscape and Seascape (BPLS) and has a Protected Area Management Board (PAMB).

The marine reserve is an area of 74 ha. It is an integrated mangrove, sea grass, and coral reef ecosystem with an oval-shaped central or core area of about 5 ha (about 7 % of the total reserve area) and a 25 ha buffer zone. Strict "no fishing" regulations are imposed inside the sanctuary, while only the gathering of shellfish, sea cucumbers, and fish from the reef flats, by hand or by use of minor implements, is allowed within the buffer zone.

12.2.2 *Field Methods and Data Analysis*

Assessment of Biodiversity and Resource Uses The assessment of the biodiversity and physical conditions of the various ecosystems in Danao Bay includes repeated dive surveys of coral and reef fish communities and assessment of the contiguous habitats of sea grass beds and mangrove forests conducted between February 2001 and July 2002, following standard monitoring protocols (English

[2] Artisanal fishers are those engaging in small-scale fishing with the use of small boats with capacity of three gross tons or less often carried out in nearshore waters (in the Philippines, they are called municipal fishers).

[3] The Pipuli Foundation is an environmental nongovernmental organization (NGO) based in Misamis Occidental, dedicated to the conservation of natural resources through the empowerment of local communities. Pipuli was founded by Mr. Neil Fraser and started off by organizing agroforestry communities in the Mt. Malindang watershed. It later expanded to coastal resource management in Danao Bay and Murcielagos Bay. "Pipuli" is an indigenous Subanen term in Mindanao, meaning "to put back" or restore. It stopped its operation in the area in 2002.

et al. 1997). Analysis of data on live coral cover and diversity, abundance, and biomass of the reef fish populations (inside the marine reserve and in non-reserve reef areas) was done to determine if significant differences exist between these parameters.

For the economic assessment of the coastal fishery surrounding the marine reserve, the fish catch per unit of effort[4] (Alcala and Russ 1990; Rodwell and Roberts 2000) and income of municipal fishers who fish adjacent to the marine reserve were used as indicators. Monitoring of fish catch and fishing effort was carried out through a combination of participatory strategies: fishers recorded their catches on prepared data forms; fish buyers (*comprador*) were given logbooks; and members of the participatory monitoring team of the Danao Bay Resource Management Organization (DB-REMO) were engaged as local field enumerators to record daily catches of fishers using different gear as they arrived from the sea. More than 200 full-time and part-time fishers (about 50 % of the fisher population) were monitored in the six coastal villages across Danao Bay. Data from catch monitoring were used to estimate total monthly and annual yields (in metric tons) from the coastal fisheries.

A survey was completed by 180 respondents to obtain relevant data on demographics, economic status, livelihood options, and resource-use patterns of the community. Interviews with key informants (community leaders, heads of community organizations and nongovernment organizations, local government officials, and national government staff) were also conducted to identify existing institutional arrangements, current fishery management strategies, and policy formulations with regard to the management of the marine reserve. Secondary data on fish catch composition and catch rates, fishing effort levels, and other pertinent data were obtained from project reports of Pipuli and DB-REMO. All these information together with data generated by fishery monitoring were used to establish the "before" and "after" components of the marine reserve establishment.

Determination of Resource Rent A survey of costs and off-the-boat prices of fish caught by each gear type was conducted on a random sample of 150 artisanal fishers; costs included investment and operation costs of fishing (Trinidad et al. 1993). Investment costs included capital costs of the boat, engine, and gear, while operating costs were classified as fixed (i.e., licenses, salaries, and depreciation costs of the boat and engine) and variable costs included running costs (fuel, oil, ice, food, etc.), shore and marketing expenses, repairs, and maintenance. Total fishing costs were calculated by multiplying the price of the quantity consumed (Trinidad et al. 1993). The cost of labor was estimated using an average value of the prevailing daily wage rate in the locality to estimate the value of salaries paid to a fishing crew or to represent the value of family labor involved in the artisanal fishery business.

---

[4] Catch per unit effort is a measure of the amount of fish catch per unit time (e.g., kg catch per fishing trip) by one fisher.

Evidence that fish biomass spillover has indeed enhanced the adjacent fishery is indicated by the generation and maintenance of positive economic rent by the resource. To demonstrate whether or not the artisanal fishery of Danao Bay is generating rent, a cost-revenue analysis was carried out, followed by estimations of the net and net present values of fishing activities. This was accomplished through a stepwise process, according to the method outlined by Trinidad et al. (1993) and Gustavson (1999). Following the work of Gustavson (1999), the net present value (NPV) is considered as the current net value associated with the use of Danao Bay waters as a fishing ground, i.e., the contribution of marine biodiversity to economic production on an annual basis, over an infinite time stream. In other words, this value is tantamount to the "resource or economic rent" that can be sustained with proper fishery management.

12.3 Findings

12.3.1 Evidence of Ecological Improvements in Danao Bay

Results show an improvement in coral diversity and the quality of coral reef habitats, both inside and outside the reserve, since the 1997 assessment by Pipuli (Fraser 2003). Average live coral cover inside the marine reserve is higher (46.3 %) than in reefs outside it (36.9 %) although the difference is not significant (see Table 12.1). The high coral cover (>50 %) in certain parts of the reef slope at 10–12 m deep in both protected and unprotected sites is presumed to be one positive impact of ceasing blast fishing activities in the marine reserve area. This is similar to the findings in other MPAs in the Philippines (Russ and Alcala 1996; Deguit and Morales 2002; Reboton 2002).

Fish diversity also increased. A total of 246 species, belonging to 36 families of reef and reef-associated fish, were identified inside the Baliangao Marine Reserve within the year surveyed (2001–2002). Earlier resource surveys conducted by Silliman University (Heinen and Laranjo 1996) reported a steady increase from 48 species in 1993 to 85 species in 1995, accompanied by increases in fish densities. Six years later the fish diversity inside the protected reefs of Danao Bay had significantly increased to 246 species (Fig. 12.1), although population densities have fluctuated since 1995. The fish communities inside the marine reserve are more diverse than in the reefs outside, but this difference in species richness is not significant. Members of the DB-REMO observed that, despite its protected status, the BMR has suffered from sporadic fishing violations. Moreover, fishing activities right along the boundary of the reserve can readily capture fish spillover from the sanctuary. On the other hand, fluctuations in fish abundance can also be influenced by season, tidal schedule, or turbulence at the time of the survey. Many fish species are often difficult to spot, while some are transient. Reef monitoring was originally scheduled by quarter to capture such variability; however, this was not accomplished due to stormy weather in some quarters.

Of particular interest is the high diversity of fish (about 199 species) found inside the small (5 ha) sanctuary core (a lagoon), which represents more than 60 % of the fish community in the whole reef system (325 species). This result indicates the importance of shelter that the sanctuary core area provides for a wide variety of organisms in Danao Bay. The unique basin-like topography of the sanctuary core makes it a natural refuge for large predators, such as emperors, snappers, and other fish that are target food, such as rabbitfish (or siganids). Fishers observe that fish aggregations move out of the reserve at high tide to feed in nearby reef and sea grass meadows. During these foraging trips, the fish may be caught by fishing gear deployed outside the MPA, thus, contributing to increased catches of fishers: a phenomenon known as a spillover effect.

No significant difference in the abundance of fish in reserve and non-reserve sites was noted (Fig. 12.2). Spatial differences in fish biomass are more apparent in significantly higher values for target species, indicator families, and major demersal groups inside the reserve ($p < 0.005$) than outside, across different time periods (Fig. 12.3). High biomass estimates (mean of 36.2 g/m^2) inside the sanctuary core area, despite the low population density, suggest that fish inside this "no-take" zone are much bigger (mean length of 25–30 cm). Average biomass (17.4 g/m^2) of target food families in the three survey periods inside the reserve is more than double that in fished areas (7.7 g/m^2). These findings provide convincing evidence that the protection of coastal ecosystems brings about significant positive changes in habitat and resource quality, such as the buildup of diversity and biomass, by either the entry of new species from surrounding areas or the return of previous expatriate or emigrant species.

12.3.2 *Evidence of Economic Impacts of the Marine Reserve*

Fishery Production Values As in many coastal villages in the Philippines, the municipal fishery of Danao Bay involves a wide variety of artisanal fishing gears that land highly diverse (154 species) catches, mostly of demersal (reef-associated) fish, dominated by parrot fish and siganids (Fig. 12.4). Fish corrals (30.6 %) and gill nets (23.5 %) contributed the biggest proportion of landed catch in 2001–2002.

The estimated total annual fish catch of Danao Bay is 285.7 mt or 14.28 mt/km^2/year. Fish production exhibits moderate seasonality, with higher catch per unit effort (CPUE) from February to May (Fig. 12.5) coinciding with the calmer summer transition between the northeast monsoon (NEM) and southwest monsoon (SWM) periods. Monsoonal winds render rougher seas, which makes fishing along the reef crest difficult and perilous. Catches of stationary gear, such as a fish corral, on the other hand, tend to increase when the water is turbid especially following heavy rains. This result is consistent with the observation made of higher catch of fish corral during the NE monsoon (Heinen 1998).

The efficiency of artisanal fishing gear varies widely with average catch rates as low as less than a kilogram (simple handlines) to 4.79 kg/unit/trip (gill nets). Daily average CPUE values are low ranging between 1.25 and 3.45 kg/fisher. An overall daily average CPUE of 2.26 kg/fisher translates to gross daily revenues ranging from less than PHP 40 to PHP 180[5] depending on kinds and sizes of fish. As a result, many fishers make daytime and nighttime fishing trips or operate a combination of gears to increase catch within a fishing day and earn higher incomes. Catch composition of landed fish indicates the increased dominance of small, poor quality, and low-value fish instead of high-value carnivores or herbivores. This common phenomenon of species replacements of higher trophic groups by lower trophic level fish in a highly exploited artisanal fishery is also known as "fishing down the marine food webs" phenomenon described by Pauly et al. (1998).

Average catch rate of siganids by fish corrals in 2001–2002 (2.29 kg/unit/day) shows only a slight improvement over its lowest point in 1997 (Fig. 12.6). The improvement is attributable to increased availability of juvenile and adult fish, presumably from increased recruitment from the protected spawning stock in the sanctuary. Visual surveys of the sanctuary core revealed an abundance of large adult (and reproductive) siganids, emperors, and snappers, which are common fish corral catches. Fish of the same species caught by fish corrals, gill nets, and fish traps were often small juvenile and subadult fish. The siganids and emperors were highly mobile and tended to move out of the sanctuary at a high frequency and are, therefore, the most likely to spill over into fishing areas outside of the reserve.

Fishing Revenues, Income, and Resource Rent Annual gross revenues, total fishing costs, and net incomes were calculated for each fisher—taking into consideration crew size and revenue sharing arrangements—and gear type. Estimated values suggest that some gear types are more profitable than others, a differential profitability that appears to be a function of CPUE and fishing costs. Gill nets, fish corrals, and handlines obtained the highest gross annual revenues. Fishing with the use of paddle boats (non-motorized canoes or *bancas*) earns positive net annual incomes (mean PHP 19,441), while fishing with motorized boats hardly earns any profit at all (mean PHP −8,694) due to large fuel costs. Estimates of net incomes by gear type show that fish corrals, bamboo fish traps, and gill nets using nonmotorized boats derive moderate profits from fishing (PHP 19,449). Spear fishing and motorized gill net and handline operations are not surprisingly unprofitable (PHP −1,540). Finally, the average monthly incomes of all gear and fishing villages, ranging from PHP 300 to PHP 2,000, are way below the poverty threshold for rural communities.

Determination of economic rent from the fishery resources of Danao Bay (Gustavson 1999) involved the derivation of total annual net operating values (ANOV) and net annual values (NAV) for each gear. Calculated ANOV and

[5] In 2001–2002, USD 1 was equivalent to PHP 44.00–48.00.

NAV (Table 12.2) show a range of estimates, including zero and negative values for motorized operation of gill nets and handline fishing, respectively. Among the gears used in Danao Bay, only fish corrals earned positive NAV estimates. The largest negative values were obtained from gill net and handline fishing. The combination of high costs of fuel, inclusion of family labor cost into total operational costs of fishing, larger crew sizes, and low CPUEs makes motorized gill net fishing around the reefs in Danao Bay generally unprofitable.

Net annual values are considered to be producer surplus (Trinidad et al. 1993; Gustavson 1999), a profit over costs that accrues to the gear owner. It also represents the annual value of contribution of the marine ecosystem (coral reefs, sea grass, and mangroves) to the artisanal fishery of the bay. Summing up the NAV for all gear results in a range of values for total NAV, with a midpoint of PHP 104,317.12 (at a 10 % discount rate). This sum indicates the amount of fishery rent generated by the bay's fishery resources. While many gears appear unprofitable, positive midpoints of total NAVs indicate that some fishery rent is being earned, although it is probably below the rent that could be earned if the fishery were managed more efficiently. Fish corrals appear to be the most profitable gear; however, if more fishers invest in this gear, then pressure on the resource would also increase. It is important that fishery management consider the bay's capacity to support various types and numbers of fishing gear.

12.3.3 Institutions and Their Role in Fishery Management

Management of the Baliangao Marine Reserve has evolved through a series of institutional arrangements from its inception as the Misom Sea Sanctuary in 1991. Although it was established as a partnership with the local government of Baliangao, the Pipuli Foundation was responsible for much of the project implementation and was the forerunner of the coastal resource management (CRM) program in Danao Bay. Apart from the Pipuli Foundation, there are three other institutions with jurisdictional interests in the marine reserve project:

1. Danao Bay Resource Management Organization (DB-REMO): A community management body formed in 1998 to undertake all future management programs in the bay. The organization maximizes community participation, ensures effective law enforcement, actively advocates and networks, conducts participatory research, monitors and evaluates the resources, and generates livelihood options.
2. The Protected Area Management Board (PAMB): A management body established under the National Integrated Protected Areas System (NIPAS) Act of 1992 (RA 7586). The PAMB is a government-led management body, with a multi-sectoral membership and the Department of Environment and Natural Resources (DENR) acting as secretariat. It provides technical and administrative support to the BPLS through its Protected Area Supervising Unit (PASU).

3. The Local Government Units (LGUs): Danao Bay is under the political jurisdiction of two municipalities, Baliangao and Plaridel, although 90 % of the area belongs to the four *barangays* of Baliangao and 10 % to two *barangays* of Plaridel. The Local Government Code of 1991 (RA 7160) endowed LGUs jurisdiction over municipal waters and the right to implement a CRM program for the sustainable development of these waters. Support of the Baliangao LGU for the coastal management program in Danao Bay might be categorized as legislative and administrative and includes the responsibility to provide some funding for the program. At least three municipal ordinances issued by the LGU of Baliangao have direct relevance to coastal resource management.

The establishment of the Baliangao Marine Reserve has led to the evolution of its management from NGO-led (Pipuli Foundation) efforts to a community-based, bay-wide coastal resource management program (Pipuli Foundation, Inc. 2001). The primary goal of establishing the Misom Sea Sanctuary was to improve the fisheries of Baliangao and the whole of Danao Bay and to restore them to their former abundance. The introduction of the marine reserve project by Pipuli Foundation in Baliangao marked the birth of a new management regime that introduced a novel concept of giving priority to local communities in managing the coastal fishery. Initial support for the project was poor, but slowly built up through time. In 1996, the Danao Bay CB-CRM program was implemented to cover a broader scope of stakeholders because Danao Bay was recognized as a complex and integrated system whose management problems required an integrated solution. Strictly protecting a portion of the bay would have had very little impact without fishery management measures also being implemented across the bay.

12.3.4 Impact of Community-Led Efforts on Coastal Fishery Management

The Baliangao Marine Reserve and the Danao Bay CB-CRM was one of the most popular[6] and long-running community-based coastal management programs in Mindanao; a wide range of programs and management measures were implemented in the area following inception of the marine reserve project by the Pipuli Foundation in 1991.

The results of this study indicate that the coastal fishery of Danao Bay is in a much better state than it was before the Baliangao Marine Reserve was established. Indications that the MPA has contributed to this improvement are provided by biological and economic data on the coastal ecosystem and fishery of the bay. The role of MPAs in enhancing coastal fisheries, however, does not depend only on the

[6] This is true at the time of data collection. In 2005, the management collapsed, and the marine reserve was laid open to a "fishing frenzy." True, the PAMB tried to rally after that, but the community-led DB-REMO lost its priority under the PAMB.

ability of the marine reserve to export biomass to fished areas (i.e., the spillover effect) but also on its ability to inspire and rally community support and involvement in fishery management.

The crucial question is whether spillover of fish biomass from the MPA in Baliangao has occurred. The present study has identified a number of indications that the coastal fishery surrounding the marine reserve in Danao Bay could be on a transition toward becoming a viable, sustainable industry. Those indicators are listed below:

(a) Improvement in the ecological conditions of the bay's ecosystem over time, namely, improved live coral, sea grass, and mangrove cover, has led to increases in fish diversity and relatively high fish biomass inside the reserve and in nearby reefs.
(b) Differential profitability of gear indicates that certain fishing gears operating in Danao Bay, such as fish corrals and bamboo fish traps, are more profitable than others, either because of the relative abundance of high-value fish species or because of reduced fishing costs and capital.
(c) The increase in CPUE for certain gear—fish corral profit increased (albeit slightly) from 1.3 kg/day in 1997 to 2.29 kg/day in 2000–2001—is an apparent reversal of the downward trend experienced by this fishery since 1945. Some fishers have also claimed increased catch from other gears, such as spears and fish traps, since the establishment of the marine reserve.
(d) Large aggregations of big adult (spawning) stock and juvenile populations of important food fishes (e.g., emperors, siganids, and snappers) inside the small sanctuary not observed in adjacent reefs. Without this protected source of fish recruits, the harvestable biomass in surrounding fished areas, particularly of large predators, would have dissipated long ago under sustained high levels of fishing effort. Community involvement and support have played an important role in ensuring the success of the reserve project. Support for the sanctuary project by fishers was very high (73.5 %), and bay-wide CRM efforts led to the complete eradication of explosive fishing techniques and other destructive gears.

12.3.5 Management Issues in Danao Bay

Despite the long list of measures that have been implemented by the various institutions, the coastal fishery of Danao Bay is still characterized by a range of management problems, as follows:

1. The coastal fishery is still an open access system with little effort put toward regulation.
2. Institutional weaknesses of the LGU and people's organizations include the lack of participation and fiscal support and weak political will.
3. "Power politics" are used in resolving resource-use conflicts.

4. Sustainability of CRM efforts amidst changing institutional arrangements (i.e., DB-REMO to PAMB) is put to risk.
5. There is potential conflict in management roles and jurisdiction over the marine reserve project and sense of "losing ownership."

Despite earlier efforts to broaden the scope of community involvement in fishery management, many fisherfolk remain unperturbed and indifferent. Enforcing marine reserve regulations has become the responsibility of a handful of DB-REMO organization members who started out as volunteers, but who now all receive token honoraria from foreign donors through the Pipuli Foundation. Problems of inefficient monitoring and enforcement of boundary regulations, continued poaching, allegations of collusion between poachers and sanctuary guards, and declining membership in people's (fishers' in this case) organizations (POs) are just a few of the many challenges confronting the management of the MPA.

12.4 Conclusions and Recommendations

This study provides biological and economic data that support the concept that marine protected areas (or marine reserves) are strategic tools for fishery management, particularly in overfished coastal ecosystems. Biological and economic data indicate that the Baliangao Marine Reserve has, to some extent, contributed to improving the conditions of the coastal fishery of Danao Bay. On the other hand, there are obvious signs that the fishery is still in distress: average CPUE is still low and, thus, so are daily gross incomes; high fishing costs result in too small or marginal profits; and the lack of viable livelihood options do not encourage part-time fishers to exit the fishery in order to reduce fishing pressure on the bay's resources.

The MPA, in providing a protected habitat for adult target predatory fish, allows buildup of fish biomass inside the reserve, which spills over into neighboring fishing areas, allowing differential profitability of gear, especially gear with low capital and operational costs. Economic analyses show that small fishery rent is being earned; without the MPA, economic rent would have been driven to zero.

That a spillover of biomass from the MPA to surrounding fished areas has really occurred cannot be proven definitively, given the limited data. Advocates of MPAs admit that spillover of benefits from a protected area cannot be easily demonstrated within a short period of time. Periodic investigations in the Apo Island marine reserve point to evidence that the marine reserve is enhancing the fisheries in adjacent areas through the migration of adult fish biomass from the "no-take" reserve to fished areas after 9–11 years since its establishment (Russ and Alcala 1996).

Coastal fisheries can be complex management systems, subject to stochastic dynamics that are difficult to predict with a high degree of certainty. The current state of the coastal fishery in Danao Bay may be described as being in transition toward sustainability. It is aided in part by local initiatives to regulate fishing effort and eradicate destructive fishing methods and by establishing protected areas. There is growing evidence that the "no-take" Baliangao Marine Reserve can help sustain the coastal fishery by maintaining high fish biomass inside the sanctuary and migration of both adult fish to surrounding fishing grounds and juvenile fish to repopulate nearby coral reefs.

The MPA as a sole management tool can neither restore fishery health nor ensure sustainability of a coastal fishery. Other policy instruments and fishery management strategies are needed, such as reducing the overall fishing effort through licensing and exclusion of nonresident fishers, enforcing stiffer penalties (disincentives) for violators, implementing a gear zoning plan, and sustaining an information, education, and communication (IEC) program for Danao Bay.

Appendix

Table 12.1 Some ecological indicators of the Baliangao Marine Reserve (BMR) and adjacent unprotected areas

Parameter	Inside BMR	Outside BMR
Average live coral cover (%)	46.24	36.89
Coral diversity (number of species)	80 (2 sites)	90 (4 stations)
Fish diversity (number of species)		
Total (over three survey periods)	246	236
Mean (of diff. stations and periods)	102	72
Average fish population density		
No. of fish per 500 m^2	626	780[a]
Average fish biomass (tons per km^2)	31.93	18.09
Mangrove species richness (number of species)	20	Not assessed
Sea grass species richness (number of species)	7	10[b]

[a]Mean value for four reef sites outside BMR; higher value than inside due to very high densities of pomacentrids in two reef sites in Oct 2001 survey

[b]Includes narrow- and wide-leaf varieties of *Halodule* species

Table 12.2 Estimates of "true" net annual values and net present values (in PHP) of the artisanal fishery of Danao Bay, derived from net operating values and equivalent annual capital costs

Type of fishing gear	Number of owners	Total annual net operating values	Net annual values DR = 0.05	DR = 0.10	DR = 0.15
Fish corral	23	2045.08 to 44,334.40	1728.93 to 43,701.40	1412.78 to 43,068.40	1097.10 to 42,436.35
Gill net (nonmotor)	31	−18,045.28 to 29,219.64	−18,213.78 to 28,994.69	−18,382.28 to 28,769.74	−18,550.53 to 28,545.13
Gill net (motorized)	7	−22,330.00 to 63,677.84	−23,661 to 39,486.84	−24,992 to 37,975.84	−26,321 to 36,467.11
Bamboo fish trap	12	−146.72 to 40,997.84	−279.72 to 27,946.00	−412.72 to 27,756.00	−545.52 to 27,566.28
Spear	17	−8978.00 to 27,270.80	−9199.75 to 27,100.80	−9421.50 to 26,930.80	−9642.92 to 26,761.05
Handline (nonmotor)	26	−14,766.32 to 18,990.00	−14,879.32 to 18,783.25	−14,992.32 to 18,666.50	−15,105.15 to 18,549.93
Handline (motorized)	3	−28,843.84 to −13,565.84	−29,876.09 to −14,490.84	−30,908.34 to −15,415.84	−31,939.04 to −16,339.45
Total	119				
	Total net annual value		−62,368.36 to 502,214.18	−64,693.36 to 493,813.18	−67,014.87 to 485,424.78
	Midpoint		**109,053.62**	**104,317.12**	**99,587.72**
	Net present value		−1,247,367.20 to 10,044,283.00	−646,933.60 to 4,938,131.80	−446,765.83 to 3,236,165.17
	Midpoint		**2,181,072.31**	**1,043,171.16**	**663,918.11**

Sensitivity analysis-derived values for different discount rates (DR)
US$1 = PHP 44–48 in 2001–2002

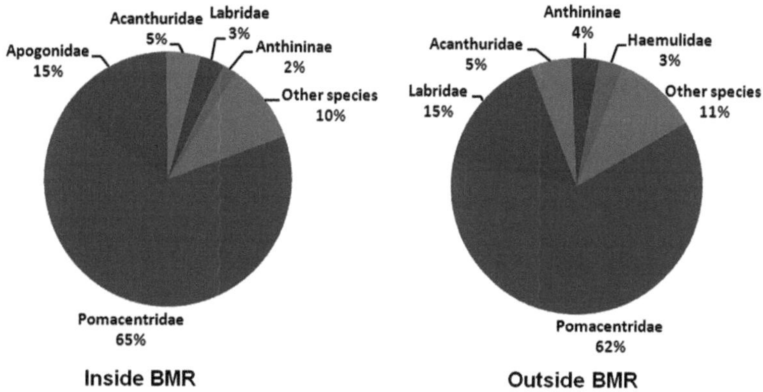

Fig. 12.1 Relative abundance of fish families inside protected (*left*) and unprotected (*right*) reefs of Danao Bay where small pomacentrids equally dominate

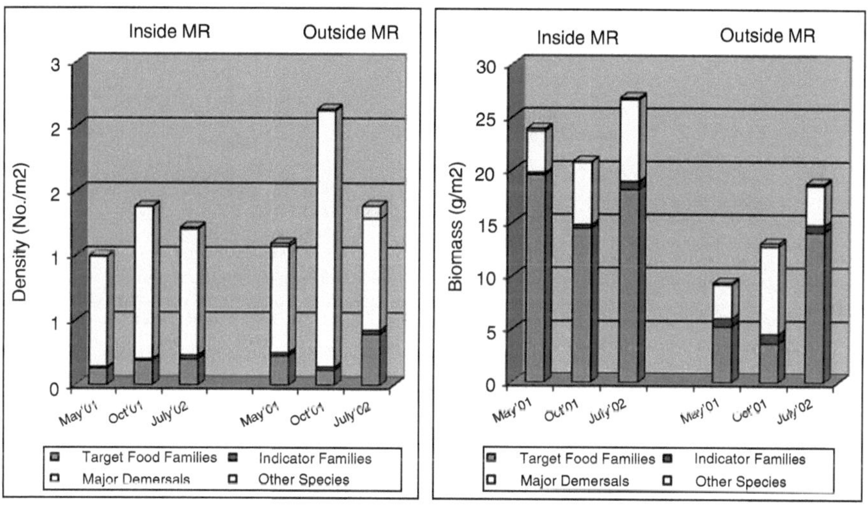

Fig. 12.2 Variations in abundance (*left*) and biomass (*right*) of different fish groups in reefs inside and outside the marine reserve across survey periods

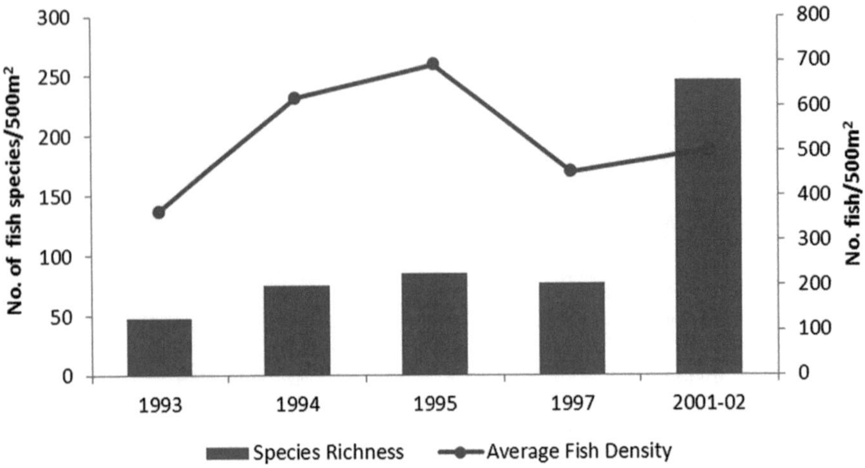

Fig. 12.3 Time series changes in species richness and abundance of reef fish inside the BMR (Data sources: 1993–1995 data from Silliman University, 1997 data from Pipuli Foundation, and 2001–2002 data from present work)

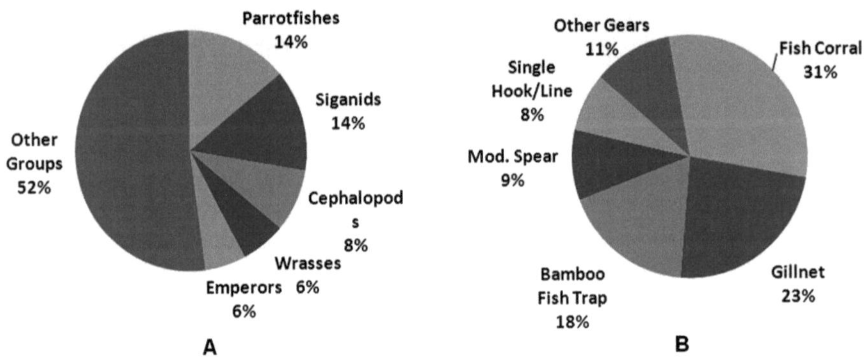

Fig. 12.4 Most abundantly caught fish groups (**a**) and contribution of major fishing gear to total landed catch in Danao Bay (**b**)

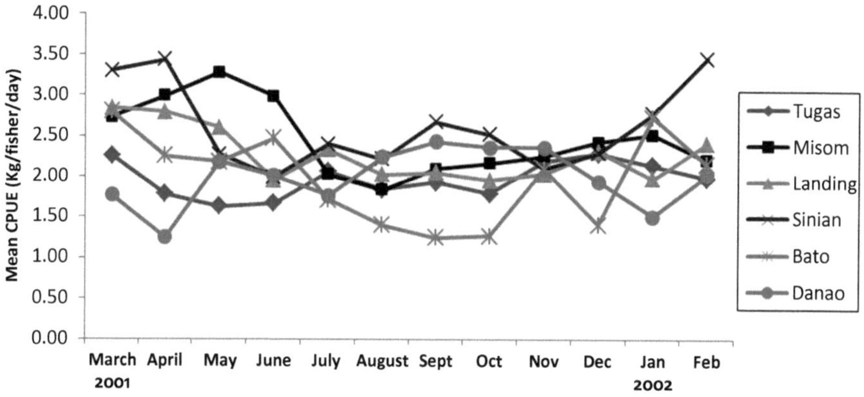

Fig. 12.5 Trends in CPUE profiles of fishers in six fishing villages of Danao Bay

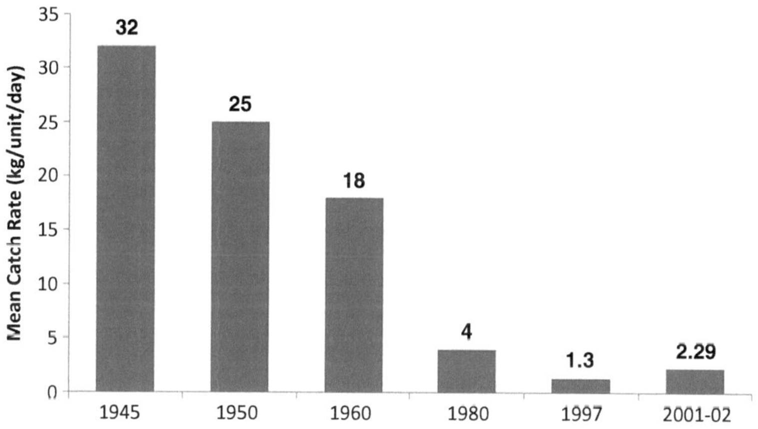

Fig. 12.6 Historical trends of average daily catch of fish corrals in Danao Bay (Data source: estimates for 1945–1980 derived from focus group discussions by Heinen (2003), for 1997 from Heinen (1998), and for 2001–2002 from this study

References

Alcala AC (1999) Fisheries enhancement effects of marine reserves. In: Pauly D, Christensen V, Coelho L (eds) Proceedings of EXPO'98 conference on ocean food webs and economic productivity. 1–3 Jul 1998. Lisbon, Portugal. ACP-EU Fisheries Research Initiative, Brussels, pp 45–46

Alcala AC, Russ GR (1990) A direct test of the effects of protective management on abundance and yield of tropical marine resources. J Cons Int Explor Mer 46:40–47

Deguit ET, Morales MN (2002) Gilutongan marine sanctuary: benefits gained from proper management. In: Campos WL, Beldia III PD, Aliño PM (eds) Formulation of a National Fish Sanctuary Strategy. Workshop proceedings of the AFMA-Marine Fishery Reserves

Program. 9–10 May 2000 (Quezon City) and 3–5 Jul 2001 (Cebu City). University of the Philippines in the Visayas, Iloilo City, pp 45–52

English SA, Wilkinson C, Baker VJ (eds) (1997) Survey manual of tropical marine resources. ASEAN-Australian Marine Science Project, Australian Institute of Marine Science, Townsville

Fraser JB (2003) Danao Bay Reef, Misamis Occidental. In: Aliño PM (ed) Philippine coral reefs through time. Workshop proceedings. 27–28 Feb 2002, Manila. Coral Reef Information Network of the Philippines (PhilReefs) and the University of the Philippines Marine Science Institute, Quezon City, pp 90–92

Gustavson H (1999) Economic production from the coral reef fisheries of Jamaica and capture ecosystem values. PhD dissertation. Department of Geography, University of Victoria, Australia

Heinen A (1998) Catches of *bungsod* in Danao Bay. Unpublished report for fisher leaders and DA personnel. Pipuli Foundation, Misamis Occidental

Heinen A (2003) Rehabilitating nearshore fisheries: theory and practice in community based coastal resource management from Danao Bay, Philippines. CB-CRM Resource Center. UP Diliman, Quezon City

Heinen A, Laranjo A (1996) Marine sanctuary establishment: the case of Baliangao wetland park in Danao Bay. In: Ferrer EM (ed) Seeds of hope. College of Social Works and Community Development, University of the Philippines, and NGO Technical Working Group for Fisheries Reform and Advocacy, Quezon City

Holland DS, Brazee RJ (1996) Marine reserves for fisheries management. Mar Resour Econ 11:157–171

Pauly D, Christensen V, Dalsgaard J, Froese R, Torres F (1998) Fishing down marine food webs. Science 279(5352):869

Pipuli Foundation Inc. (2001) Danao Bay CBCRM program annual report (January–December). Oxfam-Philippine Office, Quezon City

Reboton C (2002) Apo Island. In: Aliño PM, Miclat EB, Nañola CL, Roa-Quiaoit HA, Campos RT (eds) Atlas of the Philippine coral reefs. Philippine Coral Reef Information Network (PhilReefs). Goodwill Trading Co., Philippines, pp 130–134

Roberts CM, Polunin NVC (1993) Marine reserves: simple solutions to managing complex fisheries. Ambio 22(6):363–368

Rodwell LD, Roberts CM (2000) Economic implications of fully protected marine reserves for coral reef fisheries. In: Cesar HS (ed) Collected essays on the economics of coral reefs. CORDIO, Kalmar University, Kalmar, pp 107–124

Russ GR, Alcala AC (1996) Do marine reserves export adult fish biomass? Evidence from Apo Island, Central Philippines. Mar Ecol Prog Ser 132:1–9

Trinidad AC, Pomeroy RS, Corpus PV, Aguero M (1993) Bioeconomics of the Philippine Small Pelagics Fishery. ICLARM technical report no. 38. ICLARM, Makati City

Chapter 13
Do Institutions Affect the Performance of Marine Protected Areas? Evidences from the Philippines

Esmyra P. Javier

Abstract This report investigated the effectiveness of different management regimes in the marine protected areas (MPAs) in the Philippines. It assessed how MPAs are performing – from an environmental, economic, and social perspectives – and determined what constitutes the key features of a successful management scheme. It determined that the type of institution in charge of an MPA makes little difference to its effectiveness. Instead, good leadership, adequate manpower and source of funding, and provision of sustainable livelihoods are the keys for effective management. The report also suggested a number of ways in which community involvement in MPA management can be strengthened and highlighted the need for community requirements to be taken fully into account in any MPA action plan.

Keywords Protected area • MPA • Community-based management • Coastal resource management • Institutional arrangements

13.1 Introduction

Coastal and marine ecosystems and their diversity are continuously being threatened by both natural and anthropogenic factors. One widely recognized way to conserve the ecosystem is to establish a marine protected area (MPA).[1] The success of an MPA depends on effective program implementation, strict enforcement of established rules, effective leadership, adequate financial resources, and strong support by the coastal communities.

[1] An MPA, as described by the World Conservation Union (Resolution 17.38 (1988) and reaffirmed in Resolution 19.46 (1994) of the IUCN General Assembly), "is any area of intertidal or subtidal terrain, together with its overlying water, and associated flora, fauna, historical and cultural features, which has been reserved by law to protect part or all of the enclosed environment."

E.P. Javier (✉)
Part of an MSc Thesis, University of the Philippines, Los Banos, Laguna, Philippines
e-mail: esmyra.javier@gmail.com

N. Olewiler et al. (eds.), *Marine and Coastal Ecosystem Valuation, Institutions, and Policy in Southeast Asia*, DOI 10.1007/978-981-10-0141-3_13

The governance of MPAs in the Philippines can be headed by the Department of Environment and Natural Resources (DENR), by local communities (with support from other stakeholders) or by the local government. Nongovernmental organizations (NGOs), people's organizations (POs), local communities, and the private sector may also partner with the local governments in the implementation of MPAs.

The role of the local government as local resource manager is based on the 1991 Local Government Code (RA 7160) that decentralized governance from the national to the local level. The dominant role and success of nongovernmental organizations, often in collaboration with local governments in resource conservation, are acknowledged by many institutions (Lewis 1996). NGO contribution has reached a wide number of benefactors, often endowed with sufficient funding and technical knowledge. The role of rural communities as managers of natural resources gained momentum in the 1990s. The sense of stewardship among the members of the community has encouraged better resource management.

These governing institutions, however, also face a number of challenges, such as lack of financial and human resources, ineffective leadership skills, and a complicated bureaucratic process. State-run initiatives, such as the Protected Area Management Board (PAMB), rely heavily on a bureaucratic process that slows down the implementation of various conservation initiatives. The local governments may be ill-equipped to act as resource managers since their main mandate is to develop the local community, in terms of fund generation, infrastructure development, and projects that increase welfare. NGO management of MPAs sometimes does not have sufficient support from the community to effectively carry out the task due to difficulties with organizing its large membership – similar to POs/communities.

The Philippines is one of the most biologically diverse coastal and marine regions in the world. It is in the country's best interest to determine the best approach to conserving this diversity. It is also in the interest of the public to be able to determine which governing institution is most effective in managing marine protected areas. Research studies that aim to provide data on how these different institutional arrangements manage the natural resources and those that seek to evaluate their performances are useful to policy-makers in terms of planning and management of the country's natural resources.

The main objective of this chapter is to determine the role of governing institutions in biodiversity conservation in selected MPAs in the Philippines. It determines the overall performance of MPA given various key indicators including socioeconomic and ecological benefits.

13.2 Methodology

13.2.1 Study Sites

Eight MPAs were chosen using the following criteria: high biodiversity richness, active management committees, significance of the sites (site is among the best in the country, in terms of its biological diversity and economic value), a fair amount of data being available, ecotourism developing in the area, and sites having relatively stable funding sources. All sites also face threats of encroachment, poaching, and various levels of unsustainable resource use, which highlights the importance of maintaining an effective MPA. These sites are grouped under three types of institutional arrangement and are described below.

13.2.1.1 As DENR/NGO-led MPAs

(A) Batanes Island Protected Landscape and Seascape (BIPLAS) is located at the northernmost tip of the Philippines and has an area of 213,578 ha. It was proclaimed a protected area in February 1994 by virtue of Presidential Proclamation No. 335. The DENR is primarily responsible for managing BIPLAS through the Conservation of Priority Protected Area Project (CPPAP). Support comes from NGOs that include the Batanes Development Foundation Inc. (BDFI), the Itbayat Integrated Area Development Program (IADP), the Ivatan Heritage Foundation, and people's organizations (POs).

(B) Siargao Island Protected Landscape and Seascape (SIPLAS) has the biggest mangrove reserve in Mindanao covering 8692 ha and was declared a protected landscape and seascape on October 10, 1996 through Presidential Proclamation No. 902. The management is led by the DENR through the Conservation of Priority Protected Area Project (CPPAP). The Surigao Economic Development Foundation Inc. (SEDF) is the area's local NGO. There are presently several POs in the area that SEDF helped organize.

(C) Apo Reef Marine Natural Park (ARMNP) is one of the largest coral atolls in the country and was created on September 6, 1996 through Presidential Proclamation No. 868. The ARMNP is under the jurisdiction of the Municipality of Sablayan, Mindoro Occidental, and is a DENR-NGO-managed site, through its Conservation of Priority Protected Area Project (CPPAP), in close collaboration with the NGO for Integrated Protected Area (NIPA).

13.2.1.2 As LGU/NGO-led MPAs

(A) Sagay Marine Reserve (SMR) is situated at the northeast tip of Negros Island and declared a reserve on June 1, 1995. The management of SMR is under the National Integrated Protected Area System (NIPAS) Act and managed by a

Protected Area Management Board (PAMB). The Sagay City mayor cochairs the PAMB with the DENR Regional Executive Director (RED).

(B) Tubbataha Reef Marine National Park (TRMNP) is situated in the middle of the Sulu Sea and is the country's first National Marine Park. It is under the jurisdiction of the Municipality of Cagayancillo Island in Palawan and is an LGU-NGO-managed protected area. The PAMB, chaired by the provincial governor, is currently managing the TRMNP. Contrary to the official setup of a PAMB, the governor has been in charge of the board since its establishment. This was brought about by a Memorandum of Agreement (MOA) signed between the office of the governor and DENR during the early stages of the park's establishment.

13.2.1.3 As Community-Managed MPAs

(A) Panggangan Island (PI) was initially declared a Mangrove Swamp Forest Reserve on December 20, 1981 by Presidential Proclamation No. 2152. PI is under the jurisdiction of the Municipality of Calape, Bohol, which has a total population of 4213. With the initiative from the local PO, the Lomboy Farmers, Fishers, and Carpenters Association created two marine sanctuaries where fishing and any other related activities are strictly prohibited. The fish sanctuaries were created on March 1995.

(B) Biri-Larosa Coastal Community (BLCC) comprised of the municipalities of Biri Island, Lavezares, Rosario, and San Jose, which are located on the northern tip of Samar Province. It was declared a protected area on April 23, 2000 by Proclamation No. 291. BLCC management is a collaborative effort between the DENR through its Coastal Environmental Program (CEP), the Samar Center for Rural Education and Development (SACRED), a local NGO, and the University of the Eastern Philippines.

(C) Apo Island Marine Sanctuary (AIMS) is on a 74 ha volcanic island in the Visayan Sea and was established in 1982 as a community-based managed protected area; it is a strict marine sanctuary. The marine sanctuary, a 500 m stretch of prime reef in the southeast side, was established in 1985. On August 9, 1994, the island was proclaimed a National Integrated Protected Area System (NIPAS) site through Presidential Proclamation No. 438. The area is under the jurisdiction of the Municipality of Dauin, Negros Oriental, but has been managed by a community organization since 1985 through the marine management committee (MMC), with assistance from Silliman University.

13.2.2 Data Collection

Data collection methods include collection and assessment of data from existing literature, key informant interviews, and household surveys. Research reports conducted by the Department of Environment and Natural Resources (DENR) through the Conservation of Priority Protected Area (CPPAP) and Protected Area Management Board (PAMB) and other NGOs and other research institutions (including academia such as the Silliman University and the University of the Philippines) were used in the study. The key informants include representatives from the DENR, who were either the Protected Area Superintendent (PASu) or the Provincial Environment and Natural Resources Officer (PENRO), from the local government unit (LGU), local NGOs, officers of POs, and other PAMB members. The survey focused on fisherfolk households. The respondents were mostly married, aged 30–39 years old, with elementary level of education, living in households with four to six members or one to three children.

13.2.3 Data Analysis

The institutional arrangement matrix was used to analyze the different management regimes prevailing in the chosen protected areas. The matrix is an application of the taxonomic approach to analyzing problems that arise from common property resources. The row vectors of Table 13.1 show the five rights, including use rights (the right to withdraw resources for different purposes like for food or commercial use), exchange rights (the right to decide on how use rights are to be transferred), distribution rights (the right to allot resources for extraction and use), management rights (the right to regulate internal use patterns and transform resource through improvements), and rights of authority (the right to enforce existing rules and laws). The column vector expresses the types of property regime (DENR-NGO run property, LGU-NGO run property, and community-based run property) overseeing the MPAs.

This study uses multiple criteria analysis to generate a composite measure of accomplishments, including biophysical, socioeconomic, and institutional indicators. A "with or without MPA" approach was used in analyzing the data using the key performance indicators. The estimates were gathered from existing literature, which describes relevant data about when an area before and after it became an MPA. However, in cases where there are no data, the study relies on the recollection of the respondents and key informant interviews. A multilevel analysis was conducted on the household, community, and park areas.

Biodiversity level indicators are used to gage changes in resources, which include area-specific species (measured through listing and number/frequency); habitats, such as sea grass beds, mangroves, coral reefs, and others (measured through listing, area extent/location, species composition/density, and condition of habitats according to the percentage of damage); and level of exploitation (presence/absence of prohibited/destructive fishing methods and violations of resource use as stated in management plans). Biodiversity indicators at the household level include respondent perceptions, based on their observations and experiences as fisherfolk. For data that were not available for some aspects of the biophysical indicators, this study relies on "keystone" species for inspecting reef health and changes in biodiversity level.

In terms of socioeconomic indicators, the transaction cost includes costs generated from the promotion of the MPA's goals through information dissemination and communication, in terms of materials distributed, meetings, seminars/workshops, and management and enforcement costs. Institutional indicators include the types and lists of laws formulated and enforced in the MPAs, the types of enforcement, and the kinds of management boards. The level of liability was also assessed, through identification of violations accorded under the PA's laws and regulations.

The views and ideas of the respondents and key informants comprised one portion of the questionnaire. The questionnaire also sought opinions of the respondents on the general performance of the MPAs, in terms of biodiversity conservation and its socioeconomic benefits, as well as their perspectives on the contributing factors to the success or failure of the respective management regimes.

13.3 Results

13.3.1 Rights System and the Institutional Arrangements

The institutional arrangement matrix (Tables 13.1, 13.2, and 13.3) shows the types of right practiced by the three institutional arrangements (Cuevas 1999). The focus is on the link between property systems and environmental conservation and its sustainable use.

Use Rights Use rights are often exercised by way of economic instruments, such as license and user fees. All the institutions utilized such instruments to regulate resource use, although not at all study sites. The use of such instruments prevents the open-access regime that often leads to unsustainable use of resources; individuals tend to continue to extract resources without regard to available stock. The use of fees regulates resource extraction to a more sustainable use level, and those who pay these fees are bestowed a specified level of use rights.

The MPAs governed by the DENR/NGO employ economic instruments, while LGU/NGO and community-managed sites (TRMNP and AIMS) charge license and user fees. Although BIPLAS does not charge a user fee, it does charge researchers visiting the area at PHP 200 (USD 4)/researcher. SIPLAS charges research fees as well at PHP 40 (USD 0.8)/local/day and PHP 100 (USD 2)/foreigner/day. SIPLAS charges other fees as well for docking, filming, and sport fishing, at rates of PHP 200 (USD 4)/boat, PHP 1500 (USD 30)/day, and PHP 45 (USD 0.9)/rod, respectively. Fishing activities in the ARMNP is generally restricted unless a fisher acquires a fishing license from the local government office, which could amount to PHP 1000 (USD 20)/year. The license is applicable to the individual or to the vessel. Upon acquiring the fishing license, fisherfolk may engage in fishing activities in the protected area's sustainable use zone. As for ecotourists, leisure activities are free in PAMB-designated recreational zones, but scuba enthusiasts pay a diving fee of PHP 300 (USD 6)/visit.

For LGU/NGO sites, the subject of license fees does not apply in TRMNP since it is a no-take zone. With regard to user fees, the management board charges around PHP 2500 (USD 50)/diver/visit for foreigners and PHP 1250 (USD 25) for a local scuba diver's first visit. The fees are 50 % less for subsequent visits to Tubbataha Reef. For SMR, no user fees are charged.

For the community-managed site of AIMS, no fishing licenses are required; fisherfolk are unlimited as long as they use sustainable methods only, such as by hook and line, fishnet, and fish cage. This practice is closely monitored by the *Bantay Dagat* and almost none of the fisherfolk engage in illegal methods. User fees for AIMS are set lower at PHP 150 (USD 3)/diver, for activity within the sanctuary and PHP 75 (USD 1.5)/diver if outside the sanctuary but within the PA. The other two sites of PI and BLCC have not yet taken steps to incorporate economic instruments for resource management.

Ecotourists visiting any of the sites are allowed to enjoy the surroundings and take photographs but not to gather marine species from any designated zones. Among the eight sites, only AIMS charges a fee of PHP 50 (USD 1) for the use of an underwater camera.

It is important to note that the management institutions, with the exception of TRMNP, arbitrarily set these user fees. The user fees in TRMNP resulted out of a willingness-to-pay study by the Philippine World Wide Fund (WWF). Thus, for the other sites, it cannot be determined whether the assigned fees reflect the true value of the resource.

Exchange Rights This refers to a practice by management institutions at all sites. License fees for all sites, where applicable, are issued either to the fisher or to the vessel and cannot be used by anyone other than the person or vessel named on the license. The tourists/visitors, such as scuba divers and sightseers, pay user fees that are transferable to other users. For example, someone who initially planned to scuba dive but decided against it may transfer the purchased right to another diver who has not yet paid the user fee. This is being practiced in all of the study sites.

Distribution Rights The MPAs in this study tend to totally restrict the gathering of endangered, threatened, and rare species. In terms of methods employed for gathering species, only sustainable fishing methods are allowed, such as by hook and line and fishnet. Scientists and researchers may be allowed to gather specific species for research purposes as their results may aid in policy-making and management.

To illustrate, distribution rights for DENR/NGO sites, fisherfolk in ARMNP who do not have fishing licenses can only gather marine species in the designated *payao* area, and those with licenses may gather within the reserve as long as the species are not threatened, endangered, or rare. Those who violate such policies are subjected to penalties, namely, fines between PHP 5000 (USD 100) and PHP 100,000 (USD 2000) and/or imprisonment for 1–3 years, as specified by PAMB. For both BIPLAS and SIPLAS, a similar scheme is employed. Although these sites do not charge license fees, locals may only fish at subsistence levels. The LGU-/NGO-managed site of TRMNP, however, does not allow any form of species collection within its boundaries, even if a species is abundant. The SMR allows only hook and line fishing within the reserve. In the strict marine sanctuary of the community-based AIMS, any collection and extraction of marine species is prohibited; however, leisure activities such as scuba diving and snorkeling, with a limit of 15 divers per day, are allowed. In general, only sustainable fishing methods including fish cages are permitted outside the marine sanctuary but within the protected area. For the PI and BLCC, well-defined restrictions have yet to be set.

Management Rights While management rights are typically bestowed upon the management board of an MPA or DENR-/NGO-managed site, the DENR PAMB has the right to manage the resource and enforce the laws and regulations governing it. However, there are MPAs where host NGOs play active roles; in ARMNP, for example, NGOs for Integrated Protected Area, Inc. (NIPA) actively manage the area. The management allows licensed fisherfolk to fish only in the sustainable use zone.[2] Open and close seasons for selected species are implemented, and there are set sizes and quantities for some species. Although such limits are specified in the management plan, no mention is made as to which particular species these limits apply. In this zone, only low impact aqua sports, like scuba diving and sport fishing, are allowed.

Although TRMNP is governed by PAMB, the marine park is actively managed by the LGU, through the office of the provincial governor in partnership with the NGO WWF-Philippines; WWF also promotes alternative livelihoods, such as seaweed farming. PAMB headed by the provincial governor – instead of the traditional DENR-RED – monitors and supervises the activities in the area. Park rangers are stationed in the area's north atoll, on regular rotations. For SMR, the PAMB is also currently being led by an LGU official, namely, the municipal mayor of Sagay.

For management rights in AIMS, the members of the community have been managing the area for more than a decade through the marine management committee, a local community organization. Recently, management of the whole island was put under the Coastal Environment Program (CEP) of the DENR, and the MPA is now managed by PAMB, which tries to regulate resource use and formulate conservation management plans. For PI and BLCC, the CEP-DENR is currently managing the area, which aims to encourage community participation in sustainable resource management.

The TRMNP is a no-take zone, which could explain the community's level of hostility toward the management. This animosity manifests as uncooperative behavior by most locals. The implementation of PA rules and regulations by the local government in Cagayancillo has been very lenient. Only verbal warnings have

---

[2] Sustainable use zones are designated areas where fisherfolk can fish using sustainable fishing methods, such as hook and line. Fisherfolk who use the area require fishing permits.

been issued to fisherfolk engaged in illegal activities, and members of the community are silent about fisherfolk known to engage in illegal practices. In contrast, there was no hostility or opposition toward the management in AIMS at the time when the area was being managed by the community; however, when the PAMB took over management, members of the community started to become hostile and were not as cooperative as before. The hostility may be partly due to the 75 % share from the Integrated Protected Area Fund (IPAF)[3] that has remained unreleased for at least 2 years, at the time of this study.

At ARMNP, local fisherfolk are dissatisfied with PAMB due to the implementation of license fees for the right to fish within ARMNP waters. Fisherfolk regard the annual fee of PHP 1000 (USD 20) as too steep for their budgets and are not happy to fish in the *payao* area designated for them as it is too far from the Sablayan community.

Rights to Authority This right is often exercised by those in charge of patrolling and enforcing the laws of the MPA, such as the *Bantay Dagat* volunteers, coast guards, and park rangers. For DENR/NGO sites, the office of the Protected Area Superintendent (PASu) has the right to monitor and supervise the activities in the area; park rangers and *Bantay Dagat* also patrol the areas. Fisherfolk, scuba divers, and any individuals engaged in any form of illegal activities are subject to different penalties, including possible imprisonment.

With respect to who monitors and patrols the area, the community-based sites like AIMS employ people from the community (e.g., local fisherfolk), whereas DENR/NGO and LGU/NGO sites employ people from the Philippine National Police, Philippine Coast Guard, and park rangers as enforcers. WWF is also active in patrolling the area in TRMNP and is currently training locals at ARMNP on enforcement activities.

Even before the establishment of AIMS as an MPA, the Silliman University has always assisted the community through environmental education and provision of alternative livelihoods. When the community in Apo Island established a sanctuary, the University's participation was extended to monitoring and enforcement. This community effort became a catalyst for converting the whole Apo Island into a marine protected area. However, entry of the PAMB into the area brought problems with distribution of the benefits. The unreleased 75 % of the IPAF has drawn widespread criticism from Silliman University and uncooperative behavior from the community; one concrete manifestation is the unwillingness of the *Bantay Dagat* members to continue patrolling the area.

---

[3] IPAF is a trust fund set up under the provisions of the 1992 NIPAS Act (Sec. 16), to receive donations, endowments, and revenues generated within protected areas and disburse the same to finance projects of the National Integrated Protected Area System. IPAF could lead to a growing portion of the costs of protected area management, being covered by income.

13.3.2 Performance of Marine Protected Areas

The performances of the different management schemes were assessed based on key indicators categorized as biological, socioeconomic, and institutional (Tables 13.4 and 13.5).[4,5] In terms of biophysical indicators, the LGU-/NGO-managed sites of SMR and TRMNP have the highest number of coral genera and the highest number of fish species, respectively. In socioeconomic terms, household income is highest in BIPLAS, a DENR/NGO site, followed closely by AIMS, a community-managed site. In terms of the amount of fish catch, ARMNP leads the three sites, with an average of 35 kg/day. For institutional indicators, it is worth noting that based on the study's general assessment, the management board of TRMNP is the most active. At a glance, it seems that all sites fare well in terms of performance, but the study cannot readily judge the best MPA based on these values and listings. In order to aid assessment, a composite index was generated to systematically determine the best performing MPA. Different weights and institutional rankings of indicators were employed, based on those given by the key informants themselves.

Table 13.6 illustrates the results of the ranking and composite indices for each performance indicator. For the biophysical performance indicators, the LGU-/NGO-managed sites of SMR and TRMNP were able to obtain the highest index of 1.87, and the community-managed sites ranked second with an index of 1.05. However, for socioeconomic indicators, the DENR-/NGO-managed sites scored the highest at 1.71, and the community-managed sites were second with 1.05. In the aspects of institutional performance indicators, again, the LGU-/NGO-managed sites together with the community-managed sites were first, garnering an index value of 1.20.

The values suggest that the LGU-/NGO-managed sites of Sagay Marine Reserve and Tubbataha Reef Marine National Park are the best among the MPAs, with an average total composite index of 3.79 compared to the other study sites. DENR/NGO sites garnered an index of 3.53 and the community-managed 3.30. However, the nonstatistically significant differences between these values make it impossible

---

[4] Update: The study was conducted a decade ago; however, all areas remain protected (2013). With the popularity of and increased access to the Internet, more information is now retrievable, such as the number of visitors and tourism income for ARMNP and TRMNP. Data recording has become more efficient, and monitoring of various indicators has improved. Fees, such as user fees, have increased through the years; TRMNP fees, for one, have increased 20 %. This development verifies the effectiveness of MPAs as a conservation tool, while providing social and economic benefits, not just to its nearby communities but to the entire nation as well. A protected area (PA) database was also developed to monitor various types of data. It was launched in 2010 by UNEP, WCMC, IUCN, and WCPA. The database shows that in the Philippines only 2.47 % (1.9 % less than the global average) of marine areas are protected (16,753 km^2 out of 678,751 km^2), which highlights that despite successful cases, institutional will to extend the coverage of protected areas in the country remains weak.

[5] Detailed assessment for each of the indicators can be found on the full report on www.eepsea.net/pub/pb/132403.pdf

to rate one governing institution as better MPA manager than the other. The scores obtained for all sites seem to imply that the MPAs are performing well under all management systems, with scores close to 5.0. This result is an important indication that MPA is a good conservation tool in managing coastal resources in the country.

13.4 Conclusions

The comanaged (LGU/NGO) MPA sites are better managed than MPAs under different institutional arrangements. However, the performance measures across various sites are not statistically different from each other. This result indicates that regardless of the type of institution governing an MPA, as long as its management is efficient and effective in implementing its conservation plan, enforcing its laws and sourcing its funds, an MPA can perform well. Surprisingly, the community-managed sites obtained the lowest indices, refuting one of the study hypotheses that a CBRM scheme is more effective than other institutional arrangements. This attests to the conclusion that one cannot generalize the effectiveness of a particular governing institution: one institution may perform well at one site, while another may do better at another site.

In summary, the study shows that establishing an MPA is better than not having established one. Indeed, a protected area that is managed effectively will result in resource conservation, ecotourism enhancement, effective monitoring, and law enforcement. That the social and economic effects are not fully determined in this study (due to lack of data) does not negate the fact that MPAs can provide economic benefits to the community, specifically through ecotourism and alternative livelihood programs. The results also establish that the best governing institution for an MPA will differ depending on the situation in the field and to generalize whether one regime is better than another is not possible. This result implies that the success of MPAs relies not on any specific type of institution, but rather on the various approaches that are effective at each specific site.

13.5 Recommendations

The following are recommended:

1. Establish CBRMs in small MPAs like AIMS. For bigger areas of jurisdiction, like TRMNP and ARMNP, comanagement styles may be more effective as there are bigger pools of resources available for different protection activities.
2. Provide sustainable alternative livelihoods for the community, appropriate to the site resources and include them in the management plans.
3. Link up to a network of adjacent MPAs that are actively managed. If there are no established MPAs near the existing protected area, encourage others to set up

some. An active network of adjacent MPAs covers a larger geographic area beyond municipal waters. Thus, law enforcers pursuing violators outside their jurisdiction can be assisted by enforcers of adjacent islands or municipalities. This strategy will encourage fisherfolk to stay within their waters, without infringing on other vicinities.

4. Establish strong collaboration between the national government (through the DENR) and the local government units (LGUs). This is especially relevant to ARMNP, BIPLAS, and SIPLAS where LGU involvement is currently minimal. Although lead institutions manage the area, it is very important that a harmonious relationship is developed among these organizations.

Appendix

Table 13.1 Institutional arrangement matrix: essential management rights practiced by DENR-/NGO-managed sites

Rights	BIPLAS	SIPLAS	ARMNP
Use rights	No license fee	No license fee	Restricted fishing area/ sustainable use zone
	Research fee – PHP 200/researcher	User fees[a]:	Fishing license fee – PHP 1000/year
		Docking fee PHP 200/boat	User fee – PHP 300/diver/day
		Filming – PHP 1500/ day	
		Sport fishing – PHP 45/rod	
		Research fee – PHP 40/day for locals and PHP 100/day for foreigners	
Exchange rights	Research fees are nontransferable	User fees are transferable	Licenses are nontransferable
			User fees are transferable
Distribution rights	No gathering of species that is threatened, endangered, or rare although not strictly monitored	No gathering of species that is threatened, endangered, or rare, although not strictly monitored	No gathering of species that are threatened, endangered, or rare
Management rights	PAMB-DENR chaired by RED-DENR	PAMB-DENR chaired by RED-DENR	PAMB-DENR chaired by RED-DENR
	CPPAP site, partnership with BDFI, and PENRO as OIC	CPPAP site, partnership with SEDF	CPPAP site, partnership with NIPS, Inc., and WWF-Philippines
Rights to authority	*Bantay Dagat* volunteers	Park rangers assigned in Sohoton	PASU, park rangers, and *Bantay Dagat* control/regulate fishing and recreational activities
		Control of illegal fishing in the area	
Liability	N/A	First offense, fine of PHP 500 and/or 1 week imprisonment at discretion of court; second offense, PHP 750 and/or 2 weeks imprisonment; third offense, PHP 1000 and/or 3 weeks imprisonment	A fine of not less than PHP 5000 but not more than PHP 10,000 and/or imprisonment of not less than 1 year but not more than 3 years as implemented by PAMB

50 PHP = 1 USD

[a]User fees are only applicable to the Sohoton Blue Lagoon located in the Bucas Grande Island in the Municipality of Socorro

Table 13.2 Institutional arrangement matrix: essential management rights practiced by LGU-/NGO-managed sites

Rights	SMR	TRMNP
Use rights	No license fees	No-take zone
	No user fees	No fishing license fee
		User fees: PHP 2500/foreigner, PHP 1250/local for first visit (for diving),
		PHP 1250/foreigner, and PHP 625/local for consecutive visits
Exchange rights		User fees are transferable
Distribution rights	No gathering of species that are threatened, endangered, or rare	No gathering of any type of species both for tourists and fisherfolk, except for scientific purposes
Management rights	PAMB-DENR cochaired by RED-DENR and Sagay Municipal Mayor	PAMB-DENR chaired by Palawan governor active partnership with WWF-Philippines
Rights to authority	Park rangers patrol the area and enforce park laws	LGU and NGO through WWF-Philippines regulate, enforce, and patrol the area
Liability	Fine of not less than PHP 5000 but not more than PHP 500,000 and/or imprisonment of not less than 1 year but not more than 6 years	Fine and/or imprisonment

50 PHP = 1 USD

Table 13.3 Institutional arrangement matrix: essential management rights practiced by community-managed sites

Rights	PI	BLCC	AIMS
Use rights	No license fees	No license fees	No-take zone for sanctuary
	No user fees	No user fees	No fishing license fee
			User fees:
			PHP 150/diver/day in the sanctuary,
			PHP 75/diver/day outside sanctuary, and PHP 50/underwater camera when diving
Exchange rights			User fees are transferable
Distribution rights			No gathering of any type of species in the sanctuary, small-scale extraction for the rest of the MPA
Management rights	DENR-CENRO, aided by BIDEF	DENR-CENRO, aided by SACRED	Marine management committee till 1996
			PAMB-DENR under CEP-DENR, present
Rights to authority	CENRO, LGU support, and community volunteers	CENRO with PO support	*Bantay Dagat* volunteers from the community
Liability			Fine and/or imprisonment determined by local court; fine is exclusive of the cost of damage pursuant to DAO 25

PHP 50 = USD 1

Table 13.4 Summary of performance indicator values for each study site

Characteristics	DENR/NGO		
	BIPLAS	SIPLAS	ARMNP
Biophysical indicators			
Species composition			
Coral species (genera)			34
Fish species		106	148
Number of habitats		2	2
Kind of fishes (family)			24
Socioeconomic indicators			
HH real income (ave/month)	3872	2302	2633
Access to information/training	63 %	88 %	65 %
Amount of fish catch (kilo/day/HH)	11.5	8.2	35
Financial aid[a]	Yes	Yes	Yes
Livelihood alternative	Wildlife collection, non-timber, livestock, forestry, and farming	Livestock, poultry raising, wood gathering, trading, and	Aquaculture, seaweed, farming, livestock raising, demo farm, ice plant

		contract labor	
Institutional indicators			
Laws/regulation/ordinances	3	3	3
Management board	DENR-NGO(BDFI), PCG	DENR-NGO (SEDF), POs	PAMB-DENR(RED), 15 POs
Monitoring and enforcement	Patrol boat		Patrol boats, lighthouse, ranger station, powerhouse

[a]Finance aid refers to the presence of foreign aid/large funding from other institutions such as the World Bank

Table 13.5 Summary of performance indicator values for each study site

Characteristics	LGU/NGO		Community based		
	SMR	TRMNP	PI	BLCC	AIMS
Biophysical indicators					
Species composition	56	46			31
Coral species (genera)	58	380	68	42	126
Fish species	2	1	2	2	2
Number of habitats		31			27
Kind of fishes (family)					
Socioeconomic indicators					
HH real income (ave/month)	2279	1412	1846	3119	3835
Access to information/training	32 %	48 %	65 %	80 %	42 %
Amount of fish catch (kilo/day/HH)	12.4	5.8	5.1	21.7	13
Financial aid[a]	No	Yes	No	No	No
Livelihood alternative	Livestock raising, fish drying, grouper/squid/crab and oyster culture	Seaweed farming, agriculture, livestock, food processing	Biointensive backyard gardening, money lending, hog dispersal, plantation	Mud crab fattening, mat weaving	Mat weaving, hog dispersal, vending, resort employees

Institutional indicators					
Laws/regulation/ ordinances	3	5	2	9	3
Management board	DENR-LGU, 13 POs	PAMB-DENR (Provl. Gov.) 3 POs, NGOs	DENR-LGU-NGO (BIDEF), 8 POs	DENR-NGO (SACRED, Bankaton), 2 POs	MMC
Monitoring and enforcement	Patrol boats, basic surveil-lance equip-ment, *Bantay Dagat* (60)	Patrol boats, ranger station (PCG/PNP & WWF-Phils.)			PAMB-DENR (RED), 1 PO

[a]Financial aid refers to the presence of foreign aid/large funding from other institutions such as the World Bank

Table 13.6 Composite index ranking for each key performance indicators for all study sites

	DENR/NGO			LGU/NGO			Community-based		
Characteristics	Rank	Index	$N=9$	Rank	Index	$N=6$	Rank	Index	$N=11$
Biophysical indicators	3.00	0.9	30	4.67	1.87	40	3.00	1.05	35
Species composition	2	0.8	40	5	2	40	2	0.64	32
Number of habitats	5	1.75	35	5	2	35	5	1.75	35
Kind of fishes	2	0.5	25	4	1	25	2	0.66	33
Subtotal		3.05			4.75			3.05	
Socioeconomic indicators	3.8	1.71	45	2.4	0.72	30	3	1.05	35
Household income	4	0.92	23	2	0.32	16	4	1.28	32
Livelihood alternative	4	0.72	18	3	0.69	23	3	0.93	31
Access to information	4	1	25	1	0.35	25	3	0.42	14
Amount of fish catch	3	0.69	23	3	0.60	20	3	0.21	7
Financial aid	4	0.44	11	3	0.48	16	2	0.32	16
Subtotal		3.77			1.33			0.95	
Institutional indicators	3.67	0.917	25	4.00	1.20	30	4.00	1.20	30
Management board	4	1.28	32	5	2	45	5	1.50	30
Monitoring and	3	0.99	33	4	1	25	3	0.60	20
enforcement laws/ regulation/ ordinances	4	1.4	35	3	1	30	4	2.00	50
Subtotal		3.67	100		4.15	100		4.10	
Total		3.53	100		3.79	100		3.30	100

A composite measure of accomplishments that include biophysical, socioeconomic, and institutional indicators was derived using multi-criteria analysis (MCA). MCA entails the solution of decision problems that involve multiple (generally conflicting) objectives (Zoints 1992). It entails asking representatives from different key institutions, to rank the given performance indicators. Composite index was determined through the arbitrary ranking of selected key performance indicators as ranked according to the highest value (a scale of five) and lowest value (a scale of one)

References

Cuevas A (1999) A model for analyzing common pool situations in natural resources: institutional arrangements in fishery. Unpublished paper. University of the Philippines, Los Baños

Lewis C (1996) In: The World Conservation Union (IUCN) (ed) Managing conflicts in protected areas. IUCN, Gland

Zoints S (1992) The state of multiple criteria decision making: past, present and future. The 9th international conference: theory and applications in business, industry and government. Springer-Verlag, New York

Chapter 14
Fisheries Management Options for Visayan Sea, Philippines: The Case of Northern Iloilo

Alice Joan G. Ferrer

Abstract The paper examined the nature of Visayan Sea and identified potentially viable options to protect the fish and help fishers earn a living. The focus was on the northern Iloilo fishery, which comprises almost half of the Visayan Sea. Ten management options, identified from various sources, were presented to stakeholders (i.e., fishers, fishery scientists, and fishery managers). These options consisted of status quo, input controls (ban of commercial fishing, ban of commercial fishing with safety nets, marine protected area, closed season, reduction in the number of commercial and municipal fishers, localization, and rotational fishing regime), output control (quota), and the creation of a special management unit. The evaluation was done in two stages: Stage 1 ruled out options with no or low impact on increasing fish stocks. The options that passed stage 1 advanced to stage 2, where each option was evaluated using a set of criteria (impact on fishers, impact on the resources, feasibility, cost to the government, and impact to the community). Feedbacks from the stakeholders were obtained through focus group discussions and in-depth personal interviews. The potentially viable options (fishing bans with and without safety nets, marine protected area, reduction in the number of municipal and commercial fishers, localization and the creation of a special management unit) were discussed. This study recommends using several combinations of management options (rather than a single one) and establishing a single management body (to be pilot tested in northern Iloilo) in the implementation of any management program in all portions of the Visayan Sea and for all its fishers.

Keywords Visayan Sea • Management options • Fishery management • Northern Iloilo

A.J.G. Ferrer (✉)
Division of Social Sciences, University of the Philippines Visayas, Miagao, Iloilo, Philippines
e-mail: aj_ferrer2005@yahoo.com

N. Olewiler et al. (eds.), *Marine and Coastal Ecosystem Valuation, Institutions, and Policy in Southeast Asia*, DOI 10.1007/978-981-10-0141-3_14

14.1 Introduction

14.1.1 Background Information

The Visayan Sea in Central Philippines covers an area of about 10,000 km^2. It is located between 11° and 12° North latitude and 123° and 124° East longitude, bound by 22 municipalities in the four provinces of Iloilo, Negros Occidental (Region 6), Cebu (Region 8), and Masbate (Region 7).[1] It is relatively shallow, with water depths of 40 m or less predominating (Armada 2004). It is divided into municipal (8460.30 km^2) and commercial waters (1539.7027 km^2) (NAMRIA 2001).[2]

The Visayan Sea is exploited by both municipal and commercial fishers from local and adjacent regions.[3] The 1998 Fisheries Code (Republic Act 8550), which incorporates the provision on the use of municipal waters for municipal fishers, has limited the access of commercial fishers to waters beyond the km from the municipality's general coastline. In the past, commercial fishers had unrestricted use of municipal waters at least 7 km outward from the shoreline.

While there are regulations that separate the two fisheries, there is still an overlap between the two fishing grounds.[4] The commercial fishers, who perceive the zoning regulation as unfair and that most fish stocks are found inshore, continue to operate in the municipal waters (Siason et al. 2005). A conflicting provision in the 1998 Fisheries Code allows local government units (LGUs) to issue permits to commercial fishing vessels to operate at 10.1 km offshore. The competition over space and resources between the municipal and commercial fishers has resulted in conflicts and losses for both fisheries. Municipal fishers blame the highly efficient commercial fishing gear for declining fish stocks and low catch.

The poor enforcement of the zoning regulation exacerbates the situation. The local enforcement groups, *Bantay Dagat* (sea patrol), are beset by logistical and operational problems in effectively carrying out their enforcement functions (Ferrer and Defiesta 2005). The commercial fishers also abuse the privilege of continued

[1] Local government units in the Philippines are provinces (the largest unit), which are made up of several municipalities, which are in turn made up of several barangays (the smallest unit).

[2] The 1991 Local Government Code (Republic Act 7160) and the 1998 Philippine Fisheries Code (Republic Act 8550, Article 1, Section 16) stipulate that municipal waters (15 km from the shoreline) are under the jurisdiction of the municipal governments. Commercial waters are under the jurisdiction of national government agencies, such as the Bureau of Fisheries and Aquatic Resources, the Philippine Navy, Philippine Coast Guard, or Philippine National Police-Maritime Group.

[3] The 1998 Fisheries Code defines municipal fishing as fishing within municipal waters using fishing vessels of ≤3 gt or fishing not requiring the use of fishing vessels, while commercial fishing uses passive or active gear for trade, business, or profit beyond subsistence or sport fishing using fishing vessels greater than 3 gt.

[4] The policy on fishers' classification into municipal and commercial has been in effect since the 1960s.

access by operating in municipal waters nearer than the allowable distance of 10.1 km. Moreover, the maximum penalty of PHP 2500 (USD 52.63) for violation, set by the 1991 Local Government Code (Republic Act 7160), is too low to deter violation or encourage compliance with regulations. Therefore, the fishery is basically an open access with regulations that do not help sustain the fish stocks.

The declining fish stocks resulting from overfishing translates to low fishing incomes. The huge number of municipal fishers continues to increase yearly, with more young people entering fishing (Ferrer et al. 2005).[5] They basically fish for consumption; the income earned from any surplus is barely enough for a family to survive.

The commercial fishers, on the other hand, use highly efficient gear, although in recent years, it has been described as a "dying industry" with the number of commercial fishing vessels in decline due to operational losses.[6] For the small-scale commercial fishers (those using active gear and vessels of 3–20 gt)—the dominant players in commercial fishing—the chances of offshore fishing are poor. Their boats are too small and too poorly constructed (made of wood) and lack the necessary equipment (such as geographic positioning devices) to fish further off-shore.[7] The boats are designed for shallow fishing. In Philippine waters, beyond 15 km it is already deep and the fish are not as aggregated as in near-shore waters. Similar to the municipal fishers, the crew has low educational attainment and lacks employable skills, preventing them from moving into other means of employment.

The perception is that the Visayan Sea has not been prioritized and the current management regime fails to address its deteriorating quality. In view of these, there is a need for a more robust regulatory environment that protects the fish and the fishers, particularly since there are few alternatives to fishing. Management measures have to be improved or implemented soon given the threat to the resources, fishers, and the larger community. In the next 5 years, if no intervention is introduced to manage the Visayan Sea sustainably, most of the catch will be composed of organisms belonging to low trophic level of the food chain.[8]

Stakeholders have made some policy proposals—notably a 5-year total ban on commercial fishing, the creation of a special management unit, season closures, and others. The policy problem is to examine the nature of the fishery and to identify policies that would help the dual goals of protecting the fish and helping fishers earn their living. In this paper, an evaluation of fisheries management options for Visayan Sea, with a focus on northern Iloilo, is presented. The aim is to identify potentially viable options and to recommend a future course of action.

---

[5] No updated and accurate number of municipal and commercial fishers in the area is available. There is poor registration of municipal fishers by the LGUs and of commercial fishers by BFAR and MARINA. Armada (2004) reported 18,650 municipal fishing boats operating in the Visayan Sea and 1157 commercial fishing boats counted in different landing sites around the Visayan Sea in 1997.

[6] Interview with key informants from the LGU and national agencies

[7] From the focus group discussions with the study participants

[8] Personal communication with Dr. Nygiel Armada, a marine scientist who has works in the Visayan Sea

14.1.2 The Visayan Sea with Focus on Northern Iloilo Area

The Visayan Sea has always been home to one of the top three fishing grounds in the country (NSO 2001a; Hermes et al. 2004). About two thirds of the total fisheries production in the area comes from commercial fishing. In the early 1990s, signs of resource depletion emerged; the commercial fisheries production went down from 165,256,000 kg in 1992 to 120,267,000 kg in 1995 (NSO 2001a).

Overfishing in the Visayan Sea manifests as low catch rates and changes in species composition. A trawl survey conducted in July 2003 showed that the standing stock biomass of trawlable fisheries resources is 2.23 mt/km^2 (Armada and Campos 2004). In 1948, the trawl survey result was 6.03 mt/km^2. The equivalent estimate from the July 2003 survey is 2.58 mt/km^2 and in March 2007 was 2.06 mt/km^2. This represents a big reduction in the trawlable biomass of 57 % over a period of 55 years and 20 % over a period of 3 years (Armada and Campos 2004). There has, meanwhile, low trophic level of demersal catch, and increasing abundance of small pelagic species has been observed through the 2003 and 2007 trawl surveys been large proportions of squid, cuttlefish, jellyfish, and crab in the total catch, and a reduction in carnivores.

Perceptions of fishers of the Visayan Sea are consistent with the biological data; signs of overfishing have been reported in the forms of decline of catch, rise of low-valued species, increased number of fishers and fishing pressure, the use of highly efficient fishing gear, and the increasing conflicts between resource users (Siason et al. 2005).[9] After a review of reports and studies on Visayan Sea, Vakily (2005) concluded that the Visayan Sea is "definitely not underexploited, most probably fully exploited, and very likely overexploited."[10]

On the other hand, the Visayan Sea is said to be in better shape than the other seas in the country because of its size, shallowness, and openness.[11] Protecting the Visayan Sea has impacts on the surrounding seas because of the interrelationship of stocks. Many observers conclude that failure to properly and effectively regulate the Visayan Sea will inevitably have some negative effects on the Philippine fishing industry inside and outside of Visayan Sea.

A 2006 study on the northern Iloilo fishery found it to be on the "verge of collapsing."[12] Catch landing data at municipal fishing ports also indicate decline of

---

[9] Same perception was gathered from the study participants.

[10] This was his view as director of the Visayan Sea Coastal Resources and Fisheries Management Project (2002–2005) funded by GTZ. The Project aimed for a Joint Management Plan (JMP) to be formulated, adopted, and implemented by the 22 municipalities surrounding the Visayan Sea. The JMP and the project failed to address the main problem of land-based alternative sources of incomes, which drives people to fishing (VisSea Project Final Report 2005).

[11] Personal communication with Dr. Wilfredo Campos, a marine scientist who has works in Visayan Sea.

[12] This was the conclusion made by the Law of Nature Foundation and the Visayan Sea Squadron after conducting a "Reconnaissance Survey of Biological Marine Resources in Three Islands Around Visayan Sea" in 18 sites in northern Iloilo using the method from the Global Coral Reef Monitoring Network.

the resource.[13] Illegal fishing activities persist: Law enforcement is weak and inconsistent. The situation is expected to continue unless management intervention is introduced to reverse the decline.

Northern Iloilo has unique characteristics that challenge fishery management. First, 7 of the 22 coastal municipalities facing the Visayan Sea are in this province (Ajuy, Balasan, Batad, Carles, Concepcion, Estancia, and Carles; Table 14.1). The 15 other municipalities are in the provinces of Negros (4), Masbate (5), and Cebu (6).

Second, jurisdiction over almost half of the entire Visayan Sea falls within the waters of the seven municipalities in northern Iloilo, and the waters of the Municipality of Carles occupy a little more than a third of the entire Visayan Sea (3577 km^2). The vastness of the area makes it the most popular fishing ground for commercial fishers not only from local and neighboring municipalities but also from others in Iloilo and other provinces.

Third, the presence of 12 offshore islands further than 15 km outward from the shoreline in the municipalities of Carles and Concepcion poses a big challenge for fisheries management. Two interpretations exist regarding the point of reckoning of municipal waters: the general coastline and the farthest island.

Fourth, smaller scale commercial fishing boats using different gear are based in northern Iloilo. Table 14.2 shows a little more than half of the owners of commercial fishing boats covered in the latest commercial fishing vessel inventory conducted from January to July 2004 by the Bureau of Fisheries and Aquatic Resources (Central Office) were based in four coastal municipalities of northen Iloilo (Carles, Concepcion, Estancia, and San Dionisio).

Fifth, northern Iloilo commercial fishers are typical to the Visayan Sea. Siason et al. (2005) found that the sociodemographic and economic characteristics of the fishers of Concepcion in Iloilo Province are not so different from the fishers in Escalante of Negros Occidental Province or the fishers from Daanbantayan in Cebu Province.

Sixth, the growth rate of 2.8 (for years 1995–2000) for seven northern municipalities is higher than the national growth rate of 2.32 for the same period, making them the fastest growing municipalities in the Visayan Sea area. Moreover, they registered the second highest projected population density for 2008 at 456 persons/km^2, much higher than the national average of 301 persons/km^2 for the same year (Table 14.3).

Seventh, the seven municipalities are members of the Northern Iloilo Alliance for Coastal Development (NIACDEV) formed in 1998 and registered at the Securities and Exchange Commission (SEC) in 1999. The alliance aims to make northern Iloilo the fish, and other marine products, capital of Western Visayas by addressing coastal resource management problems. While collaboration among

[13] Copies of report from the Field Enumerator for the National Stock Assessment Project Region 6, assigned at Estancia and Fishing Port Complex, are available at the Estancia Municipal Agriculture Office and with the Coastal Resource Management Office, Municipality of Concepcion.

member-LGUs has been achieved at the policy level, very minimal has at the operational level of project implementation (Ferrer et al. 2005).

14.2 Methodology

14.2.1 Data Collection Methods

Multiple field data collection methods were used from September 2007 to April 2008 to generate information for the analysis. These include focus group discussions (FGDs), key informant interviews, secondary data collection, and observation.

FGD was the main method used to collect data from the fishers, which is neither objective nor representative. However, FGD permits the researcher to elicit a multitude of honest views that cannot be obtained as easily through individual interviews, and the participative nature of the discussion permits the researcher to explore and contrast the views of different participants. A total of 27 FGDs with 166 participants were carried out in 3 municipalities (Table 14.4). Separate FGDs were held with the boat owners (36 participants), boat captains and machinists (38 participants), and crew members (92 participants).

This was the first time for commercial fishers in the area to participate in a research project of this nature. Gatekeepers helped organize the FGDs by ensuring they were timed properly with fishing schedules and the FGDs were held close to the residences of the participants. To promote constructive group dynamics and active participation, the participants invited were from different commercial fishing vessels and were not informed in advance of the groupings and members of the discussion group. Within-group status homogeneity was achieved by separating the owners, boat captains and machinists, and crews, thereby creating conditions for open discussion. Each session lasted 2–3 h and was conducted in the local dialect; they were audio recorded and have been transcribed in full.

The information gleaned mainly includes descriptions of fishing operations, perceptions of the state of fishery resources, and perceptions of the likely impacts of the different management options on fishing activities, fishery resources, and the larger community. The FGD participants were asked to rank the management options qualitatively and indicate their acceptance or preference.

Twenty-six key informants were interviewed on various topics.[14] The purpose was to generate information on fishery management, alternative fishing grounds for Visayan Sea commercial fishers, enforcement and monitoring capabilities,

[14] They included 15 fishery managers (4 local chief executives, 2 municipal agricultural officers, 3 agriculture technician in fisheries, 1 coastal resource management officer, 2 municipal legislative council members, 2 municipal administrators, 1 municipal planning and development officer), 6 marine scientists, heads of 3 law enforcement teams (regional superintendent of the Philippine National Police Maritime Group, regional chief of the Philippine Coast Guard, and team leader of the provincial sea watch group), Bureau of Fisheries and Aquatic Resources personnel (Head of Fishery Resource Management Project, Region 6; provincial field personnel). The interviews were conducted person-to-person and through questionnaires sent through electronic mails.

readiness of commercial fishers in the Visayan Sea for offshore fishing, possible alternative jobs and livelihood for the fishers, insights on the different fisheries management options, among others. Interviews were also recorded and transcribed in full. The fishery managers ranked the management options in terms of enforceability/feasibility, susceptibility to corruption, prevalence of illegal fishing, and overall preference using the same qualitative process as in the FGDs.

Demographic observations were limited to the characteristics of the households and barangays where the commercial fishing vessel crew reside, the characteristics of commercial fishing vessels, and the fish landings at Estancia, Carles, and Concepcion fishing ports. The information derived from this method was used to validate information gathered from other methods.

To complement data from the methods mentioned above, relevant secondary data were collected from official and unofficial documents, statistical reports, reports of previous assessments and surveys, research reports, documentation of previous or ongoing projects, experiences in other areas on fish bans, among others. The sources of information include LGUs, Bureau of Agricultural Statistics, National Statistics Office, Maritime Industry Authority (MARINA), Philippine Coast Guard, Bureau of Fisheries and Aquatic Resources (main office), university libraries, and other sources.

14.2.2 The Management Options

The management options range from continuation of the present regime to complete ban of commercial fishing. These management options were identified through various sources (newspapers, previous studies, and during preliminary consultation with commercial fishers, fishery managers, and marine biologists from September to October 2007). The list of management options was presented to the fishers during the focus group discussions and to the key informants during interviews—adjusting the language and format to be appropriate for each—from December to April 2008.

The ten options include status quo (business as usual); a 5-year ban on commercial fishing; a 5-year ban on commercial fishing with safety nets for the affected fishers; establishment of marine protected areas; creation of a special management unit with the authority to manage, regulate, and enforce laws; quota/inter-tradable quota (ITQ) (limitation to total harvest allowed during a fishing season in a fishing ground, where commercial fishers are assigned quotas of fish or limits on their catch and, to allow more flexibility, the quotas can be traded among the fishers); closed season (a geographic closure requiring 4 months of no commercial fishing and opening municipal waters for the next 8 months to commercial fishing); localization (preferential access given to local fishers of a certain municipality over the defined waters of the municipality); reduction of the number of municipal and commercial fishers; or a rotational fishing regime (by dividing the municipal waters within 7 km from the shoreline into several sectors and, each year, opening a number of sectors to commercial fishing after a stock assessment finds the area sustainable for fishing).

14.2.3 Evaluation Procedure and the Criteria

The evaluation procedure involved two stages. In stage 1, the goal was to rule out options with no or low capacity to increase the fish stocks significantly. Options that passed stage 1 advanced to stage 2, at which point evaluation is done in terms of impact on fishers, impact on resources, feasibility, cost to the government, and impact on the community. This analysis is qualitative.

Impact on the Fishers An appropriate fishery management option should take full account of its likely impact on the lives of the fishers as they depend on the sea and the fishery resources for their main livelihood. Important impacts are expected to manifest in employment, income, and other fishing-related behavior. For this criterion, the main sources of information are the fishers themselves with input from local fishery technicians.

Impact on the Resources An option must also be able to address the protection, conservation, rehabilitation, and sustainability of resources. That is, it must be able to balance the needs of conservation with those of the fishing communities. The indicators include volume of catch, size of catch, and composition of catch. For this criterion, the main sources of information are fishery scientists with input from the fishery managers.

Feasibility The success of the management option hinges on the extent to which it can be implemented by the responsible agencies of the national and local government units and has the support of existing laws. Also have a fair chance at success, it must be acceptable to the fishers. In this criterion, the main sources of information are the fishery managers (local government and national agencies), the fishers, and a variety of secondary sources of information (described above).

Cost to the Government Selection of the proper management regulation should take full account of the implementation cost. Cost would include research, enforcement and monitoring, and the cost of coordination and collaboration among the local government units and between the local and national governments. These costs can be regarded as the cost of preserving the fisheries. For this criterion, the main sources of information are the fishery managers (local government and national agencies).

Impact on the Community Likely impacts on the community include impacts on poverty, peace and order, food security, unemployment, demographics, among others. For this criterion, the main sources of information are the fishers, fishery managers and technicians, and local officials.

14.2.4 *Preference Ranking of Options*

The management options were subjected to collective ranking by the fishers during 22 of the 27 FGDs and by the fishery managers during their interviews. Each FGD group or fishery manager was requested to rank the options based on their preferences, with the most preferred option given rank 1 and the least preferred option given rank 10 out of ten options. This procedure allowed the participants to provide direct ordinal judgments and placed minimal demands on their memory.

The rankings were tabulated, and the ranks given to each option were summed. Then overall ranking (the most preferred option has the lowest sum and the least preferred option has the highest sum) was determined and converted into a qualitative description using the schedule: rank 1–2, very high; rank 3–4, high; rank 5–6, moderate; rank 7–8, low; and rank 9–10, very low.

14.3 Results

In stage 1 of the evaluation, the information from personal interviews with marine scientists and past studies was used to evaluate the options for their potential impact on fish stocks. Six were identified as highly impactful for increasing fish stocks: ban of commercial fishing for 5 years; ban of commercial fishing for 5 years with safety nets; marine protected areas; localization; reduction of municipal fishing effort; and establishment of a special management unit (detailed discussion in the next subsection). Four were identified as having no or low potential for increasing fish stocks and were ruled out before advancing to the next stage of detailed evaluation: status quo, quota, rotational fisheries management, and closed season. These four options are immediately presented below.

14.3.1 *Options with No or Low Impact on Increasing the Fish Stocks*

Status Quo No improvement in the biological condition and productive capacity of stocks is expected under the status quo. Trawl surveys conducted at different times in the past show continuous decline in the biomass. Testimonies of fishers in the area and fishery managers point to the overexploited condition of the Visayan Sea and, in particular, the northern Iloilo fisheries: declining catch, smaller fish, disappearance of high-value species and dominance of low-value fish, and longer fishing times, along with increasing numbers of fishers.

Quota/ITQ Under this option, fishers would be limited to specified amounts of catch to reduce overall fishing mortality. Total Allowable Catch (TAC) is stipulated in the 1998 Fisheries Code (Section 8), but efforts toward this direction have been very slow, and many are unconvinced that it could work; generating the maximum sustainable yield (MSY) in a multispecies fisheries has been difficult to impossible. Moreover, quota is expected to have a minimal effect, much less than the fishing ban. If quotas were imposed over a long term (i.e., indefinitely), it may eventually disrupt declining trend in catch rates. How long it would take to reverse that trend would depend on the limit of the quota, but it would likely take well over 5 years. Studies also show that enforcing quotas in multispecies fisheries are impractical because of natural fluctuations in stocks, the fluidity of catch numbers, and the fluidity and dispersion of the fleet (Sprout 1997; Porter 2004)—the fishery has numerous users landing at a multitude of landing point where fish are readily sold directly to the public.

Closed Season This is different from the current closed season for sardines, herring, and mackerel observed every November 15 to March 15. A group of commercial fishers suggested that the municipal waters could be closed for commercial fishing for 4 months and open for the next 8 months. Although the closure might allow for natural restocking and regeneration of fish species, this option might also encourage wasteful expansion of efforts as fishers attempt to make the best of the open areas and seasons. The adverse consequences of the increased fishing intensity and the resulting increase in overall fishing costs could more than offset any beneficial effects the closure might have on the productivity of the stocks.

Rotational Fishing Regime Although a rotational regime is typically used with sedentary species, one municipality has decided to adopt this option for non-sedentary species after dialogue with the local chief executive and the commercial fishers. Under this option, depleted stocks are expected to rebound, but much less than they would under a total ban. However, this is only practical in municipalities with large municipal water areas like Carles and Concepcion. Similar to the quota option, the biological information required by this option is high, and the acquisition of such information is very costly.

14.3.2 Potentially Effective Options

The six options that passed stage 1 of the evaluation were then evaluated in stage 2 for their likely impacts on the fishers and resources, their feasibility, their cost to the government, and their potential impacts on the community. Table 14.5 summarizes the stage 2 evaluation of these options, and Table 14.6 presents the preference ranking of all the options by the fishers.

Ban Commercial Fishing for 5 Years Such a ban would have different impacts on the commercial fishing owners and the crews (boat captains, machinists, and utility workers), with the latter group shouldering more of the burden. For the fishing crew, the ban would likely result to labor dislocation and loss of income[15]; fishing is their main or sole source of livelihood. Their options are limited because they lack other employable skills, a result of low formal education, isolation, a preference for fishing as their way of life, and lack of knowledge or exposure to other opportunities. Alternative sources of income and chances to diversify income sources are very limited in the area. Without other options, some crewmembers expressed shifting to municipal fishing, which, according to them, could not provide enough food for their family or put their children through school.

With a ban, boat owners would be concerned about income and investment losses. The amounts they report having invested vary depending on the sizes of their boats and the gear they use. It ranges from PHP 0.25 million (USD 5.32 thousand) to PHP 2.8 million (USD 59.57 thousand) with a mode value of PHP 1.5 million (USD 31.91 thousand). They expressed a willingness to venture into other businesses or sell their boats anytime there is a buyer.

Although commercial fishing grounds are not accessible to municipal fishers, this option would involve a reallocation in favor of the municipal fishers, who may or may not employ less efficient types of gear. On the other hand, the medium (use active gear and vessels grossing 20.1–150 tons)- to large-scale commercial fishing vessels (use active gear and vessels greater than 150 gt) would likely move their operations elsewhere. The nearest alternatives are Samar Sea (111 km), Camotes Sea (114 km), Tañon Strait (134 km), and Guimaras Strait (154 km). This would mean increasing pressure in other fishing grounds.

With regard to impacts on resources, marine scientists expect a ban would arrest the continual decline in the quality of resources. Five years would be enough for the stocks and habitat to recover, but not to the levels they were before being overexploited (catches in most fishing grounds seemed to have breached sustainable levels around the mid-1970s through the 1980s). It is not certain how great the improvement would be. Catches would likely increase within the first 2 years but slow down, or even stagnate, again soon thereafter. Moreover, the improvements would likely be short-lived if municipal fishing efforts were to intensify, perhaps taking up the slack left by the banned commercial sector. If commercial fishing were to resume after 5 years, conditions would likely revert to pre-ban levels within a year or 2 if there were no measures in place to protect the rebounded resources.

Feasibility of this option is low, even with the support of the 1998 Fisheries Code (Section 128). The concurrence required of the 22 municipalities surrounding the Visayan Sea would be difficult to achieve, for it might mean the loss of political support from the commercial fishing sector. In addition, enforcement would likely be poor given the financial, staffing, and technical constraints faced by the LGUs and national agencies (BFAR, PNP-Maritime Group, and Coast Guard). It is expected that there would be more corruption in law enforcement than the current level as this option faces significant active opposition from commercial fishers (see

[15] The income of the crew would depend on the position in the vessel and the type of gear used.

Table 14.6). With a lack of other means of economic survival, even the strict implementation of a fishing ban would not stop commercial fishing.

The enforcement and monitoring costs of this option would also be high, and the resources to enforce compliance are scarce, despite the autonomy of LGUs to generate fiscal and other revenues. Ferrer and Defiesta (2005) showed large annual budgets (up to PHP 1 million or USD 21.28 thousand) are already set aside by municipalities for regular *Bantay Dagat* operations. Similarly, Campos et al. (2003) found an annual budget of about PHP 614,000 (USD 13.06 thousand) is required to ensure compliance of regulations in Lamon Bay.

Adams et al. (2009) listed the changes that have occurred in Florida, USA, after a net ban. They found declines in landings, effort, dockside values, and retired vessel prices. There were subsequent changes in income sources, business management activities, fishing patterns and species targets, and participation in assistance programs. Similarly, implementation of a ban would not only impact on the livelihoods of fishers but also several other user groups: fish middlemen (commission men, wholesalers, retailers), fishing input suppliers, fish processors, and consumers. A ban would hurt not only northern Iloilo but also the Panay Island; with low fish supplies, prices go up.

Estancia, the center of fish marketing in northern Iloilo, would be greatly affected by a ban. Most of the area's established business enterprises are related to or support the fishing industry. Farming is marginal and cannot absorb the labor that would be displaced. Similarly, economic collapse would be expected in Carles, as 95 % of the municipality is on islands where the topography makes agriculture difficult if not impossible.

With higher unemployment and loss of income, poverty is likely to worsen. This would breed further social and economic problems, thus threatening peace and order. According to the FGDs, increases in illegal fishing activities and other criminal activities, like theft and robbery, would be expected under a ban. The potential of emigration out of the area by the young people looking to find employment to support their families back home would also be high.

Ban Commercial Fishing for 5 Years with Safety Nets With safety nets, the ban may moderate the impacts on fisher income. The fishers would be able to put the assistance program to leave fishing toward other means of employment or diversify their sources of income.

Similar impacts on the resources would be expected, as with the previous option (ban without safety nets), but the scenario after the 5-year ban might be better for displaced fishers succeeding in other livelihoods. This would mean a less rapid return of commercial fishing to pre-ban levels. There would still be a tendency for conditions to return to pre-ban levels, although over a slightly longer period.

The safety nets make the option moderately feasible given its level of acceptability by some fishers. However, most are skeptical (they only moderately prefer this option) because experiences with government assistance programs have bred mistrust. They are doubtful of the government's ability to deliver the needed safety net program.

The cost of this option would be high because of the need to finance the safety net program. The realization of this safety net program would depend on the government's ability to carry off a big budget or find funding source.

Given the cushion provided by the safety nets, impacts on poverty rates might be better than under the ban only option. With successful safety nets, the model results indicated communities would be in the same situation or better, but without successful safety nets, communities would be worse off.

Marine Protected Area (MPAs) MPAs in the Philippines started in the mid-1970s (White et al. 2006). However, many LGUs have yet to be encouraged to adopt MPA as a strategy for coastal resource management and establishment; only a few are currently establishing MPA. Some LGUs do not have the technical expertise in establishing MPA. White et al. (2006) pointed out that as of 2003, out of the 54 MPA projects in the country, only 40 % were being managed sustainably.

In northern Iloilo, MPAs are found in four of the seven municipalities. The total area of the MPAs is far below the total recommended in the 1998 Fisheries Code (15 % of the municipal waters), and most are not managed sustainably. During the FGDs, the fishers shared that they do not even know the location of MPAs in their areas, or that they even exist.

MPAs pose no threat to commercial fishing operations. They are established nearshore, and commercial fishing is supposed to be offshore. It is the municipal fishers who are affected by reducing the area open for fishing in the municipal waters, and yet the fishers expect that their catches and incomes would likely increase with the establishment of an MPA (relative to the status quo). Fishers and managers understand that fish breed and grow near the shore before migrating offshore; hence, they expect higher catch, thus higher incomes with an MPA.

If properly managed, the MPAs are expected to deliver the results desired for northern Iloilo fisheries. It is expected that the impact of MPAs on the resources would be different when accompanied by other measures or alone. Together with a 5-year commercial fishing ban, and assuming the size and location of the MPAs ensure improved recruitment, there would be improvements in the productive capacity of local stocks; catches would increase shortly after implementation and would likely continue to rise, although indiscernibly. If commercial fishing were to resume after 5 years, there would likely still be a return to pre-ban conditions (overexploited), although at a slower rate.

Many experiences with MPAs in the Philippines and elsewhere have shown positive results of increased fish stocks (White et al. 2006; Hilborn et al. 2004; De Guzman 2004; Alcala et al. 2004; Adan 2004; Gell and Roberts 2002; UP-MSI 2002; Dalby and Sorensen 2002). However, Hilborn et al. (2004) warned that MPAs require careful planning and evaluation and appropriate monitoring. Without these, there is a "risk of unfulfilled expectations, the creation of disincentives, and a loss of credibility of what is potentially a valuable management tool."

Implementing an MPA entails minimal costs if the local community can be mobilized to manage it. In De Guzman's 2004 study of an MPA in Danao Bay,

Philippines, community involvement in the management of the bay played a significant role in the successful implementation of the marine reserve and in sustaining the coastal fishery. White et al. (2006) pointed out that the salient characteristic of all successful MPA projects in the country is the strong involvement of communities and the local government in planning and enforcement.

Studies also show positive social impacts of MPAs. According to White et al. (2006), the likely benefits of MPAs include food security and income from tourism. Gell and Roberts (2002) identified positive social impacts of marine reserves to include increased environmental awareness among the locals; educational opportunities for the locals; increased community harmony; reduced conflict between user groups and managers as they work together to make decisions; and venues where fishers, tourist business operators, and local politicians interact are created.

Special Management Unit Under a special management unit (SMU), only licensed fishing activities would be permitted resulting in a reduction in fishing efforts. To the displaced fishers, this would mean losses of income.

Under an SMU regime, the present decreasing trend in catch rates could be arrested within the 5-year period, which is probably the least that would happen since this would likely be the minimum goal for any management body. With improvement in the condition of stocks, catches might slowly increase and continue to do so in the long term, depending on the management interventions taken.

Although an SMU might promise a quick and straightforward solution to different problems in the Visayan Sea, its creation and management powers over the Visayan Sea would violate the 1998 Fisheries Code and the 1991 Local Government Code, both of which bestow the authority and responsibility to manage municipal fisheries upon LGUs and commercial fisheries upon BFAR. As such, it is highly likely to attract significant resistance and insufficient support from the LGUs. Failure to enforce and maintain the SMU, in the face of resistance from LGUs, would likely result in a complete lack of regulation or reversion to the current status quo, potentially for a significant period. The fishers only moderately prefer this option because they doubt the sincerity and capability of the SMU to bring about improvements in the fishery resources and their livelihood.

To function efficiently, the creation and maintenance of the SMU would require substantial budget allocation. The national government could finance this, or it could be financed through a foreign assistance project. The large budgetary requirements make study participants doubt the feasibility of this option.

The SMU might be a viable solution to the fishery problems of the Visayan Sea, in northern Iloilo particularly. While it garners the support of fishery managers and some fishers, local officials whose roles might be threatened by an SMU could resist. Experience with the VisSea Project points to the difficulty of securing support of the LGUs (VISSEA 2005). For the creation of an SMU to be taken seriously by all stakeholders, a law outlining all the necessary requirements—sufficient budget, personnel with management capability, accessible to the people, and clarity on the role, responsibilities, and functions of the SMU—is needed.

SMU models working in the county show positive results. For instance, the Palawan Council for Sustainable Development, a multi-sectoral and interdisciplinary body created in 1992 under the law (RA 7611 Strategic Environmental Plan for Palawan Act), is charged with the governance, implementation, and policy direction of the law. It is directly under the Office of the Philippine President. A second case is the Laguna Lake Development Authority established in 1966 as a quasi-government agency that leads, promotes, and accelerates sustainable development in the Laguna de Bay Region. Regulatory and law enforcement functions are carried out with respect to environmental management, particularly in water quality monitoring, conservation of natural resources, and community-based natural resource management.

Reduction in the Number of Municipal and Commercial Fishers Both the commercial and municipal fishers are the central drivers of Philippine marine habitat degradation and resource extraction (Green et al. 2003). Banning commercial fishing but allowing increases in municipal fishing effects results similar to those modeled for the status quo condition. The 2003 national project entitled "Sustainable Philippine Fisheries Agenda (SuPFA)," run by the Fisheries Resources Management Project (FRMP), recommended the reduction in the number of municipal fishers in nine of the country's bays facing overfishing. For sustainable management, its Fisheries BioEconomic (FISH-BE) simulation revealed a large reduction in the number of municipal fishers (ranging from 31 % to 84 %, or an average of 49 %) would be necessary. If it is assumed that the bays are representative of the municipal fisheries throughout the country, then there are twice as many fishers operating in coastal waters than can be supported sustainably.

A reduction in the number of municipal and commercial fishers is expected to show positive impacts within a year. Even with just a cap (i.e., no further increase), it could arrest further decreases in catch rates within 3–5 years.

Strong opposition among fishers would be expected as the displaced fishers would have nowhere to go. Given the already high rates of unemployment, the coastal communities would be further impoverished, possibly resulting in civil unrest. This option, however, has a chance if safety nets were available. The LGUs need national government support in providing safety nets to the fishers. There must be programs, such as income diversification, in place to help the affected fishers.

Localization Under this option, preferential access to the municipal waters is given only to the local fishers. Currently, the seven northern Iloilo municipalities have an agreement to allow municipal fishers to fish in their local waters, as long as they secure a permit from the municipality and use legal gear. The reason for a policy restricting access rights to outsiders is to ensure fishers of small or resource-poor municipalities have access to better fishing grounds (Ferrer et al. 2005). This privilege would not be extended to fishers from other LGUs, yet the waters of northern Iloilo are popular among fishers from nearby provinces. Local fishers blame transient fishers for the destruction of their fishing grounds.

This option would remove some fishing efforts, which would mean less pressure on the resources. This might be good for the stock overall and over the long term. While this option is highly favorable to LGUs that have large areas of municipal waters, the opposite is true for those with small fishing grounds. It is no surprise that the fishers from Concepcion and Carles very highly support this option: They have the greatest area of municipal waters, covering more than one third of the Visayan Sea, making it a popular fishing ground for transient fishers.

Compared to other options, this option is easier to enforce. For instance, color-coding the boats would help enforcement. However, the municipalities do not currently have the capacity to implement this option given the state of their local sea patrol teams. Additionally, this option would disrupt the harmonious relationship between municipalities in northern Iloilo. There is a strong culture of sharing in the area that would run into conflict with implementing this option, and the area has no tradition of tenure, which may render this option ineffective.

14.4 Conclusions and Recommendations

The current state of the Visayan Sea calls for immediate intervention to arrest further decline in the quality of fish stocks. The future of the people in the surrounding communities rests with the sustainability of the resource. The sea is an important source of livelihood, income, and food for thousands of people. Delay in intervention would only mean the difficult situation many fishers are in now will worsen, particularly if the resource collapses altogether.

The need to protect the fishery resources must be balanced with the needs of the fishers. However, no single potentially viable fisheries management option can fully satisfy the criteria. There are tradeoffs. Any fishing reduction schemes, like banning commercial fishing or reducing the number of fishers (municipal and commercial), could help the stocks recover fast, but would displace a number of fishers. Localization would protect local fishers, but compared to a fishing ban, it would take longer to be effective. An improperly managed MPA would not bring about any improvement in the condition of the habitat and fish stocks. A combination of options would likely be more beneficial than any single option.

Opposition to fishing reduction schemes is likely because they threaten the livelihoods of the locals; for most fishers, fishing is the only option. A safety net program that is responsive to the needs of the fishers and leaves them with the same level of, if not improved, welfare as before any intervention must accompany any fishing reduction option. Short-term support for the fishers could take the form of income loss compensation, credit assistance, or boat buyback programs. Any income support, however, might encourage fishers to return to fishing upon termination of the intervention, thus perpetuating the cycle.

A possible long-term solution to breaking the cycle would be through education and skills training for the young people of coastal communities to dissuade them from entering the fishing industry. Implementation of a reproductive health and population program could address the high rate of population growth in coastal areas. The realization of this safety net program would depend on the government's ability to establish a big budget. The budget could come from the 20 % of revenues from the expanded value-added tax program currently earmarked for social services programs. External funds could also be accessed for financing the program. Funds could be recuperated from the fisheries once they have successfully recovered.

The option of reducing the number of municipal and commercial fishers in the area would require improvement in the registration and licensing systems. The current number of fishers is more than what the resources can carry sustainably. The moratorium of awarding licenses for commercial and municipal fishing in Visayan Sea areas should be strictly observed.

To establish an MPA, community organization and information dissemination should be conducted in parallel with scientific biological surveys/assessments. The support of the community must be solicited to ensure smooth implementation of the MPA.

If only local fishers were able to access the resources, then fishing pressure would be reduced. Agreement by the seven municipalities to allow free mobility of their fishers in the area should be accompanied by a unified fishery code and unified enforcement of that code.

Lastly, the current management regime is clearly not responsive to the needs of the resources or the people. A special management unit (SMU) for the Visayan Sea might offer a better arrangement and could reverse the current course to self-destruction toward one of conservation and sustainable management of coastal resources. The implementation of a pilot program in northern Iloilo should be considered.

Appendix

Table 14.1 Area of municipal waters and number of islands in Visayan Sea

Province	Number of coastal municipalities facing Visayan Sea	Surface area of municipal waters (km^2)	No. of islands within 0–10 km from the shoreline	No. of islands within 10–15 km from the shoreline	No. of islands further than 15 km from the shoreline
Iloilo	7	4258[a]	41	8	12
Negros Occidental	4	900	5	2	1
Masbate	5	598[b]	2	2	1
Cebu	6	1992[c]	19	4	1
Total	22	7748[d]	67	16	15

Source of raw data: Vissea (2005)
[a]For Ajuy, Balasan, Carles, Concepcion, and Estancia. No data for Batad and San Dionisio
[b]For Cawayan, Balud, Esperanza, and Placer. No data for Milagros
[c]For Bantayan, Daanbantayan, Madridejos, Santa Fe, and San Remigio; no data for Medellin
[d]Total for those with data only

Table 14.2 Number of owners and commercial fishing vessels in provinces facing the Visayan Sea

	Iloilo[a]	Cebu[b]	Negros Occidental[c]	Masbate[d]	All
Number of owners	77	38	24	8	147
Number of vessels	130	149	91	11	381
Gross tonnage (mean)	14.07	21.99	58.90	8.13	26.86
Classification of fishing vessel (number)					
Small (3.1–20 GT)	94	98	5	7	204
Medium (20.1–150 GT)	19	44	57		120
Large (>150 GT)		1	1		2
Net tonnage (mean)	9.28	13.91	31.53		15.87
Gear (number)					
Danish seine	44	137	2	9	183
Trawl	43		47		90
Ring net	37	12			49
Purse seine	1		14		15
Bagnet				2	

Source of raw data: BFAR (2005)
[a]Carles, Concepcion, Estancia, and San Dionisio
[b]Bantayan, Daanbantayan, Madridejos, and San Remigio
[c]Cadiz City and Escalante City
[d]Balud, Cawayan, and Milagros

Table 14.3 Projected population and density for 2008 and population growth

Province	Number of municipalities facing Visayan Sea	Estimated total land area (km^2)[a]	Projected total population for 2008[b]	Projected population density for 2008 (persons/ km^2)	Population growth rate (1995–2000)
Iloilo	7	606.4	276,559	456.07	2.87[c]
Negros Occidental	4	1102.7	442,805	401.56	1.26[c]
Masbate	5	1317	194,628	147.78	1.68[e]
Cebu	6	383.6	310,857	810.37	1.69[f]

Source of basic data:
[a]NSO 2001b
[b]NSO 1996
[c]NSO 2001c
[d]NSO 2001d
[e]NSO 2001e

Table 14.4 Type and number of focus group discussions held and participants

	Owner		Boat captain/machinist		Crew		All	
Location	FGD	Participants	FGD	Participants	FGD	Participants	FGD	Participants
Carles	2	14	3	17	4	47	9	78
Concepcion	5	18	1	5	2	9	8	32
Estancia	2	4	1	16	7	36	10	56
Total	9	36	5	38	13	92	27	166

Table 14.5 Evaluation summary of the potentially viable management options

Option	Impact on the income of fishers[a]	Impact on the resources[a]	Feasibility	Cost to the government	Impact on the community[a]
Ban commercial fishing for 5 years	Low	High	Low	High	Low
Ban commercial fishing for 5 years with safety nets	Moderate	High	Moderate	High	Moderate
Marine protected area	High	High	High	Low	High
Special management unit	High	High	Low	High	High
Localization	High	Moderate	Low	high	High
Reduce municipal and commercial fishing effort	Low	High	Low	High	Low

[a]High means good

Table 14.6 Preference ranking of the options by the study participants

Very high	High	Moderate	Low	Very low
Localization	Status quo	Special management unit	Rotational fisheries management	Reduce municipal fishing effort
Marine protected area	Closed season	Ban commercial fishing for 5 years with safety nets	Quota	Ban commercial fishing for 5 years

References

Adams C, Jacob S, Smith S (2009) What happened after the net ban? University of Florida IFAS Extension Publication No. FE123. University of Florida, Gainesville

Adan WR (2004) Multiple small-scale marine sanctuaries in municipal waters: the Magsaysay example. In: In turbulent seas: the status of Philippine marine fisheries. Coastal Resource Management Project, Department of Agriculture-Bureau of Fisheries and Aquatic Resources, Cebu City, pp 232–236

Alcala AC, Russ GR, Maypa AP (2004) Evidence for fishery enhancement effects of marine reserves in Central Philippines. In: Turbulent seas: the status of Philippine marine fisheries. Coastal Resource Management Project, DA-BFAR, Cebu City, pp 215–218

Armada NB (2004) Assessment and management of small pelagic fisheries in the Visayan Sea. Unpublished research report. University of the Philippines-Visayas, Iloilo City

Armada N, Campos W (2004) A hydrobiological survey of the Visayan Sea (trawl, plankton, and oceanography): status of demersal resources. Unpublished research report. University of the Philippines in the Visayas, Iloilo City

BFAR (Bureau of Fisheries and Aquatic Resources) (2005) Inventory of Philippine commercial fishing vessels, January to July 2004. Bureau of Fisheries and Aquatic Resource, Quezon City

Campos MA, Pantoja BR, Manalili NM, Bravo MR (2003) Economic evaluation of fishery policies in Lamon Bay, Quezon, Philippines, EEPSEA research report NO. 2003-RR9. Economy and Environment Program for Southeast Asia, Singapore

Dalby J, Sorensen TK (2002) Coral reef resource management in the Philippines: with focus on marine protected areas as a management tool. , University of Copenhagen, Botanical Institute, Department of Physical Ecology

De Guzman AB (2004) A fishery in transition: impact of a community marine reserve on a coastal fishery in northern Mindanao, Philippines, EEPSEA research report No. 2004-RR6. Economy and Environment Program for Southeast Asia, Singapore

Ferrer AJG, Go J, Cainglet E (2005) Factors influencing the entry of young people into the fishery sector in the Visayan Sea. Final report. VisSea Project of the German Technical Cooperation (GTZ) and the Department of Agriculture- Bureau of Fisheries and Aquatic Resources (DA-BFAR), Quezon City

Ferrer AJG, Defiesta G (2005) Assessment of the operation of the Bantay Dagat in Iloilo Province and Sagay City, Negros Occidental. Final report. VisSea Project of the German Technical Cooperation (GTZ) and the Department of Agriculture- Bureau of Fisheries and Aquatic Resources (DA-BFAR), Quezon City

Gell FR, Roberts CM (2002) The fishery effects of marine resources and fishery closures. WWF-US, Washington, DC

Green SJ, White AT, Flores JO, Carreon MF III, Sia EA (2003) Philippine fisheries in crisis: a framework for management. Coastal Resource Management Project of the Department of Environment and Natural Resources, Cebu City

Hermes R, Armada NB, Aparri RA, Zaragoza EC, Lohmeyer U (2004) Overexploitation in the Visayan Sea: designing a project solution. In: Turbulent seas: the status of Philippine marine fisheries. Coastal Resource Management Project, DA-BFAR, Cebu City, pp 312–317

Hilborn R, Stokes K, Maguire J, Smith T, Botsford LW, Orensanz MMJ, Parma A, Rice J, Bell J, Cochrane KL, Garcia S, Hall SJ, Kirkwood GP, Sainsbury K, Stefansson G, Walters K (2004) When can marine reserves improve fisheries management? Ocean Coast Manag 47(3):197–205

NAMRIA (National Mapping and Resource Information Authority) (2001) MW series of 2001. NAMRIA, Taguig City

NSO (National Statistics Office) (1996) 1995 census-based city/municipal population projections. National Statistics Office, Metro Manila

NSO (National Statistics Office) (2001a) Philippine statistical yearbook. National Statistics Office, Metro Manila

NSO (National Statistics Office) (2001b) 2000 census of population and housing, report no. 3 -population, land area and density: 1990, 1995 and 2000. National Statistics Office, Manila

NSO (National Statistics Office) (2001c) 2000 census of population and housing, report no. 1-f (Region VI – Western Visayas) – population by province, city/municipality and barangay. National Statistics Office, Manila

NSO (National Statistics Office) (2001d) 2000 census of population and housing, report no. 1-g (Region VII – Central Visayas) – population by province, city/municipality and barangay. National Statistics Office, Manila

NSO (National Statistics Office) (2001e) 2000 census of population and housing, report no. 1-e (Region V – Bicol Region) – population by province, city/municipality and barangay. National Statistics Office, Manila

Porter G (2004) Analysing the resource impact of fisheries subsidies: a matrix approach. The Economics and Trade Branch, United Nations Environment Programme, Nairobi

Siason IM, Ferrer AJ, Monteclaro HM (2005) Philippine case study on conflict over use of municipal waters in the Visayan Sea. Final report. In: A cross-country project on fish fights over fish Rights: managing exit from the fisheries and security implications in Southeast Asia. The World Fish Center, Penang

Sprout P (1997) Management issues and quotas in the salmon fishery of British Columbia. Fraser Institute, Vancouver

UP-MSI (2002) Marine protected areas in Southeast Asia. ASEAN Regional Centre for Biodiversity Conservation, Department of Environment and Natural Resources, Los Baños

Vakily M (2005) The status of the Visayan Sea. A paper presented at the Symposium on Managing the Visayan Sea, March 8, 2005. University of the Philippines in the Visayas, Iloilo City
Vissea (Visayan Sea Coastal Resources and Fisheries Management) (2005) Project final project report. Final report. German Development Cooperation on the Management of Natural Resources, Deutsche Gesellschaft für Technische Zusammenarbeit (GTZ) GmbH and the Republic of the Philippines Department of Agriculture, Bureau of Fisheries and Aquatic Resources (BFAR), Iloilo City
White A, Meneses ABT, Ovenden M, Tesch S (2006) Sustaining marine protected areas through continued monitoring and evaluation: the MPA report guide and management rating system. Int Coral Reef Symp 10:1466–1470

Chapter 15
Response of Fishermen to Fishing Control Policies in Southern Songkhla Lake, Thailand: A Field Experiment

Kunlayanee Pornpinatepong, Pathomwat Chantarasap, Jumtip Seneerattanaprayul, Wittawat Hemtanon, and Papitchaya Saelim

Abstract This study used game theory to analyze the extraction behavior of fishermen around Southern Songkhla Lake, Southern Thailand. The field experiments were designed based on the concept of noncooperative game theory to investigate fishermen's behavior in response to four management policy options: external regulations with individual transferable quotas (ITQ) and with individual quotas (IQ) and co-management with ITQ and with IQ.

The analysis examined fishermen's responses under scenarios of high and low fish stocks, which arise due to the seasonal salinity of the lake. Results showed that higher fish stocks encouraged fishermen to increase their extraction. A co-management policy led to significantly better results than imposed external regulation in terms of reducing extraction, ensuring resource sustainability, and lessening violation behaviors. There were no significant differences between "with ITQ" and "with IQ" in terms of reduction of extraction and sustainability of resource use. However, there were significantly less violation behaviors in ITQ than in IQ. ITQ provided more flexibility for fishermen who wanted to increase their extraction while still following conservation guidelines. Therefore, this study recommends implementing ITQ but with appropriate penalties.

Keywords Songkhla Lake • Fishery • Field experiments • Fishermen's behavior • Co-management

15.1 Introduction

15.1.1 Background Information

The Southern Songkhla Lake, or Southern Lake, is one of the three lakes that make up the Songkhla Lake lagoon system. Located on the eastern coast of southern

K. Pornpinatepong (✉) • P. Chantarasap • J. Seneerattanaprayul • W. Hemtanon • P. Saelim
Faculty of Economics, Prince of Songkla University, Songkhla, Thailand
e-mail: kunlayanee.p@psu.ac.th; pathomwat.c@psu.ac.th; jumtipgift@gmail.com; wittawat.h@psu.ac.th; papitchaya.s@psu.ac.th

N. Olewiler et al. (eds.), *Marine and Coastal Ecosystem Valuation, Institutions, and Policy in Southeast Asia*, DOI 10.1007/978-981-10-0141-3_15

Thailand, this lagoon system covers an area of about 180 km^2 ranges from brackish to salt water. It connects to the Gulf of Thailand at Songkhla Town.

Although many species are fished in the lake, the shrimp fishery is the most profitable. Shrimps are available throughout the year. Historically, yearly shrimp income had been almost ten times more than the other catches combined. Thus, many people around the lake are shrimp fishers (Choonhapran et al. 1996). Intensive shrimp fishing effort, however, has contributed to a decline in shrimp populations.

In Southern Lake, the use of standing traps increased from 900 units in 1983 (Sirimontaporn et al. 1983) to 5,250 units in 1995 (Choonhapran et al. 1996). The figure continued to rise further to 8500 units in 1997 (Department of Fisheries 1997 as cited in ONEP 2005), to 29,604 in 2003 (ONEP 2005), and to 23,150 in 2010 (ONEP 2011). This means an overall annual increase of approximately 120 %.

No solution has been found to arrest the continuous decline in shrimp production. This chapter addresses this gap by providing an answer to the following question: What are the potential policy options for controlling fishing effort in Southern Lake? This study investigated fisher responses to proposed management policies that would lead to sustainable fisheries in Songkhla Lake. Specifically, this study (1) identifies viable policy options to encourage sustainable fishery resource use in the Southern Lake; (2) investigates fishers' behavior in response to suggestions that they need to consider the public good when deciding on their personal choices, in terms of their fishing efforts; (3) investigates how sustainable fishery resource use would be encouraged under five scenarios: open access as the base case, external regulation with individual transferable quotas (ITQs) and individual quotas (IQs), and co-management with ITQs and with IQs. These outcomes were used as central information inputs in the design of appropriate mechanisms for implementing potential policy options to control fishing effort in Southern Lake.

15.2 Methodology

15.2.1 Policy Options for Fishing Effort Control

Four policy options were evaluated in terms of their potential to reduce fishing effort in Southern Lake: external regulation, co-management, ITQs, and IQs on the standing traps used by shrimp fishers. These sets of policy options were set up in combinations and compared to a no-policy option. The five cases tested are (1) open access extraction as a baseline case, (2) external regulation with ITQs and a penalty system, (3) external regulation with IQs and a penalty system, (4) co-management with ITQs and a social sanction system, and (5) co-management with IQs and a social sanction system. Extraction decisions and sustainability were used as criteria to compare the different options.

15.2.2 Theoretical Framework

To choose the level of extraction, each individual fisher formed an expectation of his/her catch based on his/her experience and considering the value of the variables at a particular time. As soon as each chose what he/she proposed to catch, the overall level of extraction was revealed to everyone. As the common-pool resources (CPR) dilemma unfolds, the effect on each individual player becomes evident since each fisher's payoff depends not only on his catch and the stock of shrimp but also on the overall amount of extraction.

In order to encourage each fisher to indicate a level of shrimp catch as close as possible to the real situation in Southern Lake, the experimental process started by providing shrimp stock information as it would be, based on the salinity of any given condition.

Model Specifications The payoff function of individual fishers depends on two components: the private and public sectors (Hartwick and Olewiler 1998; Cárdenas 2004; Moreno-Sanchez and Maldonado 2008). The private (net) benefit function is f (x_i, S), where x_i is the level of extraction by individual i and S is stock of shrimp. This private benefit depends on the level of extraction by fishers and their costs in achieving that extraction. In this study, the level of extraction (x_i) of each fisher is assumed to range from 1 to 8 units of extraction.

In studying CPR as the fishery resource use in the Southern Lake, there is non-exclusion but also rivalry. Therefore, the decisions of each fisher using the resource affect the availability of the resource for other fishers, hence the function g (Σ x_i), where $i = 1$, ..., n is used to describe the public benefit function, which reflects the effect of overall fishing on individual payoff. The payoff function for an individual fisher is expressed by Eq. 15.1:

$$\pi_i = f(x_i, S) + g\left(\sum_i^n x_i\right) \tag{15.1}$$

For optimization conditions, it is assumed that $f_x > 0, f_{xx} < 0, f_s > 0, f_{ss} < 0$, and $g_x < 0$, $g_{xx} > 0$ and in the private function, there are private benefit and cost functions. The private benefit, αx_i, is the revenue that the fisher receives from his/her output. It depends on α, the revenue parameter, such as the price in a competitive market and level of extraction. On the other hand, the cost function, $\beta x_i^2/2S$, depends on his/her level of extraction, x_i, and the available shrimp stock, S. This cost is negatively related to the shrimp stock—the smaller the existing shrimp stock, the greater the per unit cost of extraction to the fisher, such as search cost—but the cost is positively related to the level of extraction. In addition, it also depends on the cost parameter, β.

The public function (CPR), γ Σ (e-x_i), $i = 1$, ..., n, implies the fisher's payoff increases if social fishing declines and vice versa. The term Σ (e-x_i) represents the benefit transferred from resource abundance to the individual; the coefficient γ

measures the proportion of the benefit transferred and e is the capacity of level of extraction, so equals 8 for this study. Rearranging the payoff function ($\gamma ne - \gamma \Sigma x_i$), fisher's behavior was analyzed using Eq. 15.2:

$$\pi_i = \alpha x_i - \frac{\beta x_i^2}{2S} + \gamma ne - \gamma \sum_i x_i \tag{15.2}$$

In setting the experiment described in the following section, individual payoff was used as a benchmark to analyze the outcomes of three scenarios: a standard, noncooperative game; a socially optimal outcome (extraction at an optimal level); and a noncooperative game with external regulation (extraction as regulated by the government or based on community agreement).

Scenario I: Standard Noncooperative Game In this scenario, each fisher is free to choose the level of extraction at which he/she considers his/her personal payoff as maximized. The objective function is

$$\max_{x_i} \left[\alpha x_i - \frac{\beta x_i^2}{2S} + \gamma ne - \gamma \sum_i x_i \right]$$

The necessary conditions for maximizing the payoff are described as follows:

$$x_i = \frac{(\alpha - \gamma)}{\beta} S \tag{15.3}$$

This scenario is a dominant strategy because it is not dependent on the level of extraction by other fishers. As all fishers follow this rule, the Nash equilibrium defines the game. The optimal level of extraction is a function of the expected shrimp stock and is negatively related to the cost parameter, β. It is also negatively related to the transferred benefit coefficient, γ, because the greater the benefit each fisher obtains from an abundant stock, the more incentive he/she has to reduce his/her personal extraction. If this coefficient is high, fishers can maintain the same revenue with a smaller investment and level of extraction.

Scenario II: Socially Optimal Outcome In order to achieve the socially optimal outcome, the social welfare function for this scenario was established by adding the individual payoff function:

$$\max_{x_i} \sum_i \left[\alpha x_i - \frac{\beta x_i^2}{2S} + \gamma ne - \gamma \sum_i x_i \right]$$

The necessary condition is

$$x^{soc} = \frac{(\alpha - \gamma n)}{\beta} S \tag{15.4}$$

As with a general noncooperative game, the socially optimal outcome is less than the individual's maximum outcome because the social level of extraction has to consider the transferred benefit for everyone, $\gamma\ n$.

Scenario III: Noncooperative Game with External Regulations To enforce external regulation, the fisher was fined if he/she extracts more than his/her share of the resource, based on the defined socially optimal level. Excess extraction is determined by the following expression: $x_i - x^{soc}$. However, even if the government chose to institute such a regulation, it faces a very large cost in monitoring compliance, and it is common for the government to try to control by fining transgressors. Moreover, fishers who are willing to break such regulations also attempt to hide their law-breaking activities. Since government oversight can never be 100 %, there is a chance fishers will benefit from their rule breaking. All of these possibilities need to be considered. Thus, two additional coefficients must be included in the payoff function, namely, value of penalty (m) and probability of fine (ρ). The objective function of the fishers then becomes

$$\max_{x_i} \sum_i \left[\alpha x_i - \frac{\beta x_i^2}{2S} + \gamma n e - \gamma \sum_i x_i - \rho m (x_i - x^{soc}) \right]$$

Considering imperfect modeling, the first order condition of external regulations is

$$x_i^r = \frac{(\alpha - \gamma - \rho m)}{\beta} S \tag{15.5}$$

This still tends toward a socially optimum outcome, as the fishers have to take the probability of a fine, $\rho\ m$, into account, which should act to reduce the overall level of extraction. Moreover, the level of the fine is a value that can be changed to test the responding behavior of the fishers.

Parameterization and the Payoff Table To compute the payoff table, the parameters of the model were defined. The level of the stock value was specified as 60 for high abundance and 30 for low abundance. The values of the revenue (α), cost (β), and transfer benefit parameters (γ) were selected to be consistent with Mahbuntham's 2002 study, while the values of ρ are drawn from the Moreno-Sanchez and Maldonado study (2008). The fine value, m, was set at 500, based on the pilot test that preceded the experiment. Mahbuntham (2002) recommended the socially optimal level to be 50–60 % of the existing level of extraction.

With these parameter assumptions, the model predicts socially optimum extraction at four units of extraction for each individual (or 20 units for overall extraction) when stock is high and two units of extraction for each individual (or ten units overall) when stock is low. In contrast, the noncooperative outcome is eight units of extraction for each individual for high stock and six units for low.

Given these parameters, the payoff function (Eq. 15.2) was used to calculate the payoff table (shown in Tables 15.2 and 15.3). This payoff table represents the major characteristic of CPR: the less others harvest, the more individual payoff increases.

Payoff is also affected by stock level. Because payoff is the difference between revenue and cost, and each individual's share of the catch is lower in a low stock situation, when costs are still high, the revenue may not cover the cost. This situation does not exist in the high stock case when high costs are less than the high revenue. These parameters and the payoff table were used to prepare the field experiment.

15.2.3 Experimental Design and Procedures

Specifications and Assumptions The field experiment conducted with the fishers followed the methodologies of Cárdenas (2004) and Moreno-Sanchez and Maldonado (2008). The details of the experiments are specified in Table 15.1. Each treatment was applied to different groups. The assumptions of the experiment were as follows:

- The experiments were conducted under a steady-state static fishery framework.
- The resource stock (S) was available at two levels, low and high abundance. The optimal catch at each level was assumed to be 10 and 20 fishing traps, respectively.
- In order to reduce the effect of information relating to the price of fishing traps on extraction behavior, this price was assumed to be constant over the experiment. Although the price per trap of fishing traps was fixed, it was still considered a variable input since the fishers could increase the number of fishing traps when they wanted a higher catch.
- Shrimp fishery was selected for this experiment.

Value of Stock and Individual Payoff Because of the varying salinity in Southern Lake, which significantly impacts shrimp stock, is an uncontrollable factor (Pornpinatepong et al. 2010), this was included as a variable in the experiment. It was assumed that the shrimp stock, S, would vary according to the salinity of the lake.

To simulate the impact of changing salinities on the stock, the game divided salinity into two variables: high and low salinity. Salinity was specified as "S" in each round and announced to the fishers before the game began; they had to decide on their level of catch.

When the experiment was conducted using the variables, a player chose his/her level of extraction (x_i) and then his/her payoff (π_i) was calculated. The payoff depended not only on his/her personal level of extraction but also on the total level of extraction by the group (Σx_i, i = 1,..., 5) and the size of the shrimp stock.

To highlight the dilemma of common-pool resources and the tradeoff between individual and social benefits in Southern Lake—that individual profits depended not only on individual decisions but also on the group decisions—the share of shrimp received by each fisher was divided into two clusters. If the sum of the level of extraction of the group was more than 10 in low salinity conditions and 20 in high salinity conditions, the shrimp share was deemed to be low due to overfishing. The shrimp share was deemed high if the sum of the levels of extraction of the group as a whole was less than 10 in the low salinity condition and 20 in the high salinity condition.

Procedure An overview of the experiment is presented in Fig. 15.1. First, the fishers were informed of the salinity level, and then the fishers chose their levels of extraction. The state of high or low catch was then calculated by adding the individual extraction levels of the five players in the group. Based on the catch, the exact value of S was specified depending on the salinity situation, which was used to calculate each fisher's payoff.

Experiments The experiments were conducted with groups of five players, in a non-finite repeated game of 20 rounds, performed in two phases. During the first phase, the players faced no rules (Fig. 15.2). During the second phase, the players operated within a framework of external regulations with ITQs (and with IQs) or co-management with ITQs (and with IQs).

In every round of the first phase, each player had to decide his/her level of extraction ranging from 1 to 8 units. These decisions generated points, convertible to monetary units that each player received. In general, the individual payoff was generated by individual extraction, with higher individual extractions resulting in higher profits. However, these profits depended not only on individual decisions but also on the decisions made by each five-person group as a whole as well as uncontrollable factors (i.e., salinity and rainfall at the study site) that relate to shrimp stock.

Following the first ten rounds, the groups played a second phase, with ten additional rounds subject to different rules. A set of rules was assigned to each group, which related to different policy options being evaluated by the players. The rules of the game were drafted based on initial information received from the focus group interviews and pretests and finalized as follows:

Steps and Assumptions Involved in the Experiment The experiment started with the moderator providing the players with general information on CPR and the optimal level of extraction (i.e., that the quantity of shrimp catch depends on three factors: the number of fishing traps used by each fisher, the total number of traps in the lake, and the shrimp stock (the optimal level of extraction is the highest level of catch received by one of the traps set at a particular time). This information was repeated after rounds 5, 10, and 15 as reminders.

1. The level of salinity (high or low), which affects the stock size of shrimp in the lake, was provided. Low salinity was indicated for rounds 1–5 and 11–15 and high salinity for rounds 6–10 and 16–20.
2. The payoff for each player was announced in special monetary units designed for the experiment, which was then converted to rice grain at the end of the experiment.
3. The total quota for extraction was limited to 20 units under high salinity conditions and ten units under low salinity conditions for the policies, external regulation, and co-management scenarios.
4. Before round 11, players were informed about the current situation in the game of shrimp fishing in the lake. The optimal level of shrimp fishing in the lake was explained (i.e., in a low stock situation, the optimal level of extraction is 10 unit traps/group; in a high stock situation, it is 20 unit traps/group). The ten rounds were then conducted by following the same steps as in phase 1.

Rules of the Game After the first ten rounds, when every group did the open access trial, they were then separated into five different groups; each group was given a different policy scenario to investigate how fishers respond to various policy options. The second phase of sessions (rounds 11–20) had the following assumptions and rules:

- *Case 1: Open access in the long term (baseline case)* The players continued playing under the same assumptions and rules as in the first phase for ten additional rounds. This enabled the assessment of baseline values through analyzing the results of doing nothing for regulating the long-term fishery resource use in the lake.
- *Case 2–5: Experimentally testing reactions to different management policies* Players continued playing under a different set of rules. Before starting the second phase, information on ITQs and IQs was provided through a script delivered by the moderator playing the role of a government representative. The rules for these sessions were as follows:

  Rule 1: Under external regulations with ITQs (and IQs) (Cases 2 and 3) Each fisher was allotted an extraction quota of two units under the low salinity situation and four units under high salinity conditions. A fisher was allowed to sell his/her quota to another fisher or to the government by auction (if ITQ but not IQ). Nobody was allowed to fish more than his/her quota or without a quota. If a fisher extracted a higher catch than the allowable quota (or any catch without a quota), he/she had to forfeit the part of his/her catch that was in excess of the quota.

To simulate a real-life situation of imperfect government enforcement of the rules, the monitoring process was randomized. Every player had a 1 in 10[1] chance of being monitored during each round (based on the method suggested by Moreno-Sanchez and Maldonado 2008). Operationally, imperfect monitoring was simulated using ten balls representing the ten players, nine white and one red. Each player blindly picked a ball from the box, and the player who picked the red ball was inspected. If the inspected individual violated the rules, he/she would pay an assigned penalty (m). This penalty was secretly subtracted from the individual payoff before the payoffs were announced. This draw was made every round during the second phase, so any player could be monitored more than once, and some not at all. All the other rules remained the same as in the first phase of open access, and decisions, as well as fines, were kept confidential throughout the course of the game. Communication among the players was confined to discussion of buying and selling extraction quotas.

Rule 2: Under co-management with ITQs (and IQs) (Cases 4 and 5) Under these rules, the group received extraction quotas of 10 units per group under the low salinity scenario and 20 units per group under high salinity. After they receive information on the system (i.e., whether with ITQ or IQ) and before starting the second phase, the group was asked to select a leader. A group discussion to define the rules of the game then followed, which lasted around 10 min and covered how the initial quotas would be distributed, the penalty system for individuals who violate the rules of extraction for their group, and how the quotas received could be transferred if fishers wish to do so (in the case of IQs, this topic was omitted).

After the group discussion, the group leader reviewed their agreement about the rules with the fishers and the next round began. Once the group members made their private decisions about their levels of extraction for the 11th round and secretly submitted them to the game moderator, the total extraction was announced and everyone was told of their payoffs. The moderator had another minute to talk to the group to confirm the rules of extraction that they had set, and then the group had an additional minute to discuss any adjustments to their rules of extraction they wanted to make before the next round began. This repeated until all rounds had been played and the game was finished. Any revenue occurring out of the fishing control process was returned to community in the game.

After the experiment, participants were asked to take part in a discussion about their perceptions about the policy options proposed, the procedure of the games they played, and their socioeconomic information.

[1] In the developing world, the probability of being monitored is one tenth of total monitoring (Moreno-Sanchez and Maldonado 2008).

15.2.4 Preparation for the Field Experiment

A preliminary study was done to pilot the experiment planned for the main study. Five presets of information were prepared:

(a) Baseline information for the first phase of every session
(b) Set scripts for particular cases during the second phase of the experiment, which were the open access situation, external regulation with ITQs (and with IQ), and co-management with ITQs (and with IQ)
(c) Rules for particular cases
(d) A payoff table
(e) A postgame interview questionnaire

To prepare realistic information, focus group discussions (FGDs) were conducted to obtain the initial information needed for information sets a, b, and c. The information included, for example, the perceptions of fishers about common-pool resources, the strategies fishers use for fishing, and the kinds of information fishers use when/if they decide to increase or decrease their takes. This information was then used to design the experimental game and its rules.

Payoff Method The payoff table (shown in Tables 15.2 and 15.3) was designed to encourage the players to behave naturally; a special monetary unit was designed and paid to the players round by round to stimulate some enthusiasm during the experiment. The participants received a payoff based on their own game results, and this special monetary unit was then exchanged for real rice grains at the end of the experiment. Rice was selected as the payoff in this experiment rather than actual cash to avoid any effects on the social customs in the community and to provide some real incentive that was good for everyone.

Questionnaire Design The questionnaire included three sections: the socioeconomic backgrounds of the fishers, evaluation of the appropriateness or difficulty they had in understanding the game they were asked to play, and perceptions about the policy options proposed in the experiment.

15.2.5 Participation in the Experiment

The economic experimental games were played by 205 fishers who were randomly selected from seven fishing villages around Southern Lake. They owned more than ten fishing traps and had more than 2 years of experience with fishing traps. The pilot testing was conducted on 35 individuals (i.e., 10 undergraduate students and 25 fishers). The number of participants in each of the five sets of experiments was six for the open access model, seven for external regulation with ITQs, seven for external regulation with IQs, seven for co-management with ITQs, and seven for co-management with IQs.

15.2.6 *Pilot Test*

The objectives of pilot testing experiment were to (1) test how understandable the procedures and rules of the game would be for the participants in the actual study, (2) test how reliably the games would help the analysis of how fishers would respond to policy options, and (3) help gauge the probable effects of various policy options on the sustainability of the resource. The performance of the players during the experiments and the discussions after each game were recorded to determine how well the players understood the game. Personal interviews were also conducted to elicit more information for full understanding of the fishers' perceptions about the games. The latter information was used as input for correcting the procedures of the actual games used in the experiment. The questionnaires and all documents needed for the experiment were finalized based on information gained through pretesting.

15.3 Results

When considering the potential policy options for managing fishing efforts, the response of the local fishers to the options proposed and tested by this study should be considered. Results show that the factors affecting fisher's decisions on resource extraction are the abundance of shrimp stock, the management options that are available, whether a fisher has a high school education, and a fisher's religion. The different violation behaviors of the players depending on the tradable and non-tradable options available were also investigated. The findings are summarized below:

Between the management methods of external regulation and co-management, the latter exhibited better results for both the sustainability of the resource and reduction of extraction. The advantages and disadvantages of the two methods are compared as follows:

- For co-management, the advantages are that it has (1) the lowest extraction level, (2) a higher percentage of reduction in extraction (relative to open access), (3) extractions that achieve optimal levels in both high and low stock situations, (4) the highest average number of rounds with high abundance (most sustainable), and (5) fewer violations because the rules are designed by the group and social sanctions tend to have a stronger effect on fisher behavior. However, there are several disadvantages to co-management: (1) the process is more complicated and needs more cooperation, (2) most fishermen are concerned that this option is impossible, and (3) these could burden community leaders.
- For external regulation, the advantages are (1) lower extraction levels compared to open access and (2) a higher average number of rounds with high abundance (higher sustainability than open access). There are more disadvantages to external regulation: (1) extraction is above optimal levels in both high and low stock

situations, (2) it is difficult to identify an optimal penalty, and (3) it requires efficient monitoring.

Under high and low fish stock scenarios (due to seasonal salinity in the lake), higher fish stocks (high salinity) encouraged fishers to increase their extraction. In a high salinity situation, the analysis showed that the rate of extraction was reduced (compared with open access) significantly lower than in the low salinity scenario. Although the significance of the extraction decisions under these two scenarios was statistically different, there was no apparent effect on sustainability (i.e., the percentage of rounds that a group reached the optimal sustainability level[2]). Moreover, the percentage of rounds the stock was fished as the optimal level was greater in the co-management situation than in the external regulation option at both stock levels.

Meanwhile, between the tradable and non-tradable options, the ITQ management results were not different from management with IQs for either sustainability or fisher extraction decisions. These results show that fishers do not behave differently with either the ITQ or IQ systems. Considering the violations or extractions without allowable quotas, IQs had a higher percentage of violations than ITQs because ITQs provide an option to buy and sell permits. However, when the price of permits grew higher than the cost of the penalty and with inefficient monitoring, the players tended to violate the rules rather than buy permits. In the co-management options, there were far fewer violations than with the regulation options because the penalties were designed by the group and partially enforced through social sanctions.

The violations occurring in the ITQ and IQ options provided important information since the fisher's behavior was significantly different in these two cases. The analysis found that there were more players extracting more than the allowable quotas in the IQ option than in the ITQ option (other variables being equal). This result confirms that the ITQ option can be expected to elicit better results as it offers an opportunity for fishers to trade their quotas, which provides more flexibility for those who wish to increase or reduce their extraction.

The information given on the optimal level of extraction influenced the decisions of many players to reduce their extraction rates. However, on its own it was not enough to ensure the optimal level would be reached; appropriate management options were needed as well.

[2] The sustainability of the resource can be measured as the proportion of rounds that a group reached optimal level (or high abundance) during a phase of the game; it ranges from 0 % to 100 %. The closer the number to 100, the more the number of the times the resource was high in abundance and, therefore, the more the sustainability in the use of the resource (Moreno-Sanchez and Maldonado (2008). In this study, therefore, the percentage of rounds that reached optimal extraction value was observed for sustainability.

15.4 Policy Recommendations

Co-management was found to have more positive benefits than external regulation. Co-management, as a game option, resulted in the lowest extractions, with the extractions often achieving optimal levels in both high and low stock conditions. Co-management also allowed for community participation in designing the fishing rules, which had the effect of creating stronger social sanctions against those who violated the rules. Therefore, co-management is recommended as the best option for fishing effort control in Songkhla Lake.

The co-management process, however, is more complex to manage than the other options, requires greater cooperation between the fishers, and burdens the community (i.e., the opportunity cost of time used for community management). Most fishers also think that although it would be the best system, it would be very difficult to implement due to potential resistance from those who wish to fish freely. Therefore, in order to apply the co-management option, there must be solid support from government agencies, particularly during the initial stages.

The external regulation option was found to be the second best option as it could at least reduce the overall extraction and improve sustainability compared to open access. To be an effective control, external regulation would require strict monitoring, particularly during the high salinity/high stock season. In addition, an appropriate fine system would have to be specified at a level high enough for effective fishing control.

When comparing tradable and non-tradable quotas, ITQs are recommended for either the external regulation or co-management options, as the analysis confirmed there were significantly fewer violations under the ITQ scenarios. ITQs provide more flexibility for the fishers who wish to increase their extraction. However, fine values need to be carefully specified and should not be lower than the value of the quotas traded in the market in they are to discourage those who would violate the rules instead of buying quotas. To apply a quota system for Songkhla Lake, further studies that examine additional factors are required (especially how to distribute quotas at the initial stage of implementation). The system would need to be fair for everyone in the community.

The analysis indicates that providing information on optimal levels could affect fisher's behavior and reduce extraction, but not necessarily to optimal levels. Additional management options would be needed for sustainable fishery resource use.

Appendix

Table 15.1 Parameters used in the model

Symbol	Definition	Value
x	Individual level of extraction	1–8 (traps)
n	Number of participants of each group	5 (persons)
e	Maximum level of extraction	8 (traps)
α	Revenue parameter	150 (units of payoff)
γ	Proportion of transfer benefit	20 (units of payoff)
β	Cost parameter	700 (units of payoff)
ρ	Probability of being fined	0.1
m	Fine value per unit extracted over quota	500 (units of payoff)

Table 15.2 Payoff table (stock in the low salinity scenario, $S = 30$)

The rest of payoff	Individual payoff							
	1	2	3	4	5	6	7	8
4	838	933	1005	1053	1078	1080	1058	1013
5	818	913	985	1033	1058	1060	1038	993
6	798	893	965	1013	1038	1040	1018	973
7	778	873	945	993	1018	1020	998	953
8	758	853	925	973	998	1000	978	933
9	738	833	905	953	978	980	958	913
10	718	813	885	933	958	960	938	893
11	698	793	865	913	938	940	918	873
12	678	773	845	893	918	920	898	853
13	658	753	825	873	898	900	878	833
14	638	733	805	853	878	880	858	813
15	618	713	785	833	858	860	838	793
16	598	693	765	813	838	840	818	773
17	578	673	745	793	818	820	798	753
18	558	653	725	773	798	800	778	733
19	538	633	705	753	778	780	758	713
20	518	613	685	733	758	760	738	693
21	498	593	665	713	738	740	718	673
22	478	573	645	693	718	720	698	653
23	458	553	625	673	698	700	678	633
24	438	533	605	653	678	680	658	613
25	418	513	585	633	658	660	638	593
26	398	493	565	613	638	640	618	573
27	378	473	545	593	618	620	598	553
28	358	453	525	573	598	600	578	533
29	338	433	505	553	578	580	558	513
30	318	413	485	533	558	560	538	493
31	298	393	465	513	538	540	518	473
32	278	373	445	493	518	520	498	453

The payoff table represents the major characteristic of CPR: the lower the total catch, the more the individual payoff increases. In each row, the individual payoff increases from left to right (increasing the extractions). In addition, the payoff of overall catch (bottom right corner) is lower than low catch (top left corner)

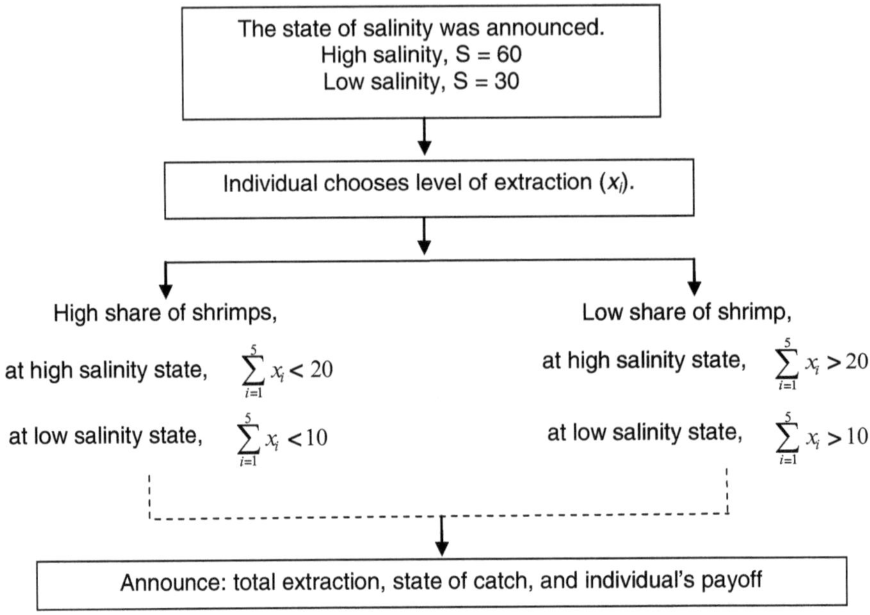

Fig. 15.1 Process of calculating the payoffs

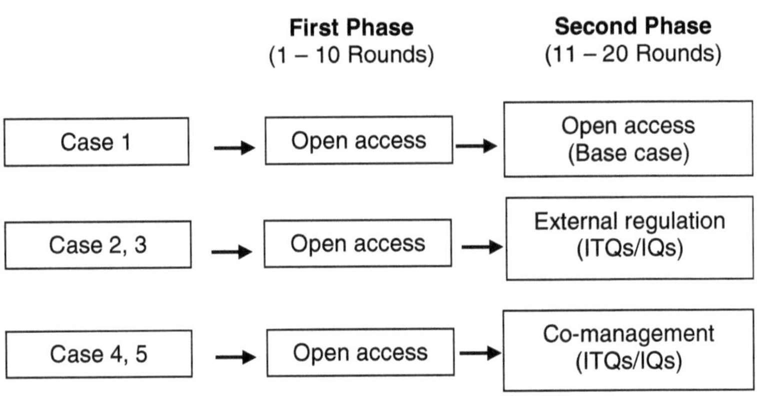

Fig. 15.2 Dynamic component of the experiments

Table 15.3 Payoff table (stock in the high salinity scenario, $S = 60$)

The rest of payoff	Individual payoff								
		1	2	3	4	5	6	7	8
	4	844	957	1058	1147	1224	1290	1344	1387
	5	824	937	1038	1127	1204	1270	1324	1367
	6	804	917	1018	1107	1184	1250	1304	1347
	7	784	897	998	1087	1164	1230	1284	1327
	8	764	877	978	1067	1144	1210	1264	1307
	9	744	857	958	1047	1124	1190	1244	1287
	10	724	837	938	1027	1104	1170	1224	1267
	11	704	817	918	1007	1084	1150	1204	1247
	12	684	797	898	987	1064	1130	1184	1227
	13	664	777	878	967	1044	1110	1164	1207
	14	644	757	858	947	1024	1090	1144	1187
	15	624	737	838	927	1004	1070	1124	1167
	16	604	717	818	907	984	1050	1104	1147
	17	584	697	798	887	964	1030	1084	1127
	18	564	677	778	867	944	1010	1064	1107
	19	544	657	758	847	924	990	1044	1087
	20	524	637	738	827	904	970	1024	1067
	21	504	617	718	807	884	950	1004	1047
	22	484	597	698	787	864	930	984	1027
	23	464	577	678	767	844	910	964	1007
	24	444	557	658	747	824	890	944	987
	25	424	537	638	727	804	870	924	967
	26	404	517	618	707	784	850	904	947
	27	384	497	598	687	764	830	884	927
	28	364	477	578	667	744	810	864	907
	29	344	457	558	647	724	790	844	887
	30	324	437	538	627	704	770	824	867
	31	304	417	518	607	684	750	804	847
	32	284	397	498	587	664	730	784	827

References

Cárdenas JC (2004) Norms from outside and from inside: an experimental analysis on the governance of local ecosystems. For Pol Econ 6:229–241

Choonhapran A, Ratanachai C, Meechookunt A (1996) Assessment of fisheries resources in Songkhla Lake during 1994–1995. Technical paper no. 4/1996. National Institute of Coastal Aquaculture, Department of Fisheries. Ministry of Agriculture and Cooperatives, Songkhla

Hartwick J, Olewiler N (1998) The economics of natural resource use, 2nd edn. Addison-Wesley, New York

Mabuntham J (2002) Species diversity and quantity of aquatic animals caught with standing traps in the outer Songkhla Lake. Master of Science thesis, Prince of Songkhla University, Songkhla

Moreno-Sanchez R, Maldonado J (2008) Can co-management strategies improve governance in a marine protected area? Lessons from experimental economic games in the Columbian Caribbean. Paper presented during the 16th annual conference of the European Association of Environmental and Resource Economists, 25–28 Jun 2008, Gothenburg
ONEP (Office of Natural Resources and Environmental Policy and Planning) (2005) Master plan for Songkhla Lake basin development project (final report). ONEP, Bangkok
ONEP (2011) Songkhla Lake basin development project: the reviewing of master plan for Songkhla Lake basin 2013–2016 (executive summary). ONEP, Bangkok
Pornpinatepong K, Kiripat S, Treewanchai S, Chongwilaikasaem S, Pornsawang C, Chantarasap P, Chandee C, Jantrakul P (2010) Pollution control and sustainable fisheries management in southern Songkhla Lake. EEPSEA research report. EEPSEA, Thailand
Sirimontaporn P, Chaiyakam K, Sae-Chu K (1983) The study of aquatic catch by standing traps in outer Songkhla Lake. Research report. No 16/1983. National Institute of Coastal Aquaculture (NICA). Department of Fisheries, Thailand

Chapter 16
The Impacts of Artificial Reefs on the Income of Artisanal Fishers in Terengganu, Malaysia

Shaufique Fahmi Sidique, Kusairi Mohd Noh, Gazi Md Nurul Islam, and Aswani Farhana Mohd Noh

Abstract The marine fishery resources in Malaysia have declined considerably over the past decades primarily due to overfishing and overcapacity. In response, the government has deployed artificial reefs (ARs) to conserve and enhance the natural fish stocks in marine waters. Specifically, AR deployment was expected to rehabilitate the degraded fish stocks and secure the livelihoods of the fishers along the coastal villages. This study examined the impact of ARs on fisher households' income in Terengganu, Malaysia. The data for this study was obtained from a survey involving 312 fisher respondents in 3 contiguous districts, namely, Besut, Setiu, and Kuala Terengganu. The results showed a difference in the value of catch of fishers fishing in an AR and those fishing in non-AR area. Fishers fishing near AR areas were found to benefit from higher monthly catch value. The regression models indicated that fishing in an AR area helps reduce the vessel operating costs. The catch value also significantly differed between fishers using an inboard-powered vessel and those fishers using outboard-powered vessels (on an average obtained MYR 5935 per month and MYR 3126 per month, respectively). Fishers using inboard-powered vessels were also less dependent on ARs as they were able to fish further to the sea.

Keywords Artificial reefs • Terengganu • Fisher's income • AR deployment • Malaysian fisheries

---

S.F. Sidique (✉) • K. Mohd Noh • A.F. Mohd Noh
Institute of Agricultural and Food Policy Studies, Universiti Putra Malaysia, Serdang, Malaysia
e-mail: shaufique@upm.edu.my; kusairimn@gmail.com; aswani85@gmail.com

G.M.N. Islam
School of Economics, Finance and Banking, College of Business, Universiti Utara Malaysia, Kedah, Malaysia
e-mail: gazinurul236@gmail.com

N. Olewiler et al. (eds.), *Marine and Coastal Ecosystem Valuation, Institutions, and Policy in Southeast Asia*, DOI 10.1007/978-981-10-0141-3_16

16.1 Introduction

The Malaysian fisheries sector plays a significant role in providing employment, alleviating poverty, contributing to export earnings, and achieving food security. Fisheries production has been increasing due to the increase in fishing capacity and technological efficiency. Talib et al. (2003) and Stobutzki et al. (2006) found that fishing vessels operating in all fishing zones have reached close to full technical efficiency and capacity utilization.

The increase in fishing activities, however, has led to overfishing and declining fishery resources. Demersal species are overexploited and the level of fishing effort is beyond the maximum sustainable yield (Talib et al. 2003). Biomass has also declined on both the east and west coasts of Peninsular Malaysia (Stobutzki et al. 2006). The trend in fisheries exploitation has affected the resources and livelihoods of small-scale fishers whose methods are less efficient relative to those of large-scale fisheries, such as trawl nets.

In curbing the problems of overfishing and depleting fish stocks, the Malaysian government has deployed artificial reefs (ARs), or *tukun tiruan*, in various locations of the marine waters. ARs refer to any material intentionally placed in the marine waters to influence physical, biological, or socioeconomic processes related to marine life (Sutton and Bushnell 2007).

The government has made significant investments in ARs amounting to MYR 131 million[1] from 1975 to 2010. The primary objectives of the AR project are to rehabilitate fish stocks (Polovina and Sakai 1989; Yodee 1994; Whitmarsh et al. 2008), to increase catch and income, and to reduce the operating costs of fishing. ARs can also deter trawlers from encroaching into inshore waters. Trawler encroachment is a major issue in the fishery where exclusive rights to fish in the waters of less than 9.26 km from the shore belong to the artisanal fishers. If the objectives of ARs are met, then it will help secure the livelihoods of artisanal fishers in coastal villages.

Studies have shown that ARs have increased catch (more than in areas without ARs) in Southern California (Johnson et al. 1994), increased fish productivity and job creation in Mexico (Kolian and Sammarco 2006), and increased tourism and recreational activities (Sutton and Bushnell 2007). Similarly, a study on the use of fish aggregating devices (FADs) in Mauritius found increases in fish productivity (Venkatasami and Mamode 1995).

In Malaysia, however, studies that examine the effectiveness of ARs in achieving their economic and socioeconomic objectives are lacking. This study addresses this gap by examining the contribution of ARs in increasing the catch and income of artisanal fishers. In the absence of baseline data to compare catch before and after AR deployments, the study examines the difference in the value of catch between fishing in an AR and a non-AR area. Other factors that affect catch value such as demographic characteristics, fishing effort, and also the general characteristics of a

[1] USD 1 = MYR 3.347.

fishing district are controlled. This study covers three fishing districts in Terengganu, a state located on the east coast of Peninsular Malaysia. The findings provide empirical evidence that ARs benefit artisanal fishers in Terengganu and serve as valuable information for policy makers in need of evidence to justify investment in ARs.

16.2 Background of Fisheries in Terengganu

16.2.1 Fisher Profile and Catch

Located on the east coast of Peninsular Malaysia, the state of Terengganu is an important fishing ground. Fishers number more than 11,000, representing about 13 % of the total number in Peninsular Malaysia (DOF 2010). The fishers are mainly artisanal fishers with low catch rates. In Terengganu, there are currently approximately 2000 licensed artisanal fishers (DOF 2010).

Artisanal fishers obtain license from the Department of Fisheries (DOF) for fishing. License fees are based on the type of gear used and are paid annually. The majority of fishing boats are powered by inboard or outboard motors. The inboard-powered boats are mostly fitted with 40 hp engines. Most of the boats are made from wood or fiberglass, with the latter being more commonly used in the artisanal fishing community. The artisanal fishers in Terengganu use a variety of fishing gear, such as hand lines, long lines, traps, gill nets, and drift nets. Each gear type has a number of different designs and sizes to catch a specific type of species, and fishers usually use more than one type at a time.

The incidence of poverty is very high among the artisanal fisher households in Terengganu. A recent study conducted by the Department of Fisheries reported that 47.6 % of the fisher households surveyed earn incomes below the poverty line (Latiff 2008). In the hopes of mitigating this problem, artisanal fishers are always a group targeted in fisheries development policies. To improve the livelihoods of artisanal fishers, the government has provided subsidies to licensed fishers for such costs as fishing boats and motors. Fishers are also given fuel subsidies and a monthly allowance of RM 200 as an incentive to fish actively. There is also an output incentive that the Fisheries Development Authority of Malaysia (LKIM) provides based on the quantity of fish sold at the jetty. However, the subsidies attract more people to fishing, creating more competition among them (Porter 2001).

Fishing is permitted year-round in Terengganu. The peak season for fishing is in the months of June–August. Fishers usually make one fishing trip daily and squid fishing is done at night. Traps are usually retrieved twice a week. Between November and January fishing activities are minimal, and shrimp catching is confined to the near shore due to strong winds and high tidal waves. Between

November and March, trawlers with relatively small boats are allowed to operate at any distance from the shoreline.

Fishing is prohibited in the coral reef areas surrounding the marine parks, and marine park rangers closely patrol these areas. Some fishers unable to afford the capital for equipment, such as nets, boat, and engine, receive support from middlemen in return for selling their catch to the middlemen at relatively low prices.

Marine landings in Terengganu accounted for around 10 % of the total landings in Peninsular Malaysia as of 2008 (DOF 2008). Total landings in Peninsular Malaysia were 8,229,000 metric tons (MT) in 1977, but had declined to 1,055,000 MT by 2008. The fish landings in Terengganu also declined from 1,286,000 MT to only 105, 000 MT in 2008 (DOF 2008). One of the possible reasons for the decline is overexploitation of fishery resources (Jahara 1988; Saharuddin et al. 2012).

16.2.2 Artificial Reefs in Terengganu

Deployment of artificial reefs (ARs) in the Terengganu coastal waters started in 1975 by the DOF and in 1980 by the LKIM. The early ARs deployed by the DOF were made of tires, derelict boats, and concrete cuboids until they were replaced in 1985 by PVC, concrete, ceramic, reef balls, and abandoned oil rigs (Ali et al. 2011). The ARs in sandy habitat are deployed within 9.26 km away from the shore, at an average water depth of 15–20 m; they are also specially designed to protect the habitat from trawl netting. On the other hand, the ARs in muddy habitat, known as *unjam-unjam*, are relatively small and placed on the seabed within 5.55 km from the shore. The LKIM *unjam-unjam* are installed at the request of artisanal fishers. They function as fish aggregating devices (FADs) that are mainly focused on increasing resource access, catches, and incomes of artisanal fishers.

The government has allocated significant funds to increasing the ARs in Terengganu. About 268 ARs of various sizes and designs have been deployed on the seabed by six districts in the state. The costs of ARs vary by size and materials. The average cost of a small AR (concrete reef balls or cuboids) is about MYR 1,000 (USD 298), and the cost of a large AR that is comprised of 20 units of concrete blocks is about MYR 300,000 (USD 89,632). The total cost of AR deployment by the DOF in Terengganu from 2006 to 2010 was approximately MYR 9.8 million (USD 2.9 million). This accounted for the largest share, at 30 % of the total amount of AR investments in Malaysia during the same period.

Most of the ARs in Terengganu have been deployed in Besut, Setiu, and Kuala Terengganu districts. A description of the characteristics and locations of ARs deployed in these areas is provided below.

In Besut. ARs made of concrete blocks were deployed in the district of Besut between April 2009 and April 2010, to a total of 107 units. Two of the ARs (AR1 and AR2) are situated in the northwestern region of the district, whereas AR3 is located in the southeastern part. Table 16.1 displays some of the features of the ARs in Besut. AR1 is the largest, with 71 structural blocks, while AR2 and AR3 have 22 and 17 units, respectively. In terms of size, however, AR3 covers the largest area (45,043 m^2), followed by AR1 (22,400 m^2), then AR2 (7146 m^2).

All units in AR1 were deployed in April 2009 and are of the soft base type. Almost all of the units in AR2 were deployed in August 2009, and they consist of a mix of soft base and cube structures. The units in AR3 are of the soft base type; half were installed in 1990 and another half were installed in April 2010.

In Setiu. AR4 (85 units) and AR5 (44 units) are located almost side by side in the northern part of the district of Setiu (Table 16.2). In terms of size, AR5 covers more than four times as much area as AR4. All units in AR4 and AR5 are cuboids and were deployed in September 2007.

In Kuala Terengganu. This district has the most number of AR units at 181 covering the largest area among the 3 districts. The majority of the units are situated in the northern-central part of the district. The units are divided into six complexes, with the following distribution (Table 16.3): AR6 (19 units), AR7 (18 units), AR8 (68 units), AR9 (26 units), AR10 (16 units), and AR11 (10 units).

The units in AR7 were installed between November and December 2007 and most are concrete. All units in AR8 are cuboids; six were deployed in 2008 and the rest were deployed in April 2010. The largest complex, AR8, consists entirely of cuboids. Almost all of the units were deployed in September and October 2007; the rest were installed later, in May 2008. All units in AR9 are cuboids and were deployed in October 2007 and May 2008. Almost all units in AR10 and AR11 consist of cuboids, and most were deployed in April and May of 2008.

16.3 Methodology

16.3.1 Data and Methods of Data Collection

The data was collected during face-to-face interviews with fishers using a structured questionnaire based on the results of focus group discussions (FGDs) and key informant interviews (KIIs). FGDs and KIIs provided input to improve and validate the questionnaire, improve the design of the survey, and gather other pertinent information related to the study. Three FGDs were conducted in the fishing community in Kuala Setiu to collect data on fishing practices, the gear used, fisher livelihood assets, and institutional arrangements. Fishers selected to participate in the FGDs use hand lines, drift nets, and traps.

The KII conducted in the DOF headquarters in Putrajaya engaged senior officers from the Licensing and Resources Development Section and also the Southeast Asian Fisheries Development Center (SEAFDEC). The KIIs in Kuala Setiu were conducted with representatives from the Area Fisherman Association and officers from the DOF. The data collected included basic AR locations, seasonal fishing calendars, and harvest levels in general. The questionnaire was piloted on

24 participants. The data collectors were officers from the DOF and LKIM who were trained in collecting data. The final questionnaire consisted of separate sections soliciting information on household characteristics, which include demographic information, information on fishing gear and values, and information on fishing locations, catch, and costs. A nautical map was used to assist the fishers identify and mark their approximate fishing grounds. Computer software was used to generate latitude and longitude coordinates of fishing locations using the fishers' markings on the maps. These fishing coordinates were used to determine distance of the fishing grounds to the AR.

16.3.2 Sampling

The final survey was conducted in Besut, Setiu, and Kuala Terengganu. The list of artisanal fishers for each district was obtained from the DOF office in Terengganu. The list of fishers contains each fisher's name, Malaysian identity card number, vessel license number, engine power, gear used, and village name.

The 312 respondents were randomly selected from a list of 926 artisanal fishers from the 3 sampling districts. The selected respondents were notified in advance of the interview with the assistance of DOF officers and representatives from the local fishers associations. The interviews were conducted late afternoon or evening when fishers would most likely be at home, during the months of September–November 2011. The final number of respondents considered in the study was 276 because 36 survey participants were found to be in commercial fishing. Sample respondents from Besut, Setiu, and Kuala Terengganu represented 32 %, 28 %, and 40 % of the total sample, respectively (Table 16.4).

16.3.3 Model

The study examines the cross-sectional variations in fishing income among fisher households. The dependent variable is household fishing income, which is the value of fish catch during different fishing seasons by the households. The independent variables are the time allocated for fishing, other effort variables such as type of vessel engine and crew size, age of household head (a demographic variable), and also other fishing site-specific variables such as AR characteristics and fishing intensity at a particular site. Fishing time is separated into two: the time spent fishing in AR locations and the time spent fishing in non-AR locations. More specifically, household fishing income function (Y) is defined as:

$$Y = f(E, D, AR, S) \tag{16.1}$$

where, E = fishing effort
D = demographic factors
AR = fishing location relative to the center of an AR area
S = other site-specific variables

Fishing effort variables are expected to have a positive relationship with income. Also, fishing in AR areas is expected to increase catch. Site-specific variables include factors such as fishing intensity and AR sizes. Fishing intensity is expected to decrease the amount of catch, and large AR areas are expected to be more productive, thereby contributing to higher catch.

Three models were developed to estimate catch value. The calculation of the dependent variable was based on the reported average monthly catch for three different fishing seasons, which include the main fishing season, shrimp season, and squid season. The three models are formally presented below:

$$VALUE_i = \beta_0 + \beta_1 AGE_i + \beta_2 INBOARD_i + \beta_3 CREW_i + \beta_4 HOUR_i + \beta_5 COST_i \tag{16.2}$$

$$VALUE_i = \beta_0 + \beta_1 AGE_i + \beta_2 INBOARD_i + \beta_3 CREW_i + \beta_4 HOUR_i + \beta_5 COST_i + \beta_6 AR_i + \beta_7 ARCOST_i \tag{16.3}$$

$$VALUE_i = \beta_0 + \beta_1 AGE_i + \beta_2 INBOARD_i + \beta_3 CREW_i + \beta_4 HOUR_i + \beta_5 COST_i + \beta_6 AR_i + \beta_7 ARCOST_i + \beta_8 FISHER_i + \beta_9 BIGAR_i \tag{16.4}$$

where subscript i represents the fisher household and the βs are the coefficients for the variables. Model 1 (Eq. 16.2) is the basic catch value model that uses only demographic and fishing effort variables as independent variables. Models 2 (Eq. 16.3) and 3 (Eq. 16.4) are extensions of the basic model. Model 2 incorporates the AR effect on the value of the catch and operating costs. Model 3 takes into account district-level effects by including fishing congestion and the number of large AR variables.

AR is a dummy variable representing fishers who are fishing in at least one AR area (Table 16.5); a fisher is considered to be fishing in an AR area if the fishing distance from the center point of an AR area is less than 20 km. *FISHER* is the number of licensed fishers in the district, and it measures the congestion due to fishing activities in a district. *BIGAR* represents the number of large AR complexes in a district; it quantifies the effect of large AR complexes within a district on fish catch. An AR area is considered large if the number of units that define it is greater than 40. There is one large AR complex in Kuala Terengganu, and there are three in Besut and two in Setiu.

16.4 Results

Table 16.6 presents the results of the three regression models of the catch value. The three models are statistically significant with an adjusted R-squared of 0.170 (Model 1), 0.177 (Model 2), and 0.178 (Model 3). All of the fishing effort variables in Model 1 are statistically significant to at least 1 %, and crew size is significant to 10 %. All coefficients in Model 2 are also statistically significant, except for crew size. The coefficients in Model 3 are statistically significant, with the exception of crew size and the district-level variables. Age was found to have a negative effect on catch value in all the three models. This means that as a fisher gets older, the value of his/her catch declines. The three models also suggest that as the number of fishing hours per trip increases, so does the value of the catch.

All three models suggest a positive relationship between operating costs per trip and the value of the catch, which confirms the expectation that an increase in fishing effort is likely to result in a higher catch. The effect of operating costs on catch is larger in Models 2 and 3 where the interaction between cost variable and fishing in the AR area *is* included. The negative sign of the interaction terms in both models implies fishers who fish in AR areas benefit from lower operating costs per trip. A plausible conclusion is that fishing in an AR area reduces the search costs of a productive fishing location.

The AR variable in both Models 2 and 3 is statistically significant. The AR coefficient in Model 2 indicates that fishing in an AR location increases the value of the catch per month by MYR 1,122 (USD 335). Model 3 has a slightly higher AR coefficient indicating that the value of the catch per month increases by MYR 1, 422 (USD 425) when a fisher operates in an AR location. Both of the district-level coefficients have the expected signs in Model 3 (although neither is statistically significant); more fishers in a particular fishing ground reduce the value of the catch, and the larger the AR complex deployed in a district, the greater the value of the catch.

There is also significant difference in catch value between fishers with outboard-powered and inboard-powered vessels. Fishers operating inboard-powered vessels, on average, earn MYR 5,935 (USD 1,773) per month, whereas fishers with outboard motors earn MYR 3,126 (USD 934). The fishers with inboard-powered vessels are less dependent on ARs as they are able to fish further to the sea. The results on the catch value of inboard fishers imply that the costs and fishing benefits are not equally distributed among the poor fishers. This result also suggests that ARs are not able to reduce the gap between inboard and outboard fishers. Therefore, this issue relating to equity needs to be addressed by other policy measures to ensure access is limited to artisanal fishers with outboard-powered vessels.

16.5 Conclusions and Recommendations

Current knowledge supports AR as an important tool for enhancement of Malaysian marine fisheries. Results of this study indicate that fishers fishing near AR areas had higher monthly catch values compared with those who fish away from AR areas.

While it is particularly true that the deployment of ARs may have attracted more fish, fishing pressure surrounding the ARs is also likely to increase in the near future. The absence of clear demarcation of AR sites—especially if access to fishing remains virtually free—makes enforcement of AR regulations difficult. This situation could lead to overfishing and a continued threat to fisheries stock. Therefore, it is important that other management measures be instituted alongside AR sites to ensure sustainable use of fisheries. This will require collaboration and coordination between government agencies with the mandate to protect the marine resources. If the AR areas were to be taken as MPAs, then they would fall under the jurisdiction of the Department of Marine Park Malaysia, Ministry of Natural Resources & Environment. Also, the support of the Malaysian Maritime Enforcement Agency (MMEA) is important in the enforcement of marine policies, including those for fisheries.

Appendix

Table 16.1 Distribution of ARs in Besut

Besut ARs	AR1	AR2	AR3
No. of units	71	22	17
Size(m^2)	22,400	7146	45,043
Density (units/m^2)	0.00317	0.00308	0.00038

Table 16.2 Distribution of ARs in Setiu

Setiu ARs	AR4	AR5
No. of units	85	44
Size(m^2)	176,350	814,200
Density (units/m^2)	0.00048	0.00005

Table 16.3 Distribution of ARs in Kuala Terengganu

Kuala Terengganu ARs	AR6	AR7	AR8	AR9	AR10	AR11
No. of units	19	18	68	26	16	10
Size(m^2)	250,681	409,194	269,912	226,467	76,472	77,622
Density (units/m^2)	0.00008	0.00004	0.00025	0.00011	0.00021	0.00013

Table 16.4 Frequency and percentage distribution of fisher population and sample by district

District	Population	Percent	Sample	Percent
Besut	296	32 %	88	32 %
Setiu	309	33 %	78	28 %
Kuala Terengganu ARs	321	35 %	110	40 %
Total	926		276	

Table 16.5 Definition of variables and summary statistics

Variable	Definition	Mean	Std. dev.
Value	Value of catch per month (MYR)	4,656	3324
Age	Age of HH	50.28	11.09
Inboard	Inboard engine (yes = 1, no = 0)	0.14	0.35
Crew	Crew size	1.70	0.61
Hour	Fishing hour per trip	9.14	3.90
Cost	Operating cost per trip (MYR)	75	103
AR	Fishing location is an AR area (yes = 1, no = 0)	0.58	0.49
ARCOST	Interaction between *AR* and *COST*	49.42	107.23
Fisher	Number of artisanal fishers in the district	322.71	10.95
	Kuala Terengganu—309		
	Besut—321		
	Setiu—337		
Bigar	Number of large AR areas in the district	1.92	0.84
	Kuala Terengganu—1		
	Besut—3		
	Setiu—2		

$n = 267$

Table 16.6 Regression models (dependent variable = *VALUE*)

	Model 1		Model 2		Model 3	
Item	Coeff.	S.E	Coeff.	S.E	Coeff.	S.E
Age	−52.36	16.68***	−48.80	16.69***	−48.05	16.70***
Inboard	1689.92	590.23***	1448.87	605.04**	1264.07	622.32**
Crew	558.49	344.06*	431.01	332.58	510.38	337.04
Hour	123.5	48.15***	129.05	48.04***	124.53	48.23***
Cost	4.95	2.05***	18.68	7.32**	18.53	7.42**
AR			1121.90	580.51*	1421.99	714.19**
ARCOST			−14.33	7.24**	−14.29	7.43*
Fisher					−28.68	20.99
Bigar					436.54	348.57
Constant	4592.76	1162.76***	3642.70	1247.98***	11789.63	6501.54*
Adjusted R-squared	0.170		0.177		0.178	
$N = 276$						

*Statistically significant at the 10 % level
**Statistically significant at the 5 % level
***Statistically significant at the 1 % level

References

Ahmad A, Raja Bidin RH, Yuttana T (2011) Enhancing management of fishery resources through intensified efforts in habitat conservation and rehabilitation, fish for the people. Southeast Asian Fisheries Development Center (SEAFDEC), 9(2):10–20. ISBN: 1685–6546

DOF (Department of Fisheries) (2008) Annual fisheries statistics, Putrajaya. Kuala Lumpur

DOF (Department of Fisheries) (2010) Annual fisheries statistics, Putrajaya. Kuala Lumpur

Jahara Y (1988) Fishery management and regulation in Peninsular Malaysia: issues and constraints. Mar Resour Econ 5(2):83–98

Johnson TD, Barnett AM, Demartini EE, Craft LL, Ambrose RF, Purcell LJ (1994) Fish production and habitat utilization on a southern California artificial reef. Bull Mar Sci 55 (2–3):709–723

Kolian S, Sammarco PW (2006) The socio-economic impacts of sustainable fisheries. Job Creation And Marine Aquaculture, EcoRigs, Louisiana

Latiff I (2008) Kajian Pendapatan Nelayan Tradisional 2008. Department of Fisheries, Putrajaya

Polovina JJ, Sakai I (1989) Impacts of artificial reefs on fishery production in Shimamaki, Japan. Bull Mar Sci 44(2):997–1003

Porter G (2001) Fisheries subsidies and overfishing: toward a structured discussion. United Nations Environment Programme (UNEP), Nairobi. UNEP website http://www.unep.ch/etu/etp/acts/capbld/rdtwo/FE_vol_1.pdf

Saharuddin AH, Ali A, Lokman MH, Salihin W (2012) Recent developments and management of artificial reefs (ARs) in Malaysia. In OCEANS, (Yeosu 2012), 2012 IEEE symposium, Kuala Lumpur, pp 1–23

Stobutzki IC, Silvestre GT, Abu Talib A, Krongprom A, Supongpan M, Khemakorn P, Armada N, Garces LR (2006) Decline of demersal coastal fisheries resources in three developing Asian countries. Fish Res 78:130–142

Sutton SG, Bushnell SL (2007) Socio-economic aspects of artificial reefs: considerations for the great barrier reef marine park. Ocean Coast Manag 50(10):829–846

Talib AA, Isa MM, Ismail MS, Yusof S (2003) Status of demersal fishery resources of Malaysia. In: Silvestre G, Garces L, Stobutzki I, Luna C, Ahmed M, Valmonte-Santos RA, Lachica-Aliño L, Munro P, Christensen V, Pauly D (eds) Assessment, management and future directions for coastal fisheries in Asian countries. WorldFish center conference proceedings No. 67. WorldFish, Penang

Venkatasami A, Sheik Mamode A (1995) Fish-aggregating devices (FADs) as a tool to enhance production of artisanal fishermen: problems and perspectives. Albion Fisheries Research Centre, Mauritius

Whitmarsh D, Santos MN, Ramos J, Monteiro CC (2008) Marine habitat modification through artificial reefs off the Algarve (southern Portugal): an economic analysis of the fisheries and the prospects for management. Ocean Coast Manag 51:463–468

Yodee K (1994) Bioeconomics of small-scale fisheries in the artificial reef areas in Ranong Province, Thailand. FAO (67):67–86

Chapter 17
Economic Evaluation of Implementing Minimum Legal Size on Blue Swimming Crab Fishery in Indonesia

Rizal Bahtiar, Nuva, Dessy Anggraeni, and Nia Kurniawati Hidayat

Abstract The blue swimming crab (BSC) or *rajungan* is one of the most valuable fish products in Indonesia. It is mostly caught by using bottom gillnets and collapsible traps and by using – to a lesser extent – the now-illegal shallow bottom trawls (baby trawls or mini trawls). Based on recent catch reports, there are indications that the average size of the landed BSC is becoming smaller (as indicated in the increase in catching effort). There are also signs that BSC is no longer producing maximum economic benefits in some regions of Indonesia. In many landing areas, even small crabs (150 crabs/kg) are being caught and harvested. In order to keep its sustainability, it is important to regulate the size of crabs to be caught by fishers. Some management options have already been established, including catch restrictions on the minimum legal size of 8 cm.

This study used a bioeconomic model to assess and compare the consequences of unrestricted fishery (no regulation) and implementation of minimum legal size on the stock condition of BSC. The study determined how much the profit loss would be due to the smaller-sized crab catch and due to the decline in fish catch (overfishing) and how much the implementation of a minimum legal size will change the productivity and income of fishermen over time. The effect of this restriction policy was evaluated using cost-benefit analysis. In the considerably depleted area, the projection showed that implementing a minimum legal size (MLS) policy would increase the stock of BSC in the early years of the projection. The model also showed that for the next 10 years, crab stock with size bigger than 8.5 cm would increase in the early years of the projection and then achieve a stable condition in the following years. Thus, MLS policy can help crab stocks to recover. Meanwhile, in the area where stock condition is considerably good, the crab stock would stay stable in the next 10 years. From the profitability point of view, MLS

R. Bahtiar (✉) • Nuva • N.K. Hidayat
Department of Resource and Environmental Economics, Faculty of Economics and Management, Bogor Agricultural University (IPB), Bogor, Indonesia
e-mail: rizal_bht@yahoo.com; nuva2@yahoo.com; niakurniawatihidayat@gmail.com

D. Anggraeni
Sustainable Fisheries Partnership (SFP), Bogor, Indonesia
e-mail: dessy.anggraeni@sustainablefish.org

N. Olewiler et al. (eds.), *Marine and Coastal Ecosystem Valuation, Institutions, and Policy in Southeast Asia*, DOI 10.1007/978-981-10-0141-3_17

would benefit those fishermen who catch the mature BSC. Likewise, implementing such policy would also benefit the miniplant industry (BSC-processing business) in the long run (assuming a 10-year period). However, their income would be slightly affected in the short term.

Keywords Bioeconomic modeling • Blue swimming crab • Indonesian fisheries • Minimum legal size • Profitability analysis

17.1 Introduction

This study examines the economic value of implementing a minimum legal size for the blue swimming crab fishery and compares that with the continued operation of an open access fishery. The goal is to contribute to the development of the Fishery Management Plan for the Blue Swimming Crab (BSC) in Indonesia and formulation of regulation for this fishery. The study was conducted in two regions: North Java (Cirebon and Dadap) and East Sumatra (Belitung). Crab stocks in North Java are considerably depleted, and only small crabs are abundant. The long-term viability of the stock and, hence, incomes of those dependent on the fishery are under threat. In East Sumatra, large crabs are of a higher share of the total catch than in North Java, so while the need for size regulation is less pressing, the goal is to sustain harvests of larger crabs.

The blue swimming crab (or locally known as "*rajungan*") is one of the important fishery export commodities in Indonesia, despite no specific management plan yet in place to regulate this fishery. Almost all of the blue swimming crabs are exported and only small portions are consumed domestically. The export of blue swimming crab started in the period 1994–1998 due to increased demand from overseas, especially from the United States. The BSC contributes to about 17 % of total Indonesian fishery products export to the US market, valued at USD 177 million in 2010, an amount that jumped by 27 % from the previous year.[1] Crab products are Indonesia's fourth largest export of fishery products following shrimp, other fish, and tuna. Total crab exports were 18,673 tons in 2009 (MMAF and JICA 2009).

The predominant fishing gear used to harvest Indonesian blue swimming crabs are bottom gillnets and collapsible traps, and to a lesser extent the now-illegal shallow bottom trawls (mini trawls or baby trawls) or locally known as "*garok.*" Harvests are mostly carried out by small-scale fishers using boats of <10 gross tons with or without motors. Catches are landed at sites scattered along the coasts. Approximately 65,000 fishers harvest the BSC. The industry also operates with middlemen who sell the fishers' catch to processing factories called "miniplants" that employ roughly 13,000 "pickers" who remove the crabmeat from the shell and the final processors/packagers who export the products.

[1] Export value is using unit price based on foreign trade data provided by NMFS http://www.st.nmfs.noaa.gov/st1/trade/index.html

There are no regulations governing harvests of the BSC; it is an open access fishery. Stakeholders have proposed the following: adoption of a minimum legal size (carapace width measured tip to tip) that begins at 8 cm and increases over time; a ban on the take of berried (egg-bearing) females; changes in gear (e.g., escape gaps in crab pots); time/area closures; a registration system for all purchasers of crabs, both fresh and processed (to improve data collection); and effort controls (e.g., licensing regime and cap numbers, specify legal gear types, minimum net length, number of pots, etc.). The proposals that would have the most immediate impact on stock size are those that reduce the harvest of small-sized crabs. Fishery scientists believe that catching undersized crabs can lead to stock depletion, since the crabs would not have a chance to reach maturity and the recruitment would not happen (some studies suggested that the age at first maturity is about 9–10 cm carapace width).

Despite the lack of government regulation, the industries have initiated efforts to improve sustainability through the sourcing policy. On March 23, 2011, the National Fisheries Institute (NFI) Crab Council in the United States launched a major initiative that recommends participating companies to add a minimum size to their sourcing policies. The Council is an industry organization that represents more than 60 % of the US market for imported blue swimming crab. The initial recommendation is that only crabs measuring at least 8 cm be purchased from the Philippines and Indonesia starting on 1 July 2011.[2]

In response to this trade policy, the Indonesia Crab Processors Association (APRI)[3] produced a letter on July 1, 2011, asking its miniplant members to agree to (1) only process BSC that has an 8 cm carapace, and (2) not accept jumbo crabs that weighed less than 2 g. The APRI consists of 11 processing companies that supply almost 90 % of total export to the US market.

In an unregulated fishery such as in this fishery, the implementation of regulations can have both positive and negative effects. In the short term, harvests will likely be lower in areas where the larger crabs represent a small proportion of the total stock, and incomes of fishers will decline. However, if the measures are successful in building up the stock and each crab harvested is bigger, then the fishery is more likely to be sustainable over time with incomes gradually rising. Gradual implementation of a rising minimum size will reduce the negative impact on fisher incomes.

[2] See www.committedtocrab.org for information about the NFI Crab Council and its commitments to promoting sustainable fish harvests in Indonesia and other countries.

[3] The APRI was formed in 2007.

17.1.1 Significance of the Study

This study investigates the potential economic effects of adopting a minimum legal size compared to maintaining the status quo (uncontrolled fishing) across the supply chain. The knowledge gained may help in the design of effective policies under a management plan. The study is part of a wider blue swimming crab fishery improvement project in Indonesia that examines the biological effects of adoption of various measures to the crab population, including holding egg-berried female, protection of spawning and nursery ground, and a hatchery project. Stock assessment will also be conducted to document the relationship between size and maturity and between length and weight.

17.1.2 Research Objectives and Scope of the Study

The overall objective of the study is *to provide information on the economic value of implementing a minimum legal size* and *comparing that with continuation of unregulated fishing (the business as usual situation).*

Management options/scenarios to be assessed:

- Scenario 1: No management (status quo) – This refers to the "business as usual" situation. This means continued unregulated fishing.
- Scenario 2: Gradual implementation of minimum size regulation, starting with smaller size (at least 8 cm) and increasing 0.5 cm a year over a period of years

Each scenario will model and assess (1) the impact on the crab stock, (2) the changes in profits to fishers and industries (miniplants), and (3) the capability of the local institutional mechanism to carry out its functions and (4) recommend further enhancements, where necessary, to the management options considered. The scenarios are applied to the two regions noted above with sampling from multiple sites within each area.

17.2 Research Methodology

17.2.1 Research Questions

The study will seek to answer the following questions:

1. How would the implementation of a minimum size restriction affect the productivity and incomes of fishers and profitability of the industries?
2. What institutional mechanisms are required to implement the minimum legal size and what capacity building may be required to achieve this?

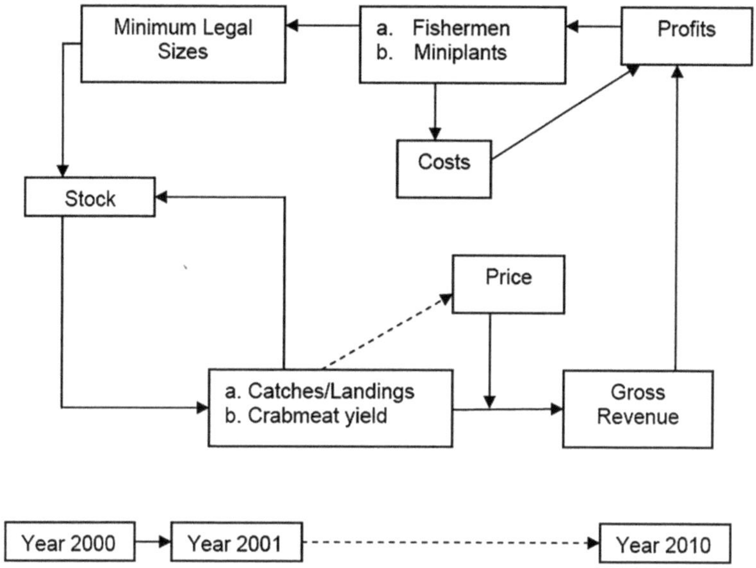

Fig. 17.1 Outline of the model

17.2.2 *Research Methodology*

We use bioeconomic modeling (the Gordon-Schaefer model), economic valuation, and institutional assessment to examine the effect of regulation on the performance of the fishery – its biological, economic, and social outcomes. A more detailed illustration of the framework is shown in Fig. 17.1. As with any modeling exercise, we can control for only some of the variables affecting the fishery. Actual outcomes are determined not only by the set of measures imposed but also by the biological, economic, social, and institutional characteristics of the fishery system and affected by influences exogenous to the fishery system (Sutinen 1999). The framework also illustrates the need for data required for management in the future (e.g., for size-at-age modeling).

Figure 17.1 illustrates the three main components of the framework:

1. An economic component which includes the relationship between quantities (catches and products) and prices (see the linkage between crab landings and price), the cost function (at both fishing boat and miniplant levels), and the determination of profits (difference between private revenues and costs)
2. An analytical biological model, which includes the fishery stocks
3. A management option where the gradual minimum legal size is implemented

17.2.3 Variables Measured

Time series data is used in the bioeconomic analysis: landing data (production) and inputs (effort), the price per unit of output (price per kilo of crab per year), consumer price index, regional gross domestic product (PDRB), and other supporting data. We do not have stock data over time, but we have crab landings and effort data over time. The study undertook a survey to estimate the standing biomass in a point in time in each area. The stocks over time were then estimated using effort and production data. Excel and Maple were used to calculate the yield-effort curve and maximum sustainable yields and to infer fishery equilibria with and without regulation.

17.2.4 Methods of Data Gathering

The data were obtained from the field data collection in the two regions and various research agencies and institutions associated to this research (for secondary data). Details are as follows:

17.2.4.1 Primary Data

Primary data was collected at three levels: fishers, landing sites (local selling stations), and processing plants. Each is now described in more detail.

Fishers Interviews with the fishers sought information about fishing gears used, harvest rate, fishing cost (e.g., price of vessels, gear, operating cost), harvests, and prices over time. A questionnaire was used to guide the interview with fishers.

Landing Sites (Local Selling Stations) Over 200 crabs were measured at each landing site surveyed (local selling stations). The crabs were randomly selected from different fishers before sorting. Each crab was also assessed for sex, and female crabs were further subdivided into berried (egg-bearing) and non-berried subclasses. The samples were individually measured and weighed for the following parameters: carapace width (notch to notch) (cm) and body weight (gram).

Processing Plants Surveys at the processing plants provided data on production cost, crabmeat production, relationship between the actual lump meat weight that comes from crabs of various sizes, composition of crabmeat type yielded over years, prices from different size categories of crabs, picking/processing time, and processing cost (labor).

We collected annual regional data from fishery offices that included official blue swimming crab landing data, number of fishers per gear type, number of processing or miniplants in the region, and total production of crabmeat in the region.

17.2.4.2 Institutional Data

Questionnaires for fishers and miniplant owners contained queries about harvests, institutional capacity, and needs assessment. The questions helped to identify the key stakeholders involved in blue swimming crab fisheries and their roles and responsibilities. We explore social dynamics among fishers regarding their catching area (spatial use, local regulation, etc.) and relationship between fishers and miniplant operators. We attended meetings with the crab exporters and government (at the local and national level) to gain information on what kind of political support would be needed to implement the policy on minimum size harvest.

17.2.5 Data Analysis

Data used in this study include:

1. Standardized fishing gear: two types of fishing gear are in common use by crab fisherman in Indonesia – collapsible traps and crab gillnets. The garoks (mini trawls and baby trawls), now illegal, were in use in some fisheries during our study period.
2. Cost per unit *standardized effort*. A concept introduced to measure the catch with an equivalent unit for each gear type.
3. Parameters for bionomic functions are estimated econometrically.
4. Stock estimates: we consulted a fishery biologist to obtain an estimation of the impact on the crab stock of continuing with no regulation versus implementation of minimum legal size. The hypothesis used in modeling is that without regulation the average size declines leading to smaller total stocks over time.

17.2.6 Simulation Analysis

Simulation analyses of the biological and economic models were conducted using Vensim software (http://www.vensim.com/) to determine the impact of minimum legal size policy on the stock and profits for fishers and miniplants.

17.2.7 Economic Analysis

For fishers and miniplants, the net present values (NPVs) and internal rates of return (IRR) over 10 years are estimated to illustrate the financial feasibility of the minimum legal size policy versus the status quo.

17.3 Description of Blue Swimming Crab Fishery in the Study Areas

17.3.1 Area Description

Cirebon Regency (North Java) Cirebon Regency is one of the coastal areas in eastern West Java, well known for its marine resources, especially shrimp and crab. Its population is 2.2 million people and the education level is relatively low. Economic activities are mainly driven by agriculture, fishing, and the trade sector. There are over 95,000 people working in fishery sector, with more than 70 % in the capture (wild) fishery, while the rest are in aquaculture and processing sector. The areas with major blue swimming crab harvests are Bondet, Muju, Gebang, and those sampled in this study.

Dadap, Jakarta Bay (North Java) Jakarta Bay is north of Jakarta City. It is a shallow bay, exposed to high pollution transported from upland region by 13 rivers. The area is facing *Kepulauan Seribu* or Thousand Islands formed by more than a hundred of small islands as long as 80 km. Dadap is a representative village on the bay and site we sampled. It has approximately 20,000 residents living in high density.

Belitung Regency (East Sumatra) Belitung Regency is an island area consisting of 98 islands in the province of Bangka and Belitung Islands. It has a population of over 166,000. Due to its location, fishing potential is great, but there are only two main blue swimming crab-producing areas, Badau and Membalong. We interviewed fishers in sub-districts – Padang Kandis, Air Saga, and Tanjung Pandan.

17.3.2 The Fishery System

Primary data was obtained through in-depth interviews with more than 30 BSC fishers in each sampled site and also from miniplant owners.

17.3.2.1 Fishers

There are two types of fishers involved in blue swimming crab fishery across Indonesia:

1. *Daily fishers*: fish overnight from late afternoon or evening and land their harvest in the morning before 12 pm. They operate close to the shore, in small boats, and typically use gillnets, traps, or trawls.

2. "*Babangan*" fishers: use boats capable of going farther offshore, usually sailing in a group containing three to five fishers. They may be fishing for 2–7 days before landing their harvests. Collapsible trap devices are commonly used in this fishing system.

This study focused on daily fishers because the Babangan harvests crabs from stocks outside our sampling sites. Most fishers in Belitung Island come from the surrounding areas, close to the landing port. In Cirebon, the fishers come from the surrounding areas and also from other areas, far away from the landing ports. In Dadap, most of the fishers are from Indramayu (West Java). Most of the fishers in the selected sites are not full-time blue swimming crab fishers. The fishers will change their target species depending on the season. Thus a potential consequence of minimum size harvest regulation may be to shift effort to other fisheries and may threaten those stocks. Only the fishers paid and funded by miniplants focus primarily on harvesting blue swimming crabs.

17.3.2.2 Fishing Boats and Fishing Gears

Fishers harvesting blue swimming crabs rarely own their own boats because they cannot afford the capital cost of the boat (IDR 22 million and IDR 40 for a 2 and 5 ton vessel, respectively).[4] The boat is usually owned by a miniplant or someone with financing ability, who become the middlemen or "*bakul*," typically owning up to 15 boats. Middlemen supply fishers with diesel fuel, kerosene fuel, bait, and capturing devices and provide maintenance for the vessel. A fisher who joins a boat owned by other fishers has to pay rent of about IDR 4000 per kilo of crab he harvests. In Cirebon Regency, there are more than 400 fishing boats using three types of fishing gear: collapsible traps, gillnets 2–4 in. in size, and the mini trawl. In Dadap, there are more than 100 fishing boats fishing for crab that use predominately gillnet from 2.5 to 3 in. and collapsible traps. In Belitung Regency, there were about 113 fishing boats and fishers use gillnets (3–3.5 in.) and collapsible traps. The most destructive fishing gear used to harvest blue swimming crab is the mini trawl, which can be dragged behind a motorboat and is more likely to be used by fishers who cannot afford a more substantial vessel. This gear type is indiscriminate; it drags along the ocean floor, catching many species including shrimp, fish, clams, and crabs and damaging local ecosystems. The smaller the net size, the larger the ecosystem damage and species bycatches and the smaller the BSC caught. The use of mini trawls has created conflict among different types of fishers, with those trying to sustain a harvest of larger and more valuable crabs thwarted by those using the mini trawls.

Table 17.1 illustrates the impact of gear on the size composition of crabs caught in the three sample areas. Of the total number of crabs harvested that are <8 cm

[4] At the time of writing, the exchange rate was 1 US dollar equals IDR 9000.

Table 17.1 Crab size distribution by fishing gears in sampled sites

Carapace width	Average size distribution					
	Cirebon (West Java)		Dadap (Jakarta Bay)		Belitung (East Sumatra)	
	Gillnet (3.5 in.)	Mini trawl	Gillnet (2 in.)	Collapsible traps	Gillnet (3.5 in.)	Collapsible traps
Size <8 cm	5 %	40 %	36 %	15 %	0 %	2 %
8 cm ≤ Size ≤9.9 cm	22 %	39 %	57 %	31 %	3 %	18 %
Size ≥10 cm	73 %	21 %	6 %	54 %	97 %	80 %

carapace width, almost 40 % are caught with mini trawls and 46 % with gillnets with mesh size <2 in. Larger mesh-sized gillnets (minimum 3.5 in.) are considerably more selective and catch bigger crabs than collapsible traps. Crab size composition also depends on the fishing grounds. The study shows that smaller crabs are caught in tidal area or <1 h trip from the shore. Tidal areas are the nursery ground for the BSC.

17.3.2.3 Trends in Harvest Rate

Crab harvests occur throughout much of the year; the actual dates depend on the location and weather. The high season for harvests typically runs from July through November (during the east monsoon); low season from March through June. Little harvesting is done from December through February due to inclement weather. Fishers interviewed in Cirebon Regency reported that average harvest rates for gillnetters range from 5–20 to 40–60 k/boat/day in high season. Mini trawl harvests range from <3 to 10 k/boat/day in low season to high season rates of 15–50 k/boat/day. The fishers and middlemen noted that harvests had declined over the last 5 years. In 2004, the gillnet fishers in Cirebon Regency could harvest about 50 k/boat/day even in low season. The decline is even larger for mini trawl fishers, falling from about 70 to 150 k/boat/day to the current levels. Fishers feel that the decline is due to overfishing (too many boats) and the use of destructive fishing gear.

In Dadap, harvest rates range from 5 to 10 k/boat/day for crab gillnet in low season to 25–30 k/boat/day in high season. Figures for collapsible traps are 15–30 k/boat/day in low season and up to 50 k/boat/day in high season. Fishers in Dadap have also noted the decline in harvests over time. The average harvest rate in Belitung, according to fishers, has been relatively stable and averages about 10 k/fisher/trip, dipping to about 5 k/fisher/trip during the low season from November to March. Fishers switch to harvesting shrimp from April to June during the west monsoon when no crabs are caught.

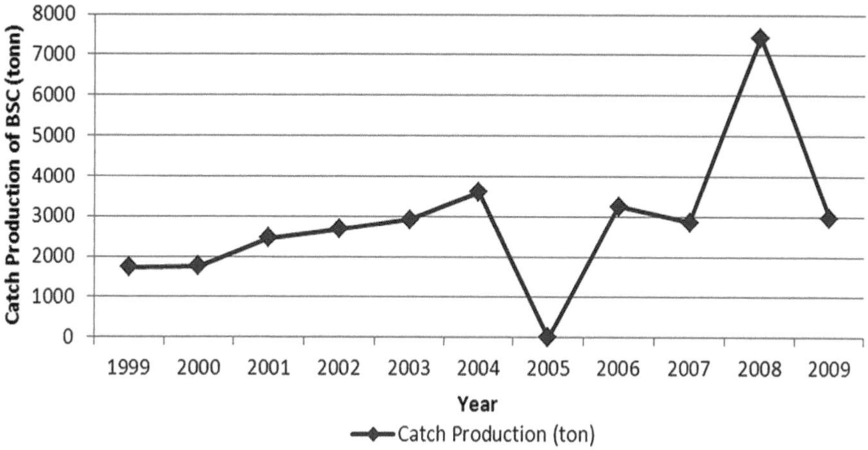

Fig. 17.2 Catch productions (landings) of blue swimming crab in Cirebon Regency in 1999–2009

17.3.2.4 Total Production

Harvests in Cirebon Regency are shown in Fig. 17.2; those from Belitung are shown in Fig. 17.3.[5] The data in Cirebon show an upward trend from 1999 to 2004. After 2005, harvests become much more unstable. The large increase in 2008 (7434 tons) was almost double than average landings and may be due to very favorable weather conditions. The 2009 harvest indicates a return to more average conditions. Landings in Belitung tend to be more stable, with the total production of about 2700 tons per year.

17.3.2.5 Trend in Prices

The landed price of crabs depends on the total supply and the size of the crab caught. Statistics from the regions show that in Cirebon, the price ranged from IDR 20,000 to 26,000/k in 2007. In April 2011, the price for raw crab increased and reached IDR 33,000 per kilo for large crabs (11–12 crabs per kilo). Prices for smaller crabs caught by the mini trawl was much lower at about IDR 8000 to IDR 15,000/k. Average prices per kilo in Dadap were about IDR 22,000 for crabs caught by gillnet, while the larger crabs caught by collapsible traps were IDR 32,000. In Belitung, the average per kilo ranges from IDR 12,000 for small crabs to IDR 15,000 for slightly larger ones. The price received by the fisher will be about 80 % of the landed price when the boats are owned by middlemen or miniplants. Recall that this is because fishers have to pay the boat owner IDR 4000 per kilo of crabs. However, in Belitung, crabs can be sold at much higher prices in local markets close

[5] In Table 17.2, harvest data for 2005 are missing, not zero.

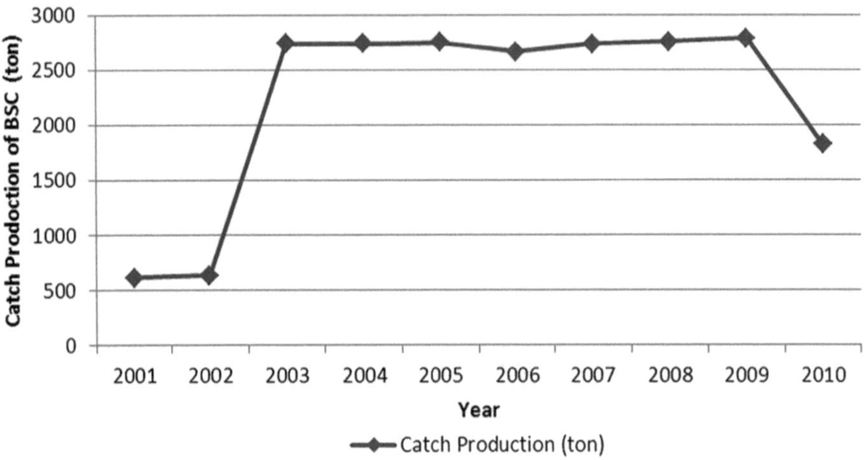

Fig. 17.3 Catch productions (landings) of blue swimming crab in Belitung Regency in 2001–2010

to the big cities (for restaurant or household consumption). Fishers can receive up to IDR 30,000/k.

17.3.3 Supply Chain

Six components comprise the blue swimming crab fishery. These are (1) fishers, (2) middlemen (known as *bakul*), (3) miniplants, (4) processors and exporters, (5) distributors, and (6) retail sectors (supermarkets, restaurants). Fishers sell the crabs to bakul who may have lent the fisher money for fishing gear or boats. The bakul sells the crab to miniplants for a price about IDR 3000–4000 above the landed price noted above. Miniplants may buy from independent bakul or employ bakul and also may own fishing boats. A typical miniplant might own between 15 and 25 boats. Vertical integration down to the boat and fisher level provides more certainty over prices but also comes with expenses related to monitoring of vessel captains and crew, scrutinizing the weighing procedures, and so on.

Miniplant workers (picker) extract the crabmeat from the carapace and sell to processing companies to be packed and exported. The net price received by the miniplant is a function of the size of the meat. Smaller crabs take more time to process, adding to labor cost, and have less of the most valuable claw meat. Table 17.2 illustrates the net yields and labor costs for different sizes of crabs. It is clearly in the miniplant's interest to see larger crabs harvested. The time between when the crab is caught and the meat enters processing plants is a key determinate of the quality of the product.[6] The ideal interval is 2 h, but it should not exceed 48 h.

[6] Foreign matter and other contaminants may also affect the quality of the crabmeat.

Table 17.2 Crabmeat yield and labor cost by crab size

Category	Average crabmeat yield	Type of product	%	Average crabmeat produced (gram)	Value (IDR)	Picking time (minutes)	Labor cost (IDR)
Large size (>10 cm)	42.50 %	Jumbo	31	450	56,750	11	2421
		Lump flower	23				
		Special	6				
		Claw meat	40				
Medium size (8.1 cm < CW <10 cm)	36.50 %	Jumbo undersize	36	380	42,332	20	2448
		Lump flower	24				
		Special	8				
		Claw meat	32				
Small size (CW <8 cm)	25 %	Super lump	32	300	24,120	35	1800
		Special	44				
		Claw meat	24				

The numbers are based on some experiments done from 2 k of crabs (raw material) from different sizes in three locations

Testing of product quality is routine and failed shipments are sent back to miniplants who in turn sell the meat to local buyers, such as the supermarket chain Carrefour and local restaurants (for rejects that are still in good quality) and to local animal farms (for meat unfit for human consumption). The local market buys about 20 % of the industry's total production; mostly the meat rejected for export.

Most of Indonesia's BSC meat is thus exported, predominately to the United States. The final two components of the supply chain are outside Indonesia and consist of distributors and the retail sector in the consuming countries. No economic modeling is done in this paper of these parts of the supply chain.

17.4 Bioeconomic Analysis

The aims of bioeconomic analysis, using a static Gordon-Schaefer fishery model, are to simulate the maximum sustainable yield (MSY), maximum economic yield (MEY), and open access (OA) conditions in the Cirebon and Belitung fisheries prior to and after the implementation of a minimum legal size regulation. Sixteen years of time series data (1994–2009), obtained from the Department of Marine and Fisheries of West Java Province and Department of Marine and Fisheries of Cirebon, includes landing data (production) and inputs (effort), the price per unit of output (price per kilo of crab per year), and other supporting data. There is no stock data available over time. Crab stocks are estimated using effort and production data. Econometric analysis is used to generate the parameters needed for the simulations from the standard Gordon-Schaefer equation:

$$d(X(t))/dt = F(X(t)) - H(t) \tag{17.1}$$

$$d(X(t))/dt = rX(1 - X/k) - qEX \tag{17.2}$$

where the left-hand side of Eq. 17.1 is the growth in the stock of fish, F(x), over time, and the right-hand side is the stock level at any time t minus the harvests at time t. Equation 17.2 substitutes in a logistic growth function for the fish biomass where r is the intrinsic growth rate of the species and a linear harvest function where q is a catchability coefficient and E the level of effort. The parameters r, k, and q can be estimated econometrically if there is data on fish stocks and effort levels. Dividing Eq. 17.2 by X gives a linear expression, with the left-hand side the rate of growth in the stock, to obtain Eq. 17.3:

$$(dX/dt)/X = r - rX/k - qE \tag{17.3}$$

An equation similar to Eq. 17.3 was estimated using secondary data for effort and computed levels of the stock over time to obtain estimates for the parameters r, k, and q.[7] In the fishery literature, r and k are typically assumed constant for a specific fish species. The parameter q is what is likely to change when a regulation that restricts harvests to larger crabs is introduced. That is, a given amount of fishing effort (people, boats, and gear) would be expected to harvest less biomass (kilograms of crab) if they cannot harvest crabs less than 8 mm in carapace.[8]

The parameter estimates along with secondary data on effort and harvests can then be used in the Gordon-Schaefer model to simulate what a fishery equilibrium under three different management regimes: open access (OA) where there is no restriction on the number of fishers or vessels operating (the current situation in the BSC fisheries), the maximum sustainable yield possible from the fishery (MSY), and the profit maximizing or maximum efficient yield (MEY) from the fishery. Effort is standardized into number of trips per year using the gear types found in each fishery.[9]

---

[7] The estimation is based on a variant to the Gordon-Schaefer model and follows the procedures developed in Hilborn (1976) for the unregulated fishery and Anna (2003) for the fishery with a minimum size regulation.

[8] The parameter k could also change if fishers increase their fishing effort by extending their fishing grounds. In the estimation for BSC done in this paper, the catchability coefficient, q, was constant, producing different levels of production with and without the minimum size regulation. Thus the carrying capacities (K) are different between the two scenarios. These conditions need to be tested again in a more robust model and after the minimum size has been implemented and enforced.

[9] There are a number of different ways to standardize effort estimates ranging in complexity. Due to data limitations, the approach taken here aggregates the number of trips using the most prevalent gear in the fishery (gillnets) plus trips using other gear.

Table 17.3 Simulated equilibrium values for MSY, MEY, and OA without minimum size limit policy, Cirebon District

	Bioeconomic simulation without minimum size limit policy		
Description	MEY	MSY	Open access
Biomass (X) in tons	2988	2268	1441
Production (H) in tons	2293	2551	2212
Effort (E) in number of annual trips	414,404	607,418	828,809
Total fishery revenue (IDR million)	96,310	107,127	92,896
Total cost of effort (IDR million)	46,448	68,082	92,896
Resource rent (π) in IDR million	49,862	39,046	0

Table 17.3 illustrates the simulation results when there is no minimum size regulation for the Cirebon fishery, and Fig. 17.4 presents a graphical illustration of the simulation.

In the simulation the MEY harvest exceeds that of OA but with far fewer fishing trips per year, resulting in higher total revenue and positive resource rent if the fishery were managed efficiently. The actual fishery is open access, and the 2009 level of effort and production in Cirebon Regency fishery illustrates how severe the open access problem is. The level of effort in 2009 was 880,419 trips, a number quite close to the predicted OA level. The total production was 1604 tons in that year, indicating adverse impacts on the stock of crabs, a situation that is unsustainable over time. This also can be seen in the decrease in catch per unit effort (CPUE) figures for Cirebon, as shown in Fig. 17.5.

Belitung Regency would have similar simulation results but fares slightly better than Cirebon in practice because while they too are an open access fishery, the supply of larger crabs has yet to decline appreciably. In 2009, actual harvests were 2784 tons using only 433,324 trips compared to a much smaller harvest in Cirebon with twice as much effort. Figure 17.6 shows Belitung's CPUE was much more stable than that of Cirebon but nonetheless trending downward indicating that here, too, regulation is needed to sustain harvests and crab stocks. Table 17.4 illustrates the three fishery equilibrium outcomes with a minimum size limit policy simulated for Cirebon.

There is not much difference between Tables 17.3 and 17.4.[10] With regulation, harvests under the MEY and OA are approximately equal, and effort is slightly lower than without regulation. There may be more positive impacts on the fish stock (biomass) over time, but the static model does not have the capacity to simulate stock effects.

[10] A more robust model and longer data series to estimate parameters may lead to greater differences in a simulation with and without size regulation.

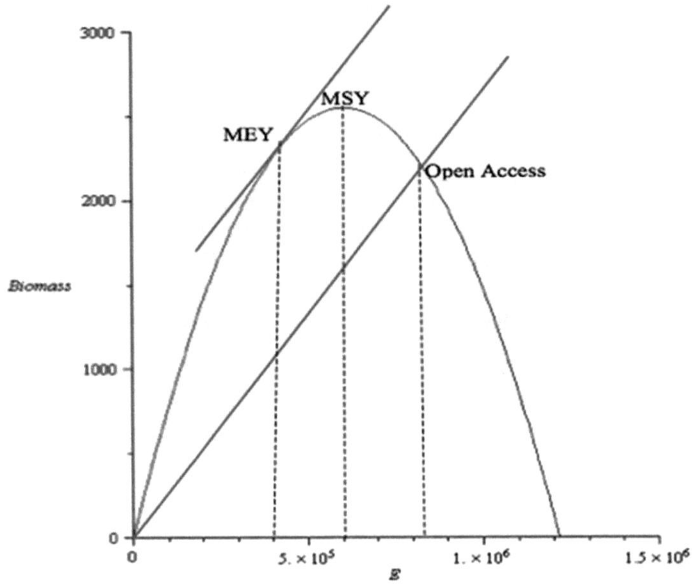

Fig. 17.4 Simulated MEY, MSY, and OA static equilibrium levels for Cirebon Regency

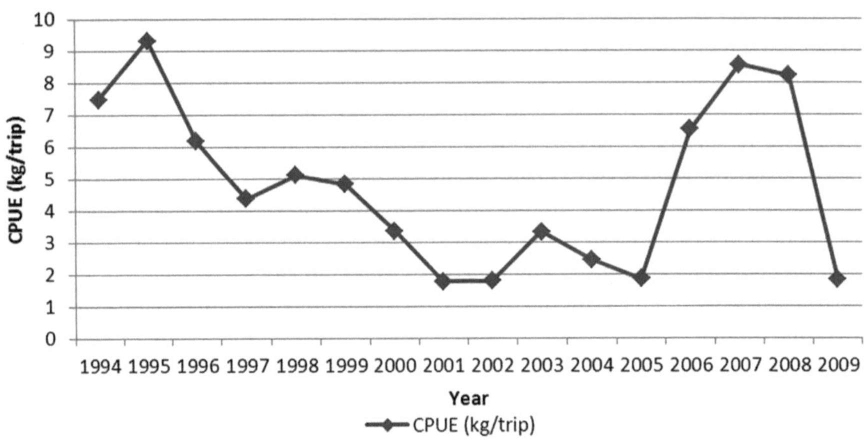

Fig. 17.5 Catch per unit effort of BSC fishery in Cirebon in 1994–2009

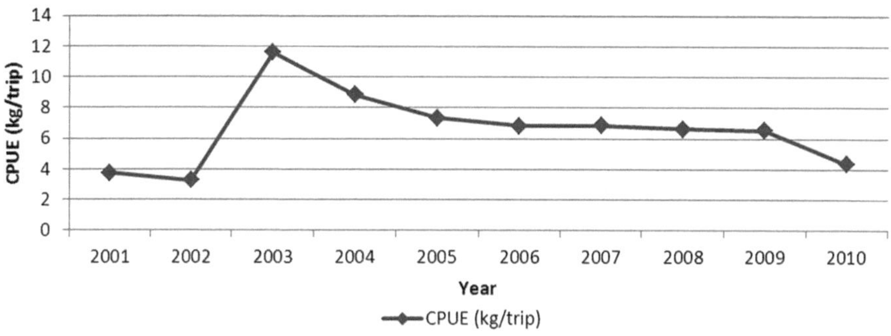

Fig. 17.6 Catch per unit effort of blue swimming crab in Belitung Regency in 2001–2010

Table 17.4 Simulated equilibrium values for MSY, MEY, and OA with minimum size limit policy, Cirebon District

	Bioeconomic simulation with minimum size limit policy		
Description	MEY	MSY	Open access
Biomass (X) in tons	2875	2154	1441
Production (H) in tons	2152	2423	2158
Effort (E) in number of trips	404,246	607,418	808,492
Total revenue (IDR million)	90,385	101,771	90,619
Total cost (IDR million)	45,309	68,082	90,619
Resource rent (π) in IDR million	45,075	33,689	0

17.5 Financial and Economic Analyses

Under the minimum legal size policy, it is assumed that in the short term there will be reduced catches and production, especially in the regions where small crabs (<8 cm) are relatively more abundant (Cirebon). The study estimates the incomes, net present value (NPV), and internal rates of return (IRR) to fishers and miniplant operations under two scenarios – with and without a minimum size regulation.

17.5.1 Profitability Analysis at the Fishers Level

17.5.1.1 Cirebon Regency

The profitability analysis at the fisher's level examines the three gear types used in this region: gillnets with 3.5 in. mesh size, collapsible traps, and mini trawls. Costs are separated into capital and operating. Capital costs are shown in Table 17.5 and include the cost of vessel, fishing gears, and machines. Operating costs – fuel, supplies, maintenance, and profit sharing – are in Table 17.6.

Table 17.5 Capital costs in Cirebon Regency by different fishing gear types (million IDR) (Data from surveys 2011)

Type of investment cost	Gillnet	Collapsible trap	Mini trawl
Boat (1 unit)	22	22	22
Machine (1 unit)	7.5	7.5	15
Fishing gear	4.8	7.0	2.7
Strings and anchor	0.22	0.60	0
Main string	0.07	4.2	0
Net floating string	0.42	0.60	0
Jerry can floating string	0	0.42	0
Floating jerry can	0.34	0.34	0
Stove	0	0.15	0
Other tools	0.15	0.20	0
Total capital cost	35.5	43.0	39.7

Table 17.6 Operating costs in Cirebon Regency per year (million IDR) (Data from surveys 2011)

Operational cost	3.5 in. gillnet	Collapsible trap	Mini trawl
Fuel	17	22.7	8.6
Consumption	7.6	8.4	0.8
Fish bait	0	18.9	0
Ice block	0	2.4	0
Boat repair	1.5	2	2
Machine repair	0.7	0.9	1
Fishing gear repair	2	2.5	3
Profit sharing (crew wage)	64.1	47.4	2.4
Total operating cost	93	105	29

Notes: (1) Maintenance costs are the costs to repair and maintain the vessels and fishing gear; (2) for gillnets, there are about four fishers per boat, and for traps approximately 5/boat; (3) fuel costs IDR 4500 per liter and fuel use is 15 L/trip for gillnet fishers and 120 L/trip for trap fishers; (4) fish bait for traps is IDR 3000/k; (5) assumes fishing occurs in 11 months per year

Table 17.7 estimates income from blue swimming crab fishing for different gear types under the current conditions (2011) and what is likely to be harvest levels if a minimum size regulation is introduced. Fishing income is given by the average total harvests from each gear type times the price of crabs, based on their size, which in turn is based on the gear used to harvest them. The estimates without regulation are based on recent data. Estimates for "with regulation" are based on current data that shows crabs caught by 3.5 in. gillnets and collapsible traps are below minimum size (8 cm). Hence, revenues fall slightly when a regulation is imposed. The major impact of the regulation is on the mini trawl fishers, whose income is forecast to fall by 40 %.

Table 17.8 provides estimates of the NPV and IRR to fishers in Cirebon without and with regulation, dependent on type of gear used. The mini trawl is the most profitable when there is no regulation due to its significantly lower operating costs.

Table 17.7 Forecasts of annual income for blue swimming crab fishers by gear type in Cirebon regency, with and without minimum size regulation

Type of fishing gear	Income per fisher per year without regulation (IDR million)	Income per fisher per year with regulation (IDR million)
3.5 in. gillnet	85.5	81.2
Collapsible trap	118.4	117.3
Mini trawl	33.0	19.8

The average price for gillnets and collapsible traps is assumed to be IDR 42,000/kg and IDR 8000–15,000/kg for mini trawls

Table 17.8 Net present value (NPV) and internal rates of return (IRR) to fishers in Cirebon Regency with and without minimum size regulation by gear type

	Fishing gear without minimum size regulation			Fishing gear with minimum size regulation			
Investment criterion		3.5 in. gillnet	Collapsible trap	Mini trawl	3.5 in. gillnet	Collapsible	Mini trawl
NPV (IDR million)	10.1	19.7	27.6	3.0	15.0	−27.2	
IRR (%)	14	17	22	9	15	–	

The discount rate is assumed to be 6.5 %, based on 2009 market rates

Collapsible traps are more profitable than gillnets with and without regulation because they harvest crabs of different sizes. Gillnets are more selective, harvesting from medium- to large-sized crabs. With regulation, harvests and hence incomes decline significantly for the mini trawl; it is no longer financially viable as a gear type. Fishers would thus likely cease using the mini trawl for harvests sold to middlemen and miniplants.[11]

17.5.1.2 Belitung Regency

The profitability analysis for Belitung Regency covers only the 3.5 in. gillnet as the majority of fishers use this type of gear for crab fishing. Fishers in this region, unlike those in Cirebon, switch to catching shrimp during the west monsoon; hence, their annual incomes depend on two types of harvest. We focus only on income from crab fishing to show the impact of a minimum size regulation.[12]

Table 17.9 provides data for the capital and operating costs in Belitung Regency and shows that these costs are roughly similar to those in Cirebon, with capital costs a bit lower and operating marginally higher. Table 17.10 presents the NPV and IRR

[11] Minimum size regulation would therefore help enforce bans on the use of mini trawls. They may still be used however for fishers selling their crabs locally or for household consumption.

[12] Data on fishing costs, incomes, and NPV from shrimp fishing in Belitung are available from the authors.

Table 17.9 Capital and operating costs in Belitung Regency for 3.5 in. gillnets

Capital cost	IDR (million)	Annual operating cost	IDR (million)
Boat	16	Fuel	19.8
Machine	5	Consumption	3.3
Gillnets	5.2	Maintenance cost	
Rope and anchor	0.25	Boat repair	1.5
Connecting rope	0.12	Machine maintaining	1.1
Floating rope	0.5	Gillnets maintaining	8.4
Jerry can	0.05	Profit sharing	
Fishing equipment	0.2	Profit sharing (crew wage)	61.9
TOTAL	27.4	TOTAL	96

Table 17.10 Net present value (NPV) and internal rates of return (IRR) to fishers in Belitung Regency using 3.5-in. gillnets

Investment criterion	3.5 in. gillnet
NPV (IDR million)	9.7
IRR (%)	16

The discount rate is assumed to be 6.5 %, based on 2011 market rates

for the fishery. Our measurement of crabs harvested in this region virtually all of the crabs harvested with gillnets were above 8 cm in carapace, so we assume that net incomes are virtually identical with and without a minimum size regulation, hence only one estimate for NPV and IRR is shown.[13] The numbers are quite comparable to those in Cirebon without the regulation. The important difference is that the NPV and IRR are not expected to change when a minimum size regulation is introduced, thus making this fishery much more profitable than that in Cirebon even without including shrimp revenue.

17.5.1.3 Profitability Analysis at the Miniplant Level

Analysis of the economics of miniplant operation assumes a 10-year life of the plant, the life of its longest durable assets. Total capital costs for a typical miniplant in Cirebon are IDR 56.6 million of which the land and the building comprise just over 50 % (land is 16 % and building 35 %).[14] Miniplants in Belitung are much larger, occupying more land and hence needing more equipment. Capital costs for a typical plant in Belitung are IDR 566.4 million. Variable costs include labor, ice blocks, fuels, and the cost of the crabs as the raw material. These costs approximate

[13] Some collapsible traps are used in Belitung. Only 1 % of crabs harvested using traps were found to be less than 8 cm carapace.

[14] Data for miniplant costs comes from interviews with plant owners.

Table 17.11 Simulated impacts of minimum size regulation on miniplant operations and financial returns for Cirebon Regency

Parameter	Without minimum legal size regulation	With minimum legal size regulation
Raw materials supplied by fishers (kilo)	24,382	23,163
Crabmeat yield	0.42	0.43
Crabmeat production (kilo)	10,240	9960
Revenue (IDR billion)	1.28	1.25
NPV (IDR million)	430	478
IRR (%)	30	34

IDR 907.5 million in Cirebon and 4.85 billion in Belitung. Average yield in Cirebon (kg/crabmeat per kg of whole crab) is 42 % and is 25 % in Belitung. The typical miniplant in Cirebon produces approximately 10,240 kg of crabmeat per year while that in Belitung produces 37,570 kg/year. Annual revenue for the miniplant is computed by multiplying average yield by the average price of crabmeat. The price received by Cirebon miniplants was IDR 125,035/kg in 2009 and IDR 135,000 for plants in Belitung. The price is higher for Belitung because the crabmeat is of higher quality due to the larger size crabs processed. Annual revenue for a typical Cirebon plant is thus IDR 1.2 billon and IDR 5.1 billion for Belitung.

The net present value of income from a miniplant operating in each region is computed over a 10-year period given the following assumptions: all costs and prices are constant over time and, given by their 2009 level (as shown above), the price of fresh whole crabs is IDR 37,000/kg, the discount rate is 6.5 %, equipment fully depreciating within the 10-year period is repurchased, and there is still some residual value for the plant at the end of the 10 years. Based on these assumptions, the NPV for Cirebon plants is just under IDR 430 million and that for Belitung is IDR 756 million.

A simulation of the impact of the minimum size regulation on the total revenues of a miniplant in Cirebon Regency is simulated assuming in the short run, total harvests will fall marginally, but as the stock rebuilds, will rise over time. Recall that the regulation is expected to have no impact in Belitung because the majority of crabs harvested there are bigger than 8 cm in carapace. Table 17.11 shows the simulated impact of the minimum size regulation. Production and, hence, revenues decline by just 5 % in the early years. Revenue still exceeds variable costs, so the miniplant would still be profitable. Over time, more large crabs will be harvested, providing more throughput and higher prices for crabmeat.[15] The NPV under regulation is thus simulated at IDR 478 million compared to IDR 430 in the current

[15] The impact on crab stocks, fishing effort, and harvests for Cirebon Regency was simulated using Vensim software. After a one-period decline in stocks and harvest, stocks rebound and harvests and effort increase. Details of the simulation can be obtained from the authors.

harvest conditions (without regulation), and the IRR rises from 30 % without regulation to 34 % with regulation.[16]

17.5.2 *Perceptions of Fishers and Miniplants Toward Minimum Legal Size Policy*

Fishers in the study areas agree on the implementation on minimum size (8 cm), based on information from our interviews. However, most of them also point out that the main problem is the fishers that are using mini trawls, catching small crabs and contributing to ecosystem degradation. In the area where bigger crabs are abundant, compared to small crabs, fishers are willing to release the small crabs (<8 cm). Some of the respondents also mentioned about the premium price for the bigger crabs. In areas such as North Java where the smaller crabs are abundant, it will be more difficult to implement the minimum legal size. Small crabs the fishers would not be able to sell, given the minimum size requirement from the miniplant, would likely be brought home and consumed by the household.

Most miniplant owners reported in the interviews that they prefer to obtain bigger crabs from the fishers because the larger crabs yield more crabmeat and are easier to pick (extract the crabmeat). However, one reason given for why the miniplants would still accept smaller crabs is that the fishers are in debt to the miniplants owners. Revenue for their crabs helps repay their debt. This may change if the miniplants cannot export the meat from the small crabs.

17.6 Conclusions and Recommendations

As noted in the introduction, the NFI US Crab Council launched a major sustainability initiative in March 2011 that recommends companies selling crabmeat to their members add a minimum size of 8 cm at the carapace to their sourcing policies as of 1 July 2011. Going forward, the Council will review the progress of this initiative and it expects to issue subsequent minimum size recommendations. In Indonesia, on 27 April 2011, the Directorate General of Fishery Products and Marketing of Ministry of Marine Affairs and Fisheries sent a memo to all heads of provincial fishery offices in 33 provinces to inform them about the minimum size requirement imposed by the NFI US Crab Council and requested they pass along the information to all crab fishers and processors with encouragement to comply with the NFI requirement. During the APRI meeting on July 1, 2011, its members signed a letter agreeing to comply with the minimum harvest size. The APRI sent its letter to approximately 600 miniplants all over Indonesia. However, one of the

[16] The internal rate of return to investment in a miniplant in Belitung Regency is 37 %.

challenges facing this industry-imposed regulation is that most of Indonesia's miniplants are independent and are not controlled by a specific processor. While the supply chain can drive some improvements, it is critical to have governmental and multilateral support to facilitate policy change that will guarantee the long-term sustainability of fisheries as follows:

1. Government needs to reinforce the NFI sustainability initiative by regulating the minimum mesh size for gillnets to be no less than 3.5 in. to prevent the harvest of smaller crabs and ensure the minimum 8 cm crab is the target of the harvest.
2. The use of mini trawls should be banned completely. A ban would help protect the fishery and reduce conflict among fishers.
3. Better monitoring and enforcement of a minimum size regulation and ban on mini trawls is essential to ensure compliance and penalize those who violate the regulations. An awareness and education program would likely improve compliance.
4. Seasonal closures are needed during the periods when there are a high proportion of berried females.

References

Anna S (2003) Model embedded: interactive economic dynamics of fisheries and pollution. Dissertation of Post Graduate. Bogor Agricultural University

Hilborn R (1976) Optimal exploitation of multiple stocks by a common fishery. Can J Fish Res Board Can 33:1–5

MMAF (Ministry of Marine Affairs and Fisheries), JICA (Japan International Cooperation Agency) (2009) Indonesian fisheries statistics index. Jakarta

Sutinen J (1999) What works well and why: evidence from fishery-management experiences in OECD countries. ICES J Mar Sci 56:1051–1058

Chapter 18
Conclusion: What We Have Learned

Nancy Olewiler

Sustaining healthy coastal ecosystems that provide for people dependent on their goods and services requires good resource management. The research presented in this book can help enhance resource management in the three theme areas, specifically in terms of providing better information on (1) how to value marine and coastal ecosystems, (2) how these values are affected by economic activity in the form of harvesting and land transformation, and (3) the design of institutions to provide for sustainable livelihoods and resources. The papers in this volume also point to research and knowledge gaps and, hence, opportunities to fill them. The book is very much a beginning rather than an end in terms of our understanding of the complexities of our natural environment and how best to ensure that it will continue to provide the essential goods and services that the planet needs to survive.

The papers in the first section provide detailed descriptions of the methods economists use to measure the value of ecosystems to society. These include a variety of forms of stated preference frameworks, i.e., contingent valuation methods and choice experiments as well as revealed preferences in the form of travel cost estimations. The papers find that there is a large variation in the willingness to pay for ecosystem goods and services, dependent primarily on two things: who is being asked to pay and for what.

The findings suggest that the more closely the respondents engage with the ecosystem in question (e.g., tourists enjoying the benefits of a beach or diving and snorkeling on a reef), the higher the willingness to pay. When local inhabitants are asked to help protect a species at risk (e.g., whale sharks or turtles), willingness to pay is not very high. These results suggest that income constraints matter, as we would expect from economic theory. People whose income levels are not high cannot think about allocating their scarce funds for the protection of species from which they may get psychic value but no consumption or commercial value.

N. Olewiler (✉)
School of Public Policy, Simon Fraser University, Vancouver, BC, Canada
e-mail: olewiler@sfu.ca

N. Olewiler et al. (eds.), *Marine and Coastal Ecosystem Valuation, Institutions, and Policy in Southeast Asia*, DOI 10.1007/978-981-10-0141-3_18

Respondents also signal that these resources are shared goods and that there needs to be a way to have a broader swath of society pay to protect them. This is one of the conundrums with willingness to pay estimation and payments for ecosystem services—*will people actually pay what they have stated? How would governments collect these amounts? And how could they be converted into payments for protecting and enhancing ecosystem services?*

The problem of providing public goods (or reducing public "bads") remains a challenging issue in economic analysis and public policy practice. Some of the papers provide examples of feasible mechanisms for collecting more of users' consumer surplus (e.g., charging higher user fees at tourist sites), but how to sustain species and their ecosystems when there is no readily identifiable user that can be charged a fee remains a broader challenge. The papers provide fulsome methodological analysis of real cases where ecosystems and livelihoods can be at risk. Future research could be directed to see what has happened in these cases—*was there any follow up to the issues explored in these papers? Were any new or incremental fees introduced? Were other policies explored in addition to fees and taxes? What sort of attention did their work receive that may lead to better management of the natural environments they investigated?*

The second section deals much more explicitly with how to improve upon the management of coastal and marine environments. While the papers are quite different, ranging from cost-benefit studies of coastal development to assessment of policies designed to sustain fisheries (both capture and aquaculture); there are cautionary tales in each. These tales provide clear examples of policy failures combined with thorough analysis of policy initiatives that could reduce environmental damages while improving local livelihoods. Viable policies are analyzed in these studies to improve water quality and, in turn, fish harvest in Songkhla Lake; reduce aquaculture stocking levels in the Philippines; improve the management of fish pond licenses and restore abandoned ponds to mangroves or other uses; and assess the impacts of coastal development on ecosystem sustainability.

Each study provides a set of pragmatic approaches that decision makers can use to identify the trade-offs between economic activities and their ecosystem impacts, for example, using cost-benefit analysis, institutional and detailed field studies, and in-depth interviews with those affected. They follow up this discovery with an analysis of policies designed to mitigate trade-offs. The studies illustrate the value of undertaking careful economic and policy analysis, not just to identify problems but also to help solve them.

The third set of papers continues with the evaluation of existing policies and design of potential policies to improve coastal resource management. These are focused on fisheries that operate in an open access environment, where there are no or few specific property rights to the harvest and where fishers are predominately artisanal and low income. Hence, policy makers have to balance the need to sustain fish stocks, which could be done by significantly reducing fishing effort, with the need to sustain the livelihoods and incomes of fishing communities.

The policies explored in these papers highlight the trade-offs. For instance, a marine protected area (MPA) can help enhance the sustainability of aquatic species and their ecosystem, but at the expense of reducing the amount of fish available for harvest by the local fishers. This creates tension between regulators and the community and therefore motivates illegal harvesting.

The findings in the three papers dealing with protected areas are similar: better outcomes with fewer trade-offs are observed the greater the role of the community in decision making with regard to regulations and their enforcement. Engagement with fishers can enhance the understanding of the goals of the policy, improve enforcement and usage of proper gear type, and assist in the management of the protected area boundaries and fishing seasons. A rich set of policies in addition to MPAs are explored—individual quotas, individual transferable quotas, artificial reefs, and the use of the power of the market, through the supply chain and the consumer demand for sustainable fisheries, to impose size restrictions on harvests. Each study illustrates how good policy design can enhance the fishers' well-being while better protecting fish stocks.

The papers also demonstrate that research quality and relevance are enhanced by going into the field to interview fishers, to go out on their vessels, and to engage them in an experiment. These help researchers understand fishers' constraints and concerns. Similar value is obtained by talking to the regulators. These studies illustrate what sorts of policies improve the likelihood of a mutually sustainable economy and environment and can serve as important examples for fisheries in many parts of the world.

The volume thus provides in-depth studies of the challenges of using our natural environment to sustain livelihoods without destroying the very ecosystems that provide the goods and services. They offer methodologies for exploring benefits to society that could be obtained with effective resource management. Some provide analysis of public policies to address the conflicts and trade-offs that arise in natural resource use. The richness of the local cases, which provide detailed descriptions of the affected communities, institutions and regulations, and consumers of the ecosystem goods and services, provides future researchers with a good base, not only for case comparisons but also for methodological treatment.

The lessons learned in these studies can be applied to other jurisdictions beyond Southeast Asia. They can help improve our understanding of how to better manage our interaction with and dependence on our natural environment. However, much more remains to be done. Some of the studies are now more than 10 years old. What has happened since these studies were first done? Did the research affect policy? What has happened to the natural resource and the people dependent upon it? More time has passed since the introduction of, for example, MPAs, gear restrictions, better accounting for fish pond licenses, and perhaps higher user fees for tourists. In our view, there aren't enough longitudinal studies of natural resource management and use. Research methodologies have advanced in some cases—more sophisticated econometric analysis and controlled experimental design to name two. Going back to a case where there is a baseline provided by the researchers in this volume could be very illuminating.

These cases are, by design, focused on the economic aspects of coastal marine management. Biological and biophysical data is always a challenge and often these studies have had to infer underlying conditions. Better data would benefit the analysis. For example, more work is needed on the interplay between exogenous (to local communities) factors such as climate change, local and global pollutants (e.g., plastics, marine "dead zones," toxic blooms, etc.), and resource management.

Finally, in updating any of these cases or exploring analogous ones in other regions, we need more fulsome policy analysis. Such analysis should explore different means of achieving sustainable resource management using criteria in addition to efficiency and equity. The cases in this volume provide sound foundation for the examination of the complexities of coastal management. The problems they identify are endemic to jurisdictions worldwide. We need ongoing analysis of how to address these problems with examples of what works and doesn't work. The marine environment is crucial to the planet's survival; we have to persist in understanding how humans can use the environment without destroying it.

Index

N. Olewiler et al. (eds.), *Marine and Coastal Ecosystem Valuation, Institutions, and Policy in Southeast Asia*, DOI 10.1007/978-981-10-0141-3

MIX
Papier aus verantwortungsvollen Quellen
Paper from responsible sources
FSC® C105338

If you have any concerns about our products,
you can contact us on
ProductSafety@springernature.com

In case Publisher is established outside the EU,
the EU authorized representative is:
Springer Nature Customer Service Center GmbH
Europaplatz 3, 69115 Heidelberg, Germany

Printed by Libri Plureos GmbH
in Hamburg, Germany